Snowbird

Snowbird

Integrative Biology and Evolutionary Diversity in the Junco

EDITED BY ELLEN D. KETTERSON
AND JONATHAN W. ATWELL

UNIVERSITY OF CHICAGO PRESS   CHICAGO AND LONDON

ELLEN D. KETTERSON is distinguished professor of biology and gender studies at Indiana University, Bloomington, a fellow of the American Academy of Arts & Sciences, and president of the American Society of Naturalists. JONATHAN W. ATWELL is a research scientist and educator at Indiana University, Bloomington. He recently produced an award-winning and widely distributed science film, *Ordinary Extraordinary Junco* (www.juncoproject.org), which serves as a complement to *Snowbird*.

The University of Chicago Press, Chicago 60637
The University of Chicago Press, Ltd., London
© 2016 by The University of Chicago
All rights reserved. Published 2016.
Printed in the United States of America

25 24 23 22 21 20 19 18 17 16   1 2 3 4 5

ISBN-13: 978-0-226-33077-8 (cloth)
ISBN-13: 978-0-226-33080-8 (e-book)
DOI: 10.7208/chicago/9780226330808.001.0001

Library of Congress Cataloging-in-Publication Data

Snowbird : integrative biology and evolutionary diversity in the junco / edited by Ellen D. Ketterson and Jonathan W. Atwell.
    pages cm
    Includes index.
    ISBN 978-0-226-33077-8 (cloth : alkaline paper) — ISBN 978-0-226-33080-8 (e-book)
  1. Juncos. I. Ketterson, E. D. (Ellen Dorcas), 1945– editor. II. Atwell, Jonathan W., editor.
    QL696.P2438S66 201
    598.8′83—dc23

                2015028037

♾ This paper meets the requirements of ANSI/NISO Z39.48–1992 (Permanence of Paper).

IN MEMORY OF VAL NOLAN JR., ORNITHOLOGIST AND TEACHER

Contents

Preface

Obtaining knowledge about how living things will respond to ever more rapidly changing environments is a pressing societal need. This volume's mission is to help meet that need by synthesizing research conducted over the past century on a common and representative songbird, the junco. This well-integrated collection of chapters written by experts from evolutionary and integrative biology will, we hope, lead to novel connections among physiological mechanisms, adaptation, and evolutionary change that can be applied more broadly.

The volume is dedicated to Val Nolan Jr. (1920–2008), who was a law professor at Indiana University and a self-taught biologist whose research interests turned increasingly towards birds as he moved through life. At age 46, Nolan was appointed professor of zoology (later biology) at Indiana University, and in the ensuing years he established a tradition of research in avian biology at Indiana, including serving as advisor to seventeen PhD students. Nolan's research focused on the behavior and ecology of natural populations of marked songbirds in the wild, particularly the prairie warbler (Nolan, Jr. 1978) and the dark-eyed junco. In 2002, Nolan was lead author on the species account of the dark-eyed junco that appeared in *The Birds of North America* (Nolan, Jr. et al. 2002). This account, which is highly detailed and scholarly, includes a complete bibliography of more than 280 references. It is not our objective to repeat the *BNA* account here, but it is an objective to make its content more accessible and to highlight connections between the account and current research themes in evolution, ecology, and organismal biology.

Support for this volume came from the National Science Foundation through a program known as Opportunities for Promoting Understanding through Synthesis (OPUS). The program welcomes proposals "aimed at synthesizing a body of related research projects conducted by a single individual or a group of investigators over an extended period" (NSF-BIO-DEB Solicitation 14-559, "Opportunities for Promoting Understanding through Synthesis" [OPUS]). We proposed to synthesize the discoveries made by a diverse and dynamic group of researchers who had all studied the junco, and when our proposal was successful, we organized a two-day workshop at Indiana University to chart the course for this book.

Organisms are the central unit of biological organization, and they can be fully understood only if seen through multiple disciplinary lenses. Individual organisms live and die. They arise from developmental processes mediating the interaction of their genes and their environments, and they adjust as individuals to changing seasons through phenotypic flexibility. As collections, they exhibit norms of reaction such that particular genotypes vary in expression according to the environment they find themselves in. Also as collections, organisms form populations that expand or shrink according to their reproductive potential in their biotic and abiotic environments. Populations vary over time owing to evolution, and they may diverge and give rise to new species. An accurate view of the organism in nature and how it will respond to environmental change requires knowledge of all these processes. The authors who contributed to this volume were able to provide the needed expertise because each had engaged with the same model organism from different disciplinary perspectives.

Thus a stated goal of the OPUS proposal, and now this volume, is to synthesize discoveries made by diverse and talented contributors who study the junco and to chart a course for future research in evolutionary organismal biology using the junco as a model system. The OPUS proposal also set two other goals. The second goal was to create a permanent archive of publications, metadata, and data related to studies of the junco. Hardly bedtime reading, this archive is intended to preserve how field research was conducted late in the last century and early in the present century. The archive can be found online at a site known as ScholarWorks (https://scholarworks.iu.edu/dspace/handle/2022/7911) and will be maintained in perpetuity by Indiana University. The third goal of the OPUS proposal was to produce a popular documentary film, *Ordinary Extraordinary Junco: Remarkable Biology from a Backyard Bird*. The film, written and produced by Atwell, Steven Burns, and Ketterson, is now available online at www.juncoproject.org and serves as counterpoint to the more scholarly content of this book.

The nineteen contributors to the volume are biologists from four countries and thirteen institutes or universities; all share a keen interest in the junco because of its role in understanding how evolution proceeds and how animals know what to do and when with respect to breeding, migrating, and overwintering. All of the contributors have conducted research on the junco, and collectively they provide diverse expertise in behavioral ecology, behavioral neuroendocrinology, birdsong, conservation biology, ecological genomics, migration biology, physiological ecology, seasonality, quantitative genetics, and more. Many of the contributors trace their interest in the junco to Val Nolan Jr. Ketterson was Nolan's student who later became his colleague, and Atwell was Ketterson's student who has since become her colleague. Many of the other contributors were also connected to Nolan and Indiana University, while still others developed their interest in the junco independently.

Intended Audience

Our primary target audience is science professionals as well as graduate and undergraduate students seeking information and insights into the biology of a representative common backyard bird and how it adapts. We also hope that the broad accessibility and inherent interest of the junco and the related documentary film will attract amateur ornithologists, birders, educators, and naturalists with a serious interest in birds. In sum, we aimed for a scholarly volume that is not too laborious to read.

Acknowledgments

We thank the National Science Foundation, Indiana University, the University of California, San Diego, and Mountain Lake Biological Station, University of Virginia. We thank all the contributing authors, the students and postdocs who have been part of the junco project over the years, and the numerous field assistants who provided their efforts and good will. We thank our editor Christopher Chung and three anonymous reviewers whose comments substantially improved the book. Some text was drawn from the Nolan et al. (2002) *BNA* account; other text was drawn from *The Princeton Guide to Evolution* (Ketterson et al. 2013) and two published papers, one from *Philosophical Transactions of the Royal Society* (Mc-

Glothlin and Ketterson 2008) and one from *Integrative and Comparative Biology* (Ketterson et al. 2009). A special thanks goes to Amanda Brothers, who helped at every stage.

Ellen D. Ketterson
Jonathan W. Atwell
Bloomington, Indiana, USA
December 25, 2014

References

Ketterson, E. D., J. W. Atwell, and J. W. McGlothlin. 2009. Phenotypic integration and independence: Hormones, performance, and response to environmental change. *Integrative and Comparative Biology* 49:365–79.
———. 2013. Evolution of Hormones and Behavior. In *The Princeton Guide to Evolution*, edited by J. B. Losos, 616–23. Princeton: Princeton University Press.
McGlothlin, J. W., and E. D. Ketterson. 2008. Hormone-mediated suites as adaptations and evolutionary constraints. *Philosophical Transactions of the Royal Society* B-Biological Sciences 363:1611–20.
Nolan, V., Jr. 1978. The ecology and behavior of the Prairie Warbler (*Dendroica discolor*). *Ornithological Monographs* 26:1–595.
Nolan, V., Jr., E. D. Ketterson, D. A. Cristol, C. M. Rogers, E. D. Clotfelter, R. C. Titus, S. J. Schoech, and E. Snajder. 2002. Dark-eyed Junco (*Junco hyemalis*). In *The Birds of North America*, edited by A. Poole, and F. Gill, no. 716. Philadelphia: The Birds of North America, Inc.

CHAPTER ONE

Introduction

Ellen D. Ketterson and Jonathan W. Atwell

Snow borne on iced winds
has the juncos dancing on
one foot then the next
—Carol Knapp

A. Introduction

The avian genus *Junco* has served as a favored subject of biological research for at least the past 100 years. One of three species, the dark-eyed junco is among the commonest and most familiar North American passerines. It occurs across the continent and from northern Alaska to Baja California, and by one estimate (Folkard and Smith 1995) has a total population of about 630 million (compare current US census figures at 321 million). The public knows the junco not so much for its ubiquity and abundance as for the tameness and conspicuousness of its ground-foraging winter flocks. Wintering juncos are found in suburbs (often at feeders), at edges of parks and similar landscaped areas, around farms, and along rural roadsides and stream edges. Their plumage is not typical of sparrows: white outer tail feathers that flash when they take flight and a gray or blackish "hood" (head, nape, throat) and dark back that contrast with a white breast and belly. Audubon (1831) stated that "there is not an individual in the Union who does not know the little Snow-bird," and to many people "snowbird" is the junco's name today[1]. The other three spe-

---

1. Text of this paragraph has been modified from the original version of the junco *BNA* account: Nolan, Jr. et al. 2002, Dark-eyed Junco (*Junco hyemalis*) in *The Birds of North America*, edited by A. Poole and F. Gill, no. 716, Philadelphia: The Birds of North America, Inc.

cies in the genus *Junco*, the yellow-eyed junco, the Guadalupe junco, and the volcano junco, are birds of Middle and Central America and are far less well known, a gap in knowledge this book will help to fill.

Books about a single species or group of species, called monographs, are often single authored works that provide a comprehensive treatment of a taxonomic group. In the world of ornithology, monographs that focus on the behavior and ecology of a bird in nature, as opposed to its evolutionary relationships, are called life histories. The junco has already been the subject of a remarkable taxonomic monograph, *Speciation in the Avian Genus* Junco, by Alden Miller (1941) (see chapter 2, this volume, for more about Miller), and the subject of life histories including two that appeared in *Birds of North America* (*BNA*), a ten-year project including all the birds that breed in North America (Eaton 1968; Nolan, Jr. et al. 2002; Sullivan 1999) (see preface, this volume). The Cornell Laboratory of Ornithology currently hosts these *BNA* accounts online on a subscription basis, employing a model for dissemination that has only recently become possible. As new information on a species becomes available, authors (original or otherwise) can add to their species accounts to keep them current.

This book is not a remake of Miller's monograph or the *BNA* accounts; interested readers are encouraged to seek out these treatises for far more detailed information about the bird. Instead our purpose is to provide the reader with a synthesis of knowledge about a bird that has long interested biologists and to demonstrate why it is poised to serve as a model system during an era of research on organismal responses to changing environments. The authors strived to draw connections among published studies that were presented as standalone projects when first published. Many of the authors have collaborated before and share a common history. We hope that the closeness of the authors and the fact that the bird acts as a hub joining many diverse research areas will provide a sense of unity not always found in edited volumes.

B. What is This Book About?

The approach taken in this book about the junco draws on research concepts, methods, and findings from two major scientific disciplines, integrative organismal biology and evolutionary biology. From an organismal perspective, it addresses how animals match their physiology and behavior to their environment by relating endocrine and timing mechanisms to an animal's ability to know when to breed, when to migrate, and how to behave. From an evolutionary perspective, it employs the junco's challenging phylogeny as it has been interpreted over time by systematists and evolutionary biologists to address questions of how populations diverge and how species form. Knowledge of the junco's mating preferences and communication systems are included to help inform both the organismal and the evolutionary perspectives.

C. Integrative Organismal Biology in the Dark-Eyed Junco

This volume includes a synthesis of the decades of research conducted by the Ketterson-Nolan research group at Indiana University on numerous aspects of junco biology including the timing of migration, the role of sex and experience in distance migrated, courtship behavior and mate choice, parental behavior, plumage variation, hormone-mediated trade-offs in life histories, and phenotypic integration.

The key conceptual insights that have emerged from the organismal aspects of research on the junco are these: (1) Hormones are environmentally mediated molecules that act systemically and interact with target tissues to give rise to attributes referred to as traits that are referred to collectively as the phenotype; (2) coordinated trait expression mediated by a single hormone is referred to as hormonal pleiotropy; (3) if selection favors, or acts against, one or more hormonally mediated traits, other hormone-mediated traits may be dragged along, at least in the short run; (4) if coexpressed traits have multiplicative effects on fitness, then selection is referred to as correlational and can result in tight phenotypic integration; (5) dependence of traits on hormones is dynamic, so that selection can also act on tissue sensitivity to a hormone, and sensitivity to a hormone can come and go leading to less integration or phenotypic independence; (6) in the short run, tight phenotypic integration can act as a constraint on adaptive evolution; (7) phenotypic integration can also enable rapid adaptation in response to environmental change because it causes traits to co-occur in individuals when recombination might otherwise be required and take longer; (8) hormone-enabled rapid evolution can also lead to population divergence; and (9) the nature of the traits that hormones influence "makes sense" in terms of trade-offs in life histories (e.g., coordinating trait values that balance viability and sexual selection or pace of life).

We highlight key findings that (1) hormones can coordinate the expression of numerous traits (hormonal pleiotropy) (Ketterson and Nolan,

Jr. 1999), potentially favoring phenotypes that are tightly integrated as a result of correlational selection (Ketterson et al. 2009; McGlothlin and Ketterson 2008); and (2) variation in hormonal pathways can predict population divergence on several time scales (Atwell et al. 2014; Bergeon Burns et al. 2013), a phenomenon with many underappreciated implications for the nature of rapid adaptation to changing environments.

D. Speciation/Population Divergence in the Dark-Eyed Junco

In the classic view, recognized groups of dark-eyed juncos diverged in geographic isolation, perhaps during glacial advances or even earlier (Klicka and Zink 1997). However, recent research from our contributors (Milá et al., chapter 8, this volume) has shown that genetic divergence in North America may have occurred as recently as the past 10,000 years (Milá et al. 2007). Not all biologists are convinced that the dark-eyed junco groups are so young or that they spent the periods of glacial advance only in Mesoamerica (Price and Hooper, chapter 9, this volume) (Klicka and Zink 1997). Regardless, whether the dark-eyed junco is one species or more, the rate of divergence has been extremely rapid, and the junco's degree of phenotypic and genetic differentiation requires explanation and bears squarely on the question of how animals respond to changing environments.

E. Merging Evolutionary and Integrative Organismal Biology

We hope this book will contribute to its objective of merging perspectives from evolution, behavior, and physiology through its focus on the importance of migratory and reproductive timing in population divergence. Traditionally, avian species were defined by morphological measures that manifested in preserved specimens (see color plate 1) (Milá et al. 2007), and modern systematists measure genetic divergence with an array of tools that have supplanted morphology. However, some of the critical traits leading to population divergence may not be gleaned from skins or sequences, and these relate to timing of population movements (i.e., migration) and reproduction (Winker 2010). In the case of the polytypic junco, some populations are sedentary, but many are migratory, exhibiting all degrees of movement from altitudinal to latitudinal, obligate to facultative, and partial to differential. The timing of migration varies geographically, and breeding dates also differ widely from north to south and along altitudinal gradients (Atwell et al. 2011).

Importantly, however, many junco groups overlap in their winter distributions, and southern populations initiate reproduction while northern populations are still present, which has a significant implication. Despite the opportunity to interbreed, timing mechanisms have apparently led to and maintain divergence. If we are to understand population divergence in this polytypic species, and other migratory species, then we need to understand not only differences in morphology and sequences but also differences in physiology. Also missing from skins and sequences is behavior, including the role that sexual selection plays in initiating and hastening divergence. Thus another goal of this book is to focus on the interaction of natural and sexual selection acting on physiology and behavior in addition to the role of hormonal pleiotropy in allowing coordinated, rapid response to environmental change.

F. New Insights Expected to Emerge from This Book

In addition to a cohesive overview of findings to date, we hope that this synthesis will extend the utility of prior findings by providing an explication that will (1) emphasize the generality of findings on the junco; (2) focus on hormones as both a proxy for genetic pleiotropy and an emergent property of pleiotropy, adding a layer of understanding to the causes and consequences of selection on correlated traits; (3) stress how hormone-mediated phenotypic plasticity, if followed by genetic assimilation, can give rise to evolutionary phenotypic divergence among populations that may—or may not—ultimately lead to speciation; and, finally, (4) explore the whole new landscape created by environmentally and hormonally driven variation in gene expression and phenotypic outcomes that relates to population divergence and will challenge and excite evolutionary and environmental biologists in the coming decade.

G. How to Use This Book

Readers will be confronted with many common and scientific names for the junco, some of which is unavoidable, and this section offers guidance.

The answer to the question of how many species of juncos there are has not been resolved, in part because modern ornithologists disagree about how to define species and about the utility of the subspecies concept (Morrison 2010). Further, despite new techniques, which have generated molecular data to bear on relatedness, the interpretation of molecular data can still be a matter of opinion (Zink and Barrowclough 2008; Zink and Davis 1999). Thus, the dark-eyed junco was once five distinct species (see chapter 2, this volume): slate-colored (*Junco hyemalis*), white-winged (*J. aikeni*), Oregon (*J. oreganus*), Guadalupe (*J. insularis*), and gray-headed junco (*J. caniceps*), and all but *J. aikeni* and *J. insularis* had recognized subspecies with their own common names. In 1983, the American Ornithologists' Union merged these five species into one (the dark-eyed junco) and ceased to recognize subspecies (American Ornithologists' Union 1983) (see chapter 2, this volume). The yellow-eyed junco, *J. phaeonotus*, has a similar complex naming history (e.g., Baird's junco, Mexican junco), which will be addressed in chapters 2 and 8. The decision we made for this volume is (1) to acknowledge the primacy of the AOU when naming species, but (2) to use the common names for subspecies found in the Clements checklist (Clements et al. 2012) because they will be useful to the authors and the reader, and (3) to use the taxonomy suggested by Milá in chapter 8, this volume, because it is the most current.

Readers will also be confronted with disciplinary terminology from avian endocrinology and evolutionary biology, much of which is also unavoidable. We provide a glossary, which deliberately strips many terms to their essentials and which was borrowed partially from the *Princeton Guide to Evolution* (Ketterson et al. 2013) and from chapter 9 (this volume). Eight pages of color figures appear in the book's center, including (1) a photo of a "mixed" wintering flock of juncos including several different subspecies; (2) a map of the breeding distributions of junco species and subspecies; (3) a rendering of the juncos' current phylogeny; (4) and (5) "head shots" of twelve different forms of the junco; (6) a cartoon portrayal of the inputs and outputs of the hypothalamic-pituitary-gonadal axis in relation to the hormone testosterone; (7) a visual overview of a contemporary junco colonization of a novel environment in San Diego, California; and (8) a junco in flight and a close up of a feather plumage ornament (tail white). Some of these figures are also referred to in specific chapters, and others are standalone.

We hope these guides and statements of our objectives will help the reader enjoy the potential for future insights to be gleaned from study of the junco.

Acknowledgments

We thank all the colleagues, postdocs, and graduate students who have been part of the junco project over the years, as well as the numerous field assistants who provided their efforts and good will. We thank our editor Christopher Chung and three anonymous reviewers whose comments substantially improved this chapter.

References

Atwell, J. W., G. C. Cardoso, D. J. Whittaker, T. D. Price, and E. D. Ketterson. 2014. Hormonal, behavioral, and life-history traits exhibit correlated shifts in relation to population establishment in a novel environment. *American Naturalist* 184:E147–60.

Atwell, J. W., D. M. O'Neal, and E. D. Ketterson. 2011. Animal migration as a moving target of conservation: Intra-species variation and responses to environmental change, as illustrated in a sometimes migratory songbird. *Environmental Law* 41:289–316.

Audubon, J. J. 1831. *Ornithological Biography*, vol. 1. Philadelphia: E. L. Carey and A. Hart.

Bergeon Burns, C. M., K. A. Rosvall, and E. D. Ketterson. 2013. Neural steroid sensitivity and aggression: Comparing individuals of two songbird subspecies. *Journal of Evolutionary Biology* 26:820–31.

Clements, J. F., J. Diamond, A. White, and J. W. Fitzpatrick. 2012. *The Clements Checklist of Birds of the World*. Ithaca: Cornell Laboratory of Ornithology.

Eaton, S. W. 1968. Dark-eyed junco (*Junco hyemalis*). *Smithsonian Institution United States National Museum Bulletin* 237 (2): 1029–43.

Folkard, N. F. G., and J. N. M. Smith. 1995. Evidence for bottom-up effects in the boreal forest: Do passerine birds respond to large-scale experimental fertilization? *Canadian Journal of Zoology* 73:2331–37.

Ketterson, E. D., J. W. Atwell, and J. W. McGlothlin. 2009. Phenotypic integration and independence: Hormones, performance, and response to environmental change. *Integrative and Comparative Biology* 49:365–79.

Ketterson, E. D., J. W. Atwell, and J. W. McGlothlin. 2013. Evolution of hormones and behavior. In *The Princeton Guide to Evolution*, edited by J. B. Losos, 161–23. Princeton, NJ: Princeton University Press.

Ketterson, E. D., and V. Nolan, Jr. 1999. Adaptation, exaptation, and constraint: A hormonal perspective. *American Naturalist* 154:S4–S25.

Klicka, J., and R. M. Zink. 1997. The importance of recent ice ages in speciation: A failed paradigm. *Science* 277:1666–69.

McGlothlin, J. W., and E. D. Ketterson. 2008. Hormone-mediated suites as adaptations and evolutionary constraints. *Philosophical Transactions of the Royal Society B-Biological Sciences* 363:1611–20.

Milá, B., J. E. McCormack, G. Castaneda, R. K. Wayne, and T. B. Smith. 2007. Recent postglacial range expansion drives the rapid diversification of a songbird lineage in the genus *Junco*. *Proceedings of the Royal Society B-Biological Sciences* 274:2653–60.

Miller, A. H. 1941. Speciation in the avian genus *Junco*. *University of California Publications in Zoology* 44:173–434.

Morrison, M. L. 2010. Avian subspecies. *Ornithological Monographs* 67.

Nolan, Jr., V., E. D. Ketterson, D. A. Cristol, C. M. Rogers, E. D. Clotfelter, R. C. Titus, S. J. Schoech, and E. Snajdr. 2002. Dark-eyed junco (*Junco hyemalis*). In *The Birds of North America*, edited by A. Poole and F. Gill, no. 716. Philadelphia: The Birds of North America, Inc.

Sullivan, K. A. 1999. Yellow-eyed junco (*Junco phaeonotus*), *The Birds of North America Online*, edited by A. Poole. Ithaca: Cornell Lab of Ornithology.

Winker, K. 2010. On the origin of species through heteropatric differentiation: A review and a model of speciation in migratory animals. *Ornithological Monographs* 69:1–30.

Zink, R. M., and G. F. Barrowclough. 2008. Mitochondrial DNA under siege in avian phylogeography. *Molecular Ecology* 17:2107–21.

Zink, R. M., and J. I. Davis. 1999. New perspectives on the nature of species. In *Proceedings of the 22nd International Ornithological Congress, Durban*, edited by N. J. Adams and R. H. Slotow, 1505–18. Johannesburg: BirdLife South Africa.

PART I

Opportunities and Challenges in Evolutionary and Integrative Biology Presented by the Avian Genus *Junco*

My (Ketterson's) first experience with the junco's fluid taxonomy came while I was still a graduate student. I had been studying migratory behavior in the slate-colored junco in eastern North America when suddenly I learned that the species I was studying was no longer a species but had been bundled with other juncos to become the dark-eyed junco. I read that the checklist committee of the American Ornithologists' Union had made this decision based largely on writings of Ernst Mayr, who concluded that if species of juncos in North America hybridized where their ranges overlapped far to the north and west of where I lived, they could not be good species. One lesson learned was that I could not afford to ignore taxonomy and that the bird I was studying had a rich history in the literature, from which I could learn a lot. Even though for my purposes, it was more important to know where a bird came from at particular times of year than to know what it was called, history matters.

Chapter 2 introduces the reader to the junco and the people who have studied it and makes the case for why the junco should be considered a model system for synthesizing integrative and evolutionary biology. The chapter introduces key historical figures such as Mayr, Alden Miller, and William Rowan, who all made seminal contributions to avian biology during the first half of the twentieth century. Their research was critical to our modern understanding of geographic variation and seasonality.

Chapter 3 describes where the junco lives today and how its appearance (morphology) and life history vary geographically. By focusing on migration this chapter lays the groundwork for how juncos may have responded to glacial cycles and other climatic changes over time. It raises critical questions about how populations that are sympatric at certain times of year may read the environment differently, such that one population migrates before it breeds and the other does not.

CHAPTER TWO

The Junco

A Common Bird and a Classic Subject for Descriptive and Experimental Studies in Evolutionary and Integrative Biology

Ellen D. Ketterson and Jonathan W. Atwell

Little mouse-bird, so soft and so gray,
You brighten our sight on a dull winter day,
So trim and so neat in your drab little coat,
So gentle and sweet is your musical note!
Junco, I'll feed you and welcome you here.
Come, stay near my home where there's nothing to fear.
—Julius King, 1934, *Birds*, Book 2 (Cleveland: Harter Publishing Company)

A. Introduction

The dark-eyed junco is both a familiar winter bird and a classic subject of scientific research, and this chapter introduces the reader to some of the researchers who studied the junco in earlier times and describes how their contributions made possible the junco's current role as a model species in avian evolutionary and organismal biology. The chapter, which builds on two species accounts published in *Birds of North America* (Sullivan 1999; Nolan, Jr. et al. 2002), concludes with a glimpse of ongoing areas of active research and debate among biologists who study the junco, particularly in the areas of speciation and seasonality.

B. The Junco as a Popular Bird and as a Subject of Research

According to Frank M. Chapman, Associate Curator of the Department of Mammalogy and Ornithology at the American Museum of Natural History in the early twentieth century, "The junco is one of the birds whose acquaintance is easily made. His suit of slaty gray with its low-cut vest of white is not worn by any other of our birds ..." Chapman concludes his treatment of the junco by writing "... There is nothing of especial interest in the Junco's habits, and only a bird-lover can imagine what a difference his presence makes in a winter landscape" (Chapman 1901). Current estimates place the number of birdwatchers in the United States at 46 million (La Rouche 2003), which indicates that there must be many who appreciate the difference a junco makes to a winter landscape.

What attributes of the junco make it attractive to people? For one, as Chapman noted, the junco is readily identifiable because its gray body and white tail feathers make it easy to distinguish from other sparrows. Because they are common and eat seeds, juncos are among the first to show up at bird feeders and eat "bird seed." In cities, juncos are relatively bold, feeding on crumbs from scones at coffee kiosks on college campuses or nesting in flowerpots along public walkways (see chapter 10). In eastern North America, where they are abundant in winter, juncos are admired for their ability to endure blizzards, ice, short days, and enforced fasting, causing people to marvel when they see juncos perched outside their windows on snowy days. Again in eastern North America, another striking habit of the junco is its migratory pattern. Unlike the many bird species that disappear in winter, juncos arrive in late autumn after locally breeding migrants have left for the south and depart northward to breed just as other migratory species return in spring. These seasonal appearances and disappearances help people tell the time and even count the passing years. We anticipate that the recent film *Ordinary, Extraordinary Junco* will only add to the numbers of people who take pleasure in being familiar with this common yet remarkable bird (http://www.juncoproject .org).

Significantly, the junco is also a prime research subject. A quick search of Web of Science using "dark-eyed junco" will turn up roughly a thousand journal articles, many of which were cited in the dark-eyed junco account in *The Birds of North America* (Nolan, Jr. et al. 2002). A web search for other species in the genus (e.g., the yellow-eyed junco or the volcano junco) will yield far fewer hits (about thirty-five and two, respectively, in 2014). Key features of the dark-eyed junco's biology that make it appealing for study include the juncos' phenotypic variability, broad distribution, highly variable migratory behavior, high abundance, and relative tameness. Dark-eyed juncos are also representative of many north-temperate songbirds in their life history: they form social groups in winter (flock) and are territorial during the summer when they nest as pairs. Like most songbirds studied to date, juncos are socially monogamous, which describes a social system in which individuals form a pair bond and both parents care for offspring, even though some of the offspring may be sired by other males. Juncos are also relatively easy to study in the lab and in the field. They thrive in captivity and do not require a complicated diet because they eat seeds and various small insects. In the field, juncos live in the understory of forests or in open areas and generally nest on the ground, so they and their nests can be readily found and followed. Owing to their diet of seeds and their use of the understory, juncos can be captured quite readily in traps and nets. Finally, juncos are a nice size to hold in the hand, and, as a side benefit, their bites do not break the skin and are not known to carry diseases that are communicable to humans.

Not surprisingly, present-day biologists are not the first to have been drawn to the junco as a research subject. Juncos have contributed considerably to our understanding of speciation and seasonality, and the sections to follow focus on past researchers who used the junco to advance knowledge in both these areas.

C. Early Contributors to Junco Taxonomy

The key early contributors to the junco's taxonomy were Jonathan Dwight, Alden Miller, and Ernst Mayr. Jonathan Dwight (1858–1929) was associated with the American Museum of Natural History. By profession he was a civil engineer and physician from a socially prominent New York family, but Dwight was also a prolific amateur in keeping with ornithology's early tradition of significant contributions by amateurs, and he left 60,000 specimens to the museum's collection (Anonymous 1929). He was a founding member of the American Ornithologists' Union and served as its president. With respect to the junco, Dwight was concerned that specimens of the junco needed standardized names if they were to be sorted into categories within the museum or to be communicated about consistently. He thought that junco nomenclature was a mess: too many names, too many types, and, in his view, the confusion was engendered in part by the effort to give names to intermediate forms that might be hybrids and by the desire for fame that induced some systematists to give their own names to forms that were better left unnamed. It was Dwight (Anonymous 1929) who made the case for classifying juncos by color, in particular by the color of specimens' heads, backs, beaks, and flanks. His reasoning was that these were traits that could be used to establish qualitative differences between forms, and they also sorted geographically. He proposed that other traits (e.g., the amount of white in the tail) were quantitatively variable and that such traits could be used to identify subspecies. Dwight articulated the basis for his classification as follows:

> Instead of accepting the presence or absence of intergradation as a guide by which to separate species from subspecies, I have endeavored to show that species may be recognized by qualitative, and subspecies by quantitative characters. Specific and subspecific characters in most of the Juncos are almost wholly confined to color and therefore by mapping the geographical distribution of color we are able to gain from a new angle a fairly distinct impression of relationships in the genus. (Dwight 1918)

Dwight's emphasis on plumage coloration led him to advise ornithologists not to concern themselves with hybrids but to focus on clearly differentiated groups, which led him to recognize five species of dark-eyed junco: white-winged junco, slate-colored junco, Oregon junco, pink-sided junco, and gray-headed junco. While Dwight firmly established the role of color-based morphology and geographic distribution in the junco's classification, his was not a biological species concept.

Alden Miller (1906–1965), Director of the Museum of Vertebrate Zoology (MVZ), University of California, Berkeley, was the true father of junco systematics, and his monograph "Speciation in the Avian Genus *Junco*" (Miller 1941) is still consulted today. He is said to have been a buttoned down, family man, who was an even-handed and talented administrator and much admired as a scientist and a man (Davis 1967; P. M. 1967; Mayr 1973; Sunderland 2012). Miller too represents a tradition in ornithology—an academic naturalist and taxonomist who was thorough and precise. Despite Miller's premature death of a heart attack at age fifty-nine, he was amazingly productive as an author of 258 articles in addition to his monograph (Mayr 1973).

FIGURE 2.1. Alden Miller, a field ornithologist in his time. Source: Museum of Vertebrate Zoology at the University of California, Berkeley's "Doing Natural History: how and why were Andean sparrows collected." website, 2011; Used with permission of the Museum of Vertebrate Zoology, University of California, Berkeley.

Miller's photo (figure 2.1) shows him wearing the field clothes of the day: tennis shoes, straw collecting basket, hat, and gun. Most ornithologists of his time worked in museums and made expeditions to the field where they collected specimens, which they then preserved as skins. A quote from one of Miller's field journals conveys life in the field:

> June 7, 1938 Left Berkeley about 9:30 and headed north toward the Shasta
> area. Ward Russell drove the Reo truck with Paul Maslin and Albert Wolfson,
> Fred Test and Donald Tappe rode in my car. Stopped at Redding for supper

and drove on to 5 miles south of Dunsmuir, camping in the Sacramento River Valley on an abandoned road.

June 8, 1938 Arose about 4:45 Hammond Flycatchers were 'singing' in the firs about us and in the black oaks and pines were Audubon and Black-throated Gray Warblers. Other prominent species were. . . .

Miller made a total of fifty-three such expeditions from 1930 to 1965, many to the outback of California and Oregon and later to more exotic sites including Columbia, where he carried out a prolonged study of *Zonotrichia capensis*, the rufous-collared sparrow. Miller prepared 12,564 specimens that reside at the University of California, Berkeley's Museum for Vertebrate Zoology (Miller 1941). As Ernst Mayr (see below) described Miller's accomplishments in an obituary, "Only he who knows how much work it is to make up a bird skin and how far into the night one has to work to prepare the yield of a successful day will appreciate the significance of this figure" (Mayr 1973). Milá et al. (chapter 8) provide a nice summary of Miller's evidence and point of view, not to be repeated here, but, briefly, Miller described twenty-one different forms of the junco, which he organized into ten species. Miller contrasted his view with Dwight's regarding the importance of hybrid populations and what they can teach us about species and speciation. He wrote:

In working upon the genus *Junco*, it became apparent to me that some rather favorable conditions existed for studying natural hybrid populations. The very hybrid complexes which Dwight [*Bulletin of the American Museum of Natural History* 38 (1918): 269–309] earlier thrust into the background as without significance taxonomically prove to be the most important groups from the standpoint of biologic theory. (Miller 1939, 211)

After Miller's death, Ernst Mayr prepared a memoir about Miller for the *Proceedings of the National Academy of Science* (Mayr 1973), in which he praised Miller's research on the junco:

Miller proved conclusively that geographic races in juncos are incipient species, and show, indeed, such a perfect transition from local race to full species that to this date (1972) there is not yet unanimity among ornithologists as to which group of populations should be considered species and which others subspecies. Miller was quite justified in concluding that "the genus *Junco* contributes a rather complete exemplification of the stages and processes that lead to the first milepost in the evolutionary path, the full species." (Mayr 1973)

As this chapter, Milá et al. (chapter 8, this volume), and Price and Hooper (chapter 9, this volume) reveal, Mayr's statement regarding lack of unanimity could have been written in 2014. Miller continues to be intriguing, as can be seen in a recent summary of his achievements by science historian M. Sunderland (Sunderland 2012).

Ernst Mayr (1904–2005), a third significant figure associated with the junco, was born and educated in Germany, emigrated to New York as a young man, and affiliated with the American Museum of Natural History (just as Dwight did) and ultimately with Harvard University's Museum of Comparative Zoology. Mayr was a leading evolutionary biologist of the twentieth century and a key contributor to the conceptual integration of systematics, genetics, and evolution now known as the modern evolutionary synthesis. He too was prolific throughout his long life, and his early synthesis of his views on evolution, *Systematics and the Origin of Species* (Mayr 1942b), was highly influential. Mayr was the developer and promoter of the biological species concept (more below) and believed that geographic isolation was a necessary step in the process of speciation.

Despite Mayr's praise of Miller above, his own view was that reproductive isolation was key to defining species and was overriding when compared to morphological differentiation. To Mayr, the forms of juncos that hybridized where their ranges overlapped could not be good species. Mayr reviewed Miller's "Speciation in the Avian Genus *Junco*" in the journal *Ecology* (Mayr 1942a) and wrote in his review, "Would it not be much simpler and biologically more nearly correct to include all the juncos in a single superspecies, with three species: (1) *vulcani*, (2) the yellow-eyed group, and (3) the brown-eyed group?" Mayr also took issue with the title of Miller's monograph, as well as with Miller's view that species can be defined based on morphology. According to Mayr, Miller did not address the key issues in speciation, "namely the establishment of isolating mechanisms, the evolution of differences in habits and ecology, and whatever else contributes toward enlarging the biological gaps between incipient species" (Mayr 1973). Mayr's influence almost certainly played a leading role in the American Ornithologists' Union (AOU)'s decision in 1973 to merge the Oregon (*J. oreganus*), slate-colored (*J. hyemalis*), and white-winged juncos (*J. aikeni*) to create the dark-eyed junco (American Ornithologists' Union 1983).

Mayr's description of Miller's species concept as entirely morphological appears not to give full credit to Miller's writings on the importance of hybrids to defining species (Miller 1939; Miller 1941; Miller 1955). Mayr's view also appears to overlook or understate the key role that phenotypic divergence (e.g., song, plumage) plays in the development of reproductive isolation because such traits influence mate choice (T. Price, pers. comm.). Regardless of Mayr's ambivalence about Miller's interpretations, it is clear that Miller has had lasting impact. Modern ornithologists interested in speciation continually return to Miller's monograph and to the collection of his specimens at the University of California's Museum of Vertebrate Zoology for data, ideas, and inspiration (Ferree 2013) (see also chapter 3, this volume).

D. How Many Kinds of Juncos Are There Currently?

Clearly the question of the number of kinds of junco has had a long history, with the changes in numbers of species recognized reflecting the different interpretations by the scientists introduced above, and those of more recent contributors to the debate, including Ned Johnson (1932–2003) (Johnson and Cicero 2004), Robert Zink (Zink 1982), George Barrowclough, and authors of later chapters in this book: Borja Milá, Pau Aleixandre, Sofía Alvarez-Nordström, and John McCormack (chapter 8); and Trevor Price and Daniel M. Hooper (chapter 9). Key to determining numbers is having a definition of species, and in the case of the junco, researchers have employed all three of the prevailing species concepts, the morphological species concept, the biological species concept, and the phylogenetic species concept, which give primacy to phenotypic differences, presence/absence of hybridization, and genetic distance, respectively (Coyne and Orr 2004).

Since 1886, the AOU has issued a total of seven checklists at intervals. The fourth edition (American Ornithologists' Union 1931) recognized six species of what is now the dark-eyed junco, the fifth edition (American Ornithologists' Union 1957) recognized five, and the sixth edition (American Ornithologists' Union 1983) recognized one, a determination that was maintained in the seventh edition (American Ornithologists' Union 1998). In 2014, the AOU took the Guadalupe junco, which had been bundled with the dark-eyed junco, and described it as another species, *Junco insularis*, based on data reviewed in chapter 8. (Figure 2.2 tracks the AOU's treatment of *Junco hyemalis* over time.)

The number of dark-eyed junco subspecies has also varied widely. The AOU checklist does not currently recognize subspecies, although its stand on this issue is controversial (Nolan, Jr. et al. 2002; Winker and Haig

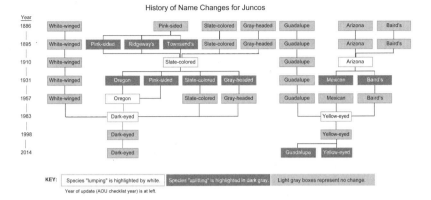

FIGURE 2.2. History of name changes in the genus *Junco*. Adapted from source image at Cornell Laboratory of Ornithology's FeederWatch Program website; used with permission.

2010). The American Birding Association, an alternative authority directed more towards birders, follows the Clements Checklist, which does recognize subspecies. In his sixth edition (Clements et al. 2012), Clements included fifteen subspecies of the dark-eyed junco. The Cornell Laboratory of Ornithology refers to the AOU as its ultimate authority but nevertheless posts Clements's subspecies, thus maintaining discrepancies between the two authorities. Similarly, the *BNA* account of the dark-eyed junco reports fifteen sub-species (Nolan, Jr. et al. 2002).

Lack of congruence among authorities in junco classification is not confined to the dark-eyed junco. The AOU currently treats the yellow-eyed junco as one species; the Clements Checklist treats it as one species with five subspecies; and Milá et al. (chapter 8) propose that the yellow-eyed junco consists of three species, the Baird's junco (*J. bairdi*), the Guatemala junco (*J. alticola*), and the yellow-eyed junco (*J. phaeonotus*). Milá et al. show that the newly recognized Guadalupe junco (*J. insularis*) is more closely aligned genetically with *J. phaeonotus, J. bairdi,* and *J. alticola* than it is with *J. hyemalis*. All authorities recognize the volcano junco (*J. vulcani*) of Costa Rica as a species, and no subspecies have been described. While the back and forth in terms of numbers of species might suggest that the topic is best ignored, the fact is that the issues raised in determining the proper number of species that comprise the genus *Junco* have helped to frame the continuing debate among evolutionary biologists about how species should be defined in general and how new species form (Price 2008).

The simplest reason for the instability in the dark-eyed junco taxonomy is that dark-eyed juncos look different from place to place. Thus the
junco's traditional classification was based on morphological features related to plumage coloration and geographic distribution. The distributions of these taxa are largely nonoverlapping during the breeding season,
which, along with the differences in their appearance, suggested distinct
species, as was seen by Dwight (1918). These morphological differences
are the way people, and probably the birds too, typically recognize different species.

The way most ornithologists recognize species, however, is based on the
biological species concept, which recognizes organisms as species when
they interbreed only with each other and not with members of other such
groups (Mayr 1963; American Ornithologists' Union 1998; Price 2008).
This definition has been used to allow for only one North American junco
(*J. hyemalis*) because where the ranges of the species recognized by early
taxonomists abut we find mixed pairs and zones of intermediate plumages indicating hybridization (Nolan, Jr. et al. 2002) (but see Johnson and
Cicero 2004). Miller (1955) provided quantitative data on the frequency
of junco hybrids as judged from samples taken in the field where pairings were observed and hybrids known to be fertile. For *Junco oreganus
x aikeni* in Montana, he reported 7 percent hybrids (five of seventy-three
samples); for *J. hyemalis x J. oreganus* in Alberta, he reported 20 percent
(nine of forty-four samples) (reinterpreted from Miller 1941); and for *J. o.
mearnsi x J. o. montanus* in California, he reported 71 percent (seventeen
"hybrids" or mixed-character birds out of twenty-four samples) (Miller
1941). Hybridization between the different forms of dark-eyed junco is
consistent with the findings of Milá et al. (2007; and chapter 8) who show
that, based on current methods, these groups are genetically extremely
similar, which makes it likely that they would freely interbreed if they all
occurred together in one place and thus, according to Price and Hooper
(chapter 9) might collapse back into a single form, in a process called reverse speciation (Seehausen 2006; Kleindorfer et al. 2014).

A limitation of the biological species concept is that it can be applied
only to taxa with sympatric (overlapping) or parapatric (adjacent) distributions and which thus have the opportunity to interbreed and not to
allopatric populations whose geographic distributions do not overlap.
However, Winker (2010) recently highlighted a fourth kind of distribution, heteropatry, which is found in migratory or incipient species that are
sympatric at one time of year but allopatric at another. Junco subspecies

or morphologically recognizable populations often differ in their migratory behavior and present examples of heteropatry. In the eastern United States, migratory slate-colored (*J. hyemalis hyemalis*) juncos are present during winter only, and in some regions (e.g., the Appalachian Mountains), they coexist with the more sedentary Carolina junco (*J. hyemalis carolinensis*). In spring, Carolina juncos begin to breed while migrants delay breeding until after they migrate. Thus hybridization is possible, yet differences in timing appear to interfere with pair formation across subspecies (see also chapter 3, this volume), and at least in ecological time, the populations remain distinguishable in morphology and behavior. Similar junco examples exist in the southwestern United States. Perhaps these junco forms are on a trajectory toward speciation or perhaps impending climate changes will reduce their level of divergence.

The challenge to the biological species concept posed by allopatry is particularly relevant in juncos found outside of North America. Guadalupe juncos are island dwelling and live about 400 kilometers off the coast of Baja Mexico, Baird's juncos live isolated on a "sky island" at the southern tip of the Baja peninsula, and Guatemala juncos are separated from the juncos of mainland Mexico by tropical habitats that are totally unsuitable for juncos and unlikely to be crossed given the sedentary nature of yellow-eyed junco populations in general. These disjunct geographic distributions mean that we cannot know whether these species of *Junco* proposed by Milá et al. (chapter 8) would hybridize naturally, and the distributions leave no option for drawing species boundaries but to assess genetic distance using molecular techniques as Milá et al. have done.

Species based on genetic structure and distance are phylogenetic species, which many prefer to the biological species concept because it can be used to recognize genetically distinct groups as species regardless of their morphology and geography (Zink and Davis 1999). However, while genetic divergence might seem an ideal solution to uncertain relationships, the issue of how genetically distinct groups must be to attain species status remains unresolved. In the genus *Junco*, we see that over time all three species concepts have been employed and led to different conclusions. Even today, some experts employ a combination of concepts. For example, Milá et al. (chapter 8, this volume) propose that the yellow-eyed junco *J. phaeonotus* remain a distinct species, despite the fact that *J. phaeonotus* will hybridize with *J. hyemalis* in the southwestern United States (McCarthy 2006) and clusters with *J. hyemalis* when genetic data are considered. What might be preventing Milá et al. from merging

J. phaeonotus and *J. hyemalis*? For the moment at least, the plumage and behavioral differences (e.g., song structure; see chapter 13, this volume) separating these two taxa are so striking, the molecular data still so tentative, and the examples of hybridization so few, that Milá et al. conclude that merging the two taxa as one species would be premature.

Finally we might ask whether it matters how many species of junco, or any other bird, there are. To some people it matters a great deal—for example, to a conservation biologist concerned with species preservation under the Endangered Species Act or to a serious birder who is compiling a life list. The contributors to this volume are probably most interested in what the junco can still teach us about the process of speciation because for us the most interesting question is not how many species there are but how long ago they diverged, whether they diverged in allopatry or sympatry, what the role of local adaptation is in population divergence, and whether there have been cycles of divergence and collapse over time.

The immediacy of these questions can be felt in chapters 8 and 9 in this volume. Going forward evolutionary biologists will also seek to determine the location and stability of the zones of hybridization that lie at the borders between the different groups of juncos and the degree of backcrossing and the fitness of hybrids in comparison to nonhybrids (Poelstra et al. 2014). More knowledge is also needed about the nature of the genetic mechanisms that account for why juncos look so different from place to place, yet apparently hybridize so readily. What is it about the junco's distribution, timing mechanisms, and mating preferences that have allowed them to diverge so strongly in appearance compared to closely related genera? Evolutionary biologists with an ecological perspective will also want to learn what accounts for the junco's consistency in its preference for cool environments, combined with its flexibility in relation to moisture. How does the junco succeed in such a wide variety of habitats? Additional questions seeking answers will include the role of disease in the junco's evolutionary history, as well as its current distribution and abundance. Clearly there is still much to learn.

E. Seasonality

A second area in which the junco played a critical historical role relates to the timing of key events in the avian annual cycle, including migration, reproduction, and molt (Nolan, Jr. et al. 2002). It is common knowledge that temperate-zone songbirds respond to lengthening days in spring by preparing to reproduce, and in males that preparation is often accompanied by the onset of song. Beneath the surface, the gonads grow and regress on a seasonal basis as well. The gonads enlarge in anticipation of the time when the climate is most benign for reproduction and regress as reproduction ends (Dawson et al. 2001). Molt often follows reproduction, and autumn migration, if it occurs, follows molt. The winter season is the least studied, but in contrast to many mammals, birds remain active throughout winter, and juncos exhibit an array of behavioral and energy-related adaptations that promote overwinter survival, including fattening and joining flocks to find food (Helms et al. 1967; Nolan, Jr. and Ketterson 1983; Lima 1988; Rogers et al. 1993; Goodson et al. 2012).

Bird species differ in the timing of migration, reproduction, and molt, so clearly they also differ in the cues they use to time these events or in how they weigh such cues (Wingfield and Farner 1993). Integrative biologists are actively exploring the many ways in which environmental cues are employed by birds to generate a seasonally "flexible phenotype" (Piersma and van Gils 2011). In fact, as the climate changes, the number of integrative biologists interested in how the environment influences timing of events of the annual cycle has grown markedly, giving rise to a subdisciplinary area known as "seasonality."

The first person to address experimentally the question of how birds know when to migrate and breed was a colorful character named William J. Rowan (1891–1957). Rowan's book, *The Riddle of Migration* (Rowan 1931), became a classic, and his life is captured in a splendid biography, *Restless Energy* by Marianne Ainley (Ainley 1993), and in several memorials (E. O. H. 1958; Anonymous 2003). Rowan's first study organism in this field was the junco, but the importance of Rowan's work far transcends the junco, for he was the first to demonstrate experimentally the role of day length (photoperiod) in the induction of seasonally appropriate behavior in vertebrate animals.

Rowan conducted his research in Alberta, Canada during the first third of the twentieth century. He held slate-colored juncos (*J. h. hyemalis*) outdoors in the harsh Canadian winter and exposed some to artificially lengthened days and others to natural days. What he showed was that the experimental, light-exposed males grew enlarged gonads (testes) prematurely and began to sing sooner in comparison to the control juncos held on natural days (Rowan 1929; Alberta 1988).

Subsequent to his groundbreaking research on the junco, Rowan switched his focus to American crows (*Corvus brachyrhynchos*), which he also studied experimentally by treating some with long days and others with natural days. He captured his crows in August, held them until November, and released them near Edmonton, Alberta. His description of the crows' diet in captivity is memorable (Rowan 1930):

> Food for both lots was identical and consisted in the main of rotten eggs, dogs and cats from the city pound, an occasional horse, green stuff of various sorts, grain and chicken mash, stale bread and buttermilk, etc. This seems to have constituted a well balanced diet. The birds could not have been in better shape at the time of liberation.

To determine whether only the light-treated group would migrate, Rowan marked the crows' tails with yellow paint and used a new medium, the radio, to ask the citizens of Alberta to help him track his birds. This approach echoes to the present when professional biologists are using the Internet to ask citizen scientists to report their observations at bird feeders (see Project FeederWatch at the Cornell Laboratory of Ornithology [http://birds.cornell.edu/pfw/]). With the help of the people of Alberta, Rowan found that the light-treated crows had traveled farther from the release point when they were sighted than had the control crows, and they had also tended to move north as predicted, convincing Rowan that light and gonadal growth were key to stimulating migration (Rowan 1931).

It was Alden Miller who reviewed Rowan's bird migration studies in the *Condor.* Miller wrote:

> In essence, Rowan's conclusions are as follows: The north and south migrations of small fringillids [now known as emberizids] and possibly of all birds are dependent upon the fall retrogression and spring recrudescence of the gonads. One of the important histological changes in the gonads at times of retrogression and recrudescence is the appearance of interstitial cells which presumably produce a hormone serving to instigate migration. The physiological rhythm of the gonads is timed and regulated by photoperiodism, or the effect of seasonal variation in length of day. Temperature and other weather conditions have either limited effect or have no effect on the gonad-migration rhythm ... (Miller 1930a)

Looking back, it is fascinating to see that Miller also had some reservations. In Miller's words:

> This alternative explanation of the experimental results seems not to be dis-
> proved so far as I can determine by any of the experimental data. . . . The sup-
> position that a hormone from the interstitial tissue of the gonads controls
> migration, probable as it may seem, is admittedly a matter of speculation on
> Rowan's part. (Miller 1930b)

Thus, while Rowan was a pioneer, he was not right about everything. Rowan's contention that migration is stimulated by secretions of the gonads
released during gonadal growth was challenged by later findings that captive castrates of some species exhibit migratory behavior (summarized by
Wingfield 1990) and this topic is still contentious (Tonra et al. 2011). Also
challenged were Rowan's ideas regarding how light induced migration
(see Wolfson below). Further, Rowan's samples were small, and inevita-
bly in this kind of work, many birds scored as migrants were unaccounted
for (compare Ketterson and Nolan 1986; Sniegowski et al. 1988). Nevertheless, Rowan's place in the history of the study of seasonality is secure.
Frustratingly, however, the endocrinological basis of bird migration remains elusive to this day (Ramenofsky 2010).

Albert Wolfson of the University of California, Berkeley, was a student
of Alden Miller's who took up Rowan's questions. Wolfson was also interested in the role of day length in regulating migration and reproduction,
and he chose to study the junco (Wolfson 1941). Wolfson, however, lived
in California, so he had at his disposal an array of junco subspecies to
study, including long-distance, short-distance, and sedentary groups. Wolfson repeated Rowan's light manipulations by advancing day length prematurely and comparing the behavior of light-advanced birds after release to the wild to control birds that experienced the days advancing in
length as they normally would. In migratory subspecies of junco, Wolfson found that light-advanced birds were more likely to disappear after
release than were controls, supporting the role of day length in promoting migration. However, he also studied experimental juncos belonging to
resident subspecies, and they too grew their gonads and sang earlier than
controls, but they were not more likely to disappear after release. This
contrast provided evidence that early gonadal recrudescence alone did
not lead to migration in the junco (Wolfson 1940; Wolfson 1942).

Wolfson also addressed Rowan's ideas about how light affected migration. Rowan attributed his initial findings of premature migrations by
light-treated juncos as the result of greater locomotor activity. Under longer days, Rowan noted that birds slept less and hopped around more. He

proposed that longer days led to gonadal secretions that triggered height-
ened activity that in turn triggered migration. He thus predicted that if he
forced juncos to exercise, they would migrate, even if they had not been
exposed to artificially lengthened days, an idea that came to be known
as the "exercise hypothesis" (Rowan 1929). Rowan reported that his attempts to induce gonadal growth with exercise alone were successful, but
Wolfson found otherwise (Wolfson 1941). A review by Donald Farner and
Richard Mewaldt summarizes much of the literature regarding the exercise hypothesis, and it provides very interesting reading about the science
of the times. They concluded that the exercise hypothesis lacked support
(Farner and Mewaldt 1955).

In the time since Rowan and Wolfson, the field of biological rhythms
has developed enormously as researchers continue to probe the role of
day length and other environmental cues in the timing of events of the
annual cycle. Three twentieth-century scientists who were highly influential were Donald Farner in the United States and Eberhard Gwinner
and Peter Berthold in Germany. Farner and colleagues focused on the
white-crowned sparrow, a close relative of the junco, and Gwinner, Berthold, and colleagues focused on old-world warblers of the family Sylvi-
idae. Echoing Rowan, all three were interested in the timing of reproduction and migration. However, Farner and colleagues focused on the role
of day length in regulating gonadal growth, whereas Gwinner and Berthold focused on how birds knew when to migrate by comparing species
and populations for the seasonal timing and duration of gonadal growth,
migratory fattening, and nocturnal restlessness (Zugunruhe).

A debate emerged about whether annual rhythms of gonadal growth
and migration were "driven" by annual changes in day length or were endogenous (self-sustaining) under constant conditions. If annual rhythms
are driven by changes in day length then they should stop if day length
does not change. If annual rhythms are inherently self-sustaining, then
they should continue under constant conditions. The findings were that
annual cycles were not self-sustaining in the temperate-zone sparrow but
they were in the case of the old-world warblers. That is, captive sparrows
under long days exhibited only one cycle of gonadal growth and regression in captivity. Unless long days were followed by a period of short days,
the sparrows became and remained unresponsive to light and thus nonreproductive. On the other hand, when old-world warblers were held on
days and nights of equal length for several years, they exhibited repeated
cycles of gonadal growth, molt, and nocturnal restlessness (i.e., their an-

nual rhythms persisted). Because the approaches differed, as did the evolutionary history and the ecology of the study species, a temporary resolution was to conclude that species differ (Dawson et al. 2001). Perhaps temperate-zone species were incapable of self-sustaining rhythms because they live in environments where day lengths vary greatly and provide reliable information about when to breed and when to migrate, whereas tropical wintering species winter where winter day lengths do not vary and may thus have evolved self-sustaining internal rhythms. In their case, day length need not drive the annual cycle, though it could entrain it.

The conclusion that temperate-zone birds lacked endogenous annual rhythms was challenged when Holberton and Able asked whether annual rhythms would persist in the temperate-zone junco if juncos were held under *truly* constant conditions (Holberton and Able 1992). Borrowing a method from the study of circadian (daily) rhythms, they reasoned that if annual rhythms of gonadal growth and other events of the annual cycle are truly endogenous, then they should continue to cycle in the absence of any day-night cues. They found that juncos held under constant dim light exhibited cycles of gonadal growth and regression that persisted for more than two years. In many ways, this was an unexpected result. The junco is a classic north-temperate bird that experiences significant changes in day length over the course of the year, so the junco might have been expected to depend on changes in day length to "drive" its annual cycle, as in the white-crowned sparrow. Instead juncos were shown to exhibit endogenous cycling. The current view is that annual cycles in birds probably consist of some components that are endogenous and others that are driven by photoperiod, but the mechanisms underlying endogenous circannual rhythms are still not known (Dawson et al. 2001).

Current researchers now focus on the brain, and significant progress is being made regarding the role of neuropeptides in the brain that act on the hypothalamus to stimulate reproduction. The junco is among the species under study; other models include the European starling and the white-crowned sparrow (Deviche 1995; Osugi et al. 2004; Dawson 2005; Meddle et al. 2006; Stevenson et al. 2013; Ubuka et al. 2013). While knowledge of the regulation of the timing of the junco's reproduction is still incomplete, later chapters in this volume will describe the endocrine axes that regulate reproduction and reproductive behavior in detail in relation to the junco (chapter 7, chapter 11).

F. How Far to Migrate?

Juncos have contributed not only to our knowledge of how birds know whether and when to migrate but also to how they know how far to fly, a topic addressed in detail in chapter 3. Partial migration, in which some individuals migrate and others do not, and differential migration, in which individuals and groups differ in how far they migrate, have been related to a migrant's sex or age. The junco was among the first species in North America to be studied in detail with respect to differential migration. By comparing population structure (sex- and age-ratios) of free-living populations by latitude in the eastern United States, and by examining museum skins amassed by early collectors, Ketterson and Nolan (1976) demonstrated that female juncos were more numerous than males in portions of the winter range that were farther from the breeding grounds, indicating that they migrate longer distances than males (see also Ketterson and Nolan, Jr. 1983). They outlined hypotheses that might account for the evolution of longer migrations by females, for example, sex differences in body size (body size hypothesis), the benefits of early return to the breeding grounds (arrival time hypothesis), and social dominance (dominance hypothesis). Subsequent research elaborated on migration patterns in the junco, and other researchers have shown that differential migration by sex is a common phenomenon (see chapter 3) (Terrill 1987; Rogers et al. 1989; Holberton 1993; Cristol et al. 1999; Maggini and Bairlein 2012). One of the newest contributions to knowledge of the junco's migration addresses the impact of climate change on differential migration by showing that the relative abundance of females at higher latitudes has increased over the past thirty years (chapter 3, this volume) (O'Neal et al., in review)

Consideration of the distance birds migrate raises the question of the regulatory mechanisms related to how long they remain in a state of physiological readiness to migrate or molt. Because caged birds exhibit migratory fattening (hyperphagia) and *Zugunruhe* (hyperactivity) at times when they would be migrating if they were free-living, comparisons of species and populations for the seasonal duration of restlessness provided strong evidence of an endogenous component to the duration of migratory readiness as described above. The evidence for sex differences in the duration of *Zugunruhe* in the junco has, however, been mixed. In one study, males and females did not differ in the duration of restlessness; in another they did (Ketterson and Nolan 1985; Holberton 1993). A recent comparison of population differences in *Zugunruhe* and fattening in sedentary and migratory forms demonstrated greater restlessness and fattening in the migratory form (Atwell et al., in review) (chapters 3 and 10, this volume).

Juncos have also been used to explore the role of prior experience in determining whether to migrate, by asking whether the autumn migratory state would be suppressed if juncos were held over the summer on their wintering range. The answer was yes, but the basis for the suppression was not resolved (Ketterson and Nolan 1986; Ketterson and Nolan 1987; Nolan, Jr. and Ketterson 1990). Others have investigated neuroanatomical and neuroendocrine factors in the junco's use of space. For example, the volume of a region of the brain associated with spatial memory, the hippocampus, is relatively larger in a long-distance migratory form of the dark-eyed junco, the slate-colored junco (*J. h. hyemalis*) than it is in an altitudinal migrant the Carolina junco (*J. h. carolinensis*) (Cristol et al. 2003). Recent blooming interest in animal cognition has caused investigators to rediscover a study of seasonal variation in habitat preference as assessed by time spent by juncos perching near images that were typical of winter and summer habitats (Roberts, Jr. and Weigl 1984). A comparison of the duration of restlessness in birds exposed to photographic images of a site they are bonded to could prove quite interesting.

G. Conclusion

Historically the junco has played an important role in our understanding of geographic variation, how new species are formed, and how animals move through flexible phenotypes across the seasons. Key players in the early accumulation of knowledge were Alden Miller and William Rowan, but the junco continues to provoke curiosity and ongoing research.

Acknowledgments

We thank all the scientists recalled here, living and dead, along with the students and postdocs who have been part of the junco project over the years, as well as the numerous field assistants who provided their efforts and good will. We thank our editor Christopher Chung and three anonymous reviewers whose comments substantially improved this chapter.

References

Ainley, M. G. 1993, *Restless Energy: A Biography of William Rowan, 1891–1957*. Montreal: Véhicule Press.

Alberta, U. o. 1988. *William Rowan Finding Aid*. University of Alberta Archives.

American Ornithologists' Union. 1931. *Checklist of North American Birds*, 4th edition. Lancaster, PA: American Ornithologists' Union.

———. 1957. *Checklist of North American Birds*, 5th edition. Baltimore: American Orinthologists' Union.

———. 1983. *Checklist of North American Birds*, 6th edition. Lawrence, KS: Ameircan Ornithologists' Union.

———. 1998. *Checklist of North American Birds*, 7th edition. Lawrence, KS: Allen Press, Inc.

Anonymous. 1929. "Obituary: Jonathan Dwight Bird Expert, Dies," *New York Times*.

———. 2003. The memorable William Rowan. *History Trails*. University of Alberta, University of Alberta Alumni Association.

Atwell, J. W., R. J. Rice, and E. D. Ketterson. In review. Rapid loss of migratory behavior associated with recent colonization of an urban environment. Manuscript.

Chapman, F. M. 1901. *Bird-Life: A Guide to the Study of Our Common Birds*. New York, D. Appelton and Company.

Clements, J. F., J. Diamond, A. White, and J. W. Fitzpatrick. 2012. *The Clements Checklist of Birds of the World*. Ithaca: Cornell Laboratory of Ornithology.

Coyne, J. A., and H. A. Orr. 2004. *Speciation*. Sunderland, MA: Sinauer Associates.

Cristol, D. A., M. B. Baker, and C. Carbone. 1999. Differential migration revisited: Latitudinal segregation by age and sex classes. In *Current Ornithology*, edited by V. Nolan, Jr., and E. D. Ketterson, 33–88. New York: Plenum Press.

Cristol, D. A., E. B. Reynolds, J. E. LeClerc, A. H. Donner, C. F. Farabaugh, and C. W. S. Ziegenfus. 2003. Migratory dark-eyed juncos (*Junco hyemalis*) have better spatial memory and denser hippocampal neurons than non-migratory conspecifics. *Animal Behaviour* 66:317–28.

Davis, J. 1967. In memoriam: Alden Holmes Miller. *Auk* 84:192–202.

Dawson, A. 2005. The effect of temperature on photoperiodically regulated gonadal maturation, regression and moult in starlings: Potential consequences of climate change. *Functional Ecology* 19:995–1000.

Dawson, A., V. M. King, G. E. Bentley, and G. F. Ball. 2001. Photoperiodic control of seasonality in birds. *Journal of Biological Rhythms* 16:365–80.

Deviche, P. 1995. Androgen regulation of avian premigratory hyperphagia and fattening: From eco-physiology to neuroendocrinolgy. *American Zoologist* 35: 234–45.

Dwight, J. M. D. 1918. The geographical distribution of color and other variable characters in the genus *Junco*: A new aspect of specific and subspecific values. *Bulletin of the American Museum of Natural History* 38:269–309.

E. O. H. 1958. Obiturary: William Rowan. *Ibis* 100:120–21.

Farner, D. S., and L. R. Mewaldt. 1955. Is increased activity or wakefulness an essential element in the mechanism of the photoperiodic responses of avian gonads? *Northwest Science* 29:53–65.

Ferree, E. 2013. Geographic variation in dark-eyed junco morphology and implications for population divergence. *Wilson Journal of Ornithology* 125: 454–70.

Goodson, J. L., L. C. Wilson, and S. E. Schrock. 2012. To flock or fight: Neurochemical signatures of divergent life histories in sparrows. *Proceedings of the National Academy of Sciences of the United States of America* 109:10685–92.

Helms, C., W. Aussiker, E. Bower, and S. D. Fretwell. 1967. A biometric study of major body components of the slate-colored junco, *Junco hyemalis*. *Condor* 69: 560–78.

Holberton, R. L. 1993. An endogenous basis for differential migration in the dark-eyed junco. *Condor* 95:580–87.

Holberton, R. L., and K. P. Able. 1992. Persistence of circannual cycles in a migratory bird held in constant dim light. *Journal of Comparative Physiology* A 171: 477–81.

Johnson, N. K., and C. Cicero. 2004. New mitochondrial DNA data affirm the importance of Pleistocene speciation in North American birds. *Evolution* 58: 1122–30.

Ketterson, E. D., and V. Nolan, Jr. 1976. Geographic variation and its climatic correlates in sex-ratio of eastern-wintering dark-eyed juncos (*Junco hyemalis hyemalis*). *Ecology* 57:679–93.

———. 1983. The evolution of differential bird migration. In *Current Ornithology*, edited by R. F. Johnston, 357–402. New York: Plenum Publishing Corporation.

———. 1985. Intraspecific variation in avian migration: Evolutionary and regulatory aspects. In *Migration: Mechanisms and Adaptive Significance*, edited by M. A. Rankin, 553–79. Port Aransas, TX: Marine Science Institute, the University of Texas at Austin.

———. 1986. A possible role for experience in the regulation of the timing of bird migration. *Proceedings of the International Ornithological Congress* 19: 2169–79.

———. 1987. Spring and summer confinement of dark-eyed juncos at autumn migratory destination suppresses normal autumn behavior. *Animal Behaviour* 35: 1744–53.

Kleindorfer, S., J. A. O'Connor, R. Y. Dudaniec, S. A. Myers, J. Robertson, and F. J. Sulloway. 2014. Species collapse via hybridization in Darwin's tree finches. *American Naturalist* 183:325–41.

La Rouche, G. P. 2003. *Birding in the United States: A Demographic and Economic Analysis.* Washington, DC: USFWS Division of Federal Aid.

Lima, S. L. 1988. Vigilance during the initiation of daily feeding in dark-eyed juncos. *Oikos* 53:12–16.

Maggini, I., and F. Bairlein. 2012. Innate sex differences in the timing of spring migration in a songbird. *PLoS ONE* 7:e31271.

Mayr, E. 1942a. Speciation in the junco. Review of "Speciation in the avian genus *Junco*" by Alden H. Miller. *Ecology* 23:378–79.

———. 1942b. *Systematics and the Origin of Species*. Cambridge: Harvard University Press.

———. 1963. *Animal Species and Evolution*. Cambridge: Harvard University Press.

———. 1973. *Alden Holmes Miller (1906–1965): A biographical memoir*. National Academy of Sciences: Washington, DC.

McCarthy, E. M. 2006. *Handbook of Avian Hybrids of the World*. New York: Oxford University Press.

Meddle, S. L., J. C. Wingfield, R. P. Millar, and P. J. Deviche. 2006. Hypothalamic GnRH-I and its precursor during photorefractoriness onset in free-living male dark-eyed juncos (*Junco hyemalis*) of different year classes. *General and Comparative Endocrinology* 145:148–56.

Miller, A. H. 1930a. Review of "Experiments in bird migration. I. Manipulation of the reproductive cycle: Seasonal histological changes in the gonads" by William Rowan. *Condor* 32:166–68.

———. 1930b. Review of "Experiments in bird migration II. Reversed migration" by William Rowan. *Condor* 32:166–68.

———. 1939. Analysis of some hybrid populations of juncos. *Condor* 41:211–14.

———. 1941. Speciation in the avian genus *Junco*. *University of California Publications in Zoology* 44:173–434.

———. 1955. Concepts and problems of avian systematics in relation to evolutionary processes. In *Recent Studies in Avian Biology*, edited by A. Wolfson, 1–45. Urbana: University of Illinois Press.

Nolan, Jr., V., and E. D. Ketterson. 1983. An analysis of body mass, wing length, and visible fat deposits of dark-eyed juncos wintering at different latitudes. *Wilson Bulletin* 95:603–20.

———. 1990. Effect of long days on molt and migratory state of site-faithful dark-eyed juncos held at their winter sites. *Wilson Bulletin* 102:469–79.

Nolan, Jr., V., E. D. Ketterson, D. A. Cristol, C. M. Rogers, E. D. Clotfelter, R. C. Titus, S. J. Schoech, and E. Snajdr. 2002. Dark-eyed junco (*Junco hyemalis*). In *The Birds of North America*, edited by A Poole and F. Gill, no. 716. Philadelphia: The Birds of North America, Inc.

O'Neal, D. M., and E. D. Ketterson. In review. Climate change and differential migration: Relaxation of sexual segregation in a songbird. Manuscript.

Osugi, T., K. Ukena, G. E. Bentley, S. O'Brien, I. T. Moore, J. C. Wingfield, and K. Tsutsui. 2004. Gonadotropin-inhibitory hormone in Gambel's white-crowned sparrow (*Zonotrichia leucophrys gambelii*): cDNA identification, transcript localization and functional effects in laboratory and field experiments. *Journal of Endocrinology* 182:33–42.

P. M. 1967. Obituary Alden H. Miller. *Ibis* 109:280.

Piersma, T., and J. A. van Gils. 2011. *The Flexible Phenotype: A Body-Centered Integration of Ecology, Physiology, and Behaviour.* New York: Oxford University Press.

Poelstra, J. W., N. Vijay, C. M. Bossu, H. Lantz, B. Ryll, I. Müller, V. Baglione, et al. 2014. The genomic landscape underlying phenotypic integrity in the face of gene flow in crows. *Science* 344:1410–14.

Price, T. 2008. *Speciation in Birds.* Greenwood Village, CO: Roberts and Company.

Ramenofsky, M. 2010. Hormones in migration and reproductive cycles of birds. In *Birds*, vol. 4 of *Hormones and Reproduction of Vertebrates*, edited by D. O. Norris and K. H. Lopez. London: Academic Press.

Roberts, Jr., E. P., and P. D. Weigl. 1984. Habitat preference in the dark-eyed junco (*Junco hyemalis*): The role of photoperiod and dominance. *Animal Behaviour* 32:709–14.

Rogers, C. M., V. Nolan, Jr., and E. D. Ketterson. 1993. Geographic variation in winter fat of dark-eyed juncos: Displacement to a common environment. *Ecology* 74:1183–90.

Rogers, C. M., T. L. Theimer, V. Nolan, Jr., and E. D. Ketterson. 1989. Does dominance determine how far dark-eyed juncos, *Junco hyemalis*, migrate into their winter range? *Animal Behaviour* 37:498–506.

Rowan, W. 1929. Experiments in bird migration. I. Manipulation of the reproductive cycle: Seasonal histological changes in the gonads. *Proceedings of the Boston Society of Natural History* 39:151–208.

———. 1930. Experiments in bird migration II. Reversed migration. *Proceedings of the National Academy of Sciences of the United States of America* 16: 520–25.

———. 1931. *The Riddle of Migration.* Baltimore: Williams & Wilkins Co.

Seehausen, O. 2006. Conservation: Losing biodiversity by reverse speciation. *Current Biology* 16:R334–R337.

Sniegowski, P. D., E. D. Ketterson, and V. Nolan, Jr. 1988. Can experience alter the avian annual cycle? Results of migration experiments with indigo buntings. *Ethology* 79:333–41.

Stevenson, T. J., D. J. Bernard, M. M. McCarthy, and G. F. Ball. 2013. Photoperiod-dependent regulation of gonadotropin-releasing hormone 1 messenger ribonucleic acid levels in the songbird brain. *General and Comparative Endocrinology* 190:81–87.

Sullivan, K. A. 1999. Yellow-eyed Junco (*Junco phaeonotus*). In *The Birds of North America Online*, edited by A. Poole. Ithaca: Cornell Lab of Ornithology.

Sunderland, M. E. 2012. Collections-based research at Berkeley's Museum of Vertebrate Zoology. *Historical Studies in the Natural Sciences* 42:83–113.

Terrill, S. B. 1987. Social dominance and migratory restlessness in the dark-eyed junco (*Junco hyemalis*). *Behavioral Ecology and Sociobiology* 21:1–11.

Tonra, C. M., P. P. Marra, and R. L. Holberton. 2011. Early elevation of testosterone advances migratory preparation in a songbird. *Journal of Experimental Biology* 214:2761–67.

Ubuka, T., G. E. Bentley, and K. Tsutsui. 2013. Neuroendocrine regulation of gonadotropin secretion in seasonally breeding birds. *Frontiers in Neuroscience* 7:38.

Wingfield, J. C., and D. S. Farner. 1993. Endocrinology of reproduction in wild species. In *Avian Biology*, edited by D. S. Farner, J. R. King, and K. C. Parkes, 164–327. London: Academic Press.

Winker, K. 2010. On the origin of species through heteropatric differentiation: A review and a model of speciation in migratory animals. *Ornithological Monographs* 69:1–30.

Winker, K., and S. M. Haig. 2010. Avian subspecies. *Ornithological Monographs* 67.

Wolfson, A. 1940. A preliminary report on some experiments on bird migration. *Condor* 42:93–99.

———. 1941. Light versus activity in the regulation of the sexual cycles of birds: The role of the hypothalamus. *Condor* 43:125–36.

———. 1942. Regulation of spring migration in juncos. *Condor* 44:237–63.

Zink, R. M. 1982. Patterns of genic and morphologic variation among sparrows in the genera *Zonotrichia, Melospiza, Junco,* and *Passerella*. *Auk* 99:632–49.

Zink, R. M., and J. I. Davis. 1999. New perspectives on the nature of species. *Proceedings of the International Ornithological Congress* 22:1505–16.

CHAPTER THREE

Axes of Biogeographic Variation in the Avian Genus *Junco*

Habitat, Morphology, Migration, and Seasonal Timing, with Implications for Diversification under Heteropatry

Jonathan W. Atwell, Dawn M. O'Neal, and Ellen D. Ketterson

The presence or absence of migratory movements in birds belongs as much to the genetic characteristics of a race or species as does size or color patterns.
—Ernst Mayr, 1942, *Systematics and the Origin of Species, from the Viewpoint of a Zoologist*

A. Introduction

The discipline of biogeography employs concepts and methods from ecology, physiology, and evolution to describe and understand the processes underlying the past, present, and future geographic distributions of species. Classic examples of biogeographic rules include Bergmann's rule and Gloger's rule, which predict that organisms tend to be larger at higher latitudes, or paler in color in drier habitats, respectively (Millien et al. 2006). Modern day biogeographers increasingly study behavior and physiology, in addition to morphology, and use their research to predict the impact of environmental change on species distributions.

Comparative biology is a subdiscipline of integrative biology that seeks to characterize and explain phenotypic variation among or within groups of organisms. Comparative biology is a prerequisite for understanding the patterns and processes of evolutionary diversification and adaptation that give rise to biogeographic patterns. As such, it is no surprise that biologists, both historical and modern from across these subdisciplines and others, have been attracted to broadly distributed, highly variable, yet closely related groups of organisms.

One such group is the avian genus *Junco* in general, and the dark-eyed junco (*Junco hyemalis*) in particular. Juncos exhibit considerable biogeographic variation in plumage coloration (see central color plates), which has long attracted biologists with an interest in understanding patterns of speciation and diversification, both in the past (e.g., Dwight, Miller, and Mayr, chapter 2, this volume) and in the present (e.g., Milá, McCormack, and Price, chapters 8 and 9, this volume). Integrative and comparative biologists have also been drawn to the junco to study variation in migration, seasonality, behavior, ecology, and underlying physiology, in order to better understand the ecological and evolutionary processes that underlie diversification (chapters 10–14, this volume).

Searching for the evolutionary mechanisms that underlie the *Junco* radiation is an important research goal, and although significant progress is being made to understand the *patterns* of evolutionary relationships and the phylogenetic history of junco groups using advancing genetic and genomic tools (chapter 8, this volume), several open questions remain regarding the evolutionary *processes* that have resulted in the diversity of physical and behavioral characters evident across highly variable lineages such as junco. Chapters 8 and 9 (this volume) emphasize alternative (though not mutually exclusive) evolutionary scenarios, as they explore possible modes of diversification of feather plumage coloration across the genus. Characterizing and considering the full scope of biogeographic variation among *Junco* subspecies and populations is foundational for ongoing and future attempts to determine evolutionary histories for additional traits across the genus and for evaluating hypotheses related to understanding the organismal and evolutionary processes that generate or maintain phenotypic and genetic diversity. It is also essential for predicting population-level responses to environmental change.

Considering a photo of multiple junco subspecies foraging together in a wintering flock (figure 3.1) conveys at least a thousand words about what makes the *process* of junco diversification so interesting *beyond* simply the extensive and diagnostic plumage color differentiation among closely (genetically) related groups. As the days grow longer, these different subspecies and races, pictured here flocking together in the foothills of Colorado's Rocky Mountains (figure 3.1), will each integrate mostly identical environmental cues, yet respond differentially across space and time as they proceed through the annual cycle—embarking in different migra-

FIGURE 3.1. Migratory junco populations with different breeding ranges often share wintering habitats in common with one another or with sedentary (nonmigratory) populations ("heteropatry," Winker 2010a). Pictured here is a wintering flock of juncos in the foothills of the Rocky Mountains, Colorado, USA, that includes at least four subspecies/races, including pink-sided, slate-colored, Oregon, and gray-headed forms. In spring, despite exposure to identical environmental stimuli, these different cohorts will follow divergent strategies with respect to the seasonal timing and geography of their migratory journeys and breeding activities. Photo: Peter Pereira. See also color plate.

tory directions, migrating variable distances (or not at all), heading back towards ecologically distinct breeding habitats (or remaining nearby to breed)—with each cohort following its own seasonal pacing.

Despite being very closely related in evolutionary terms, having likely diversified within the last 10,000 to 20,000 years (chapter 8, this volume), the dark-eyed junco forms pictured in figure 3.1 have diverged, not only in plumage coloration but also in a whole host of other morphological, physiological, and behavioral characteristics that likely underlie their adaptation to different breeding habitats. Thus, the stunning feather plumage variation in such a mixed flock provides an impetus for fully considering the scope of the several different axes of biogeographic variation evident among junco groups—including differences in habitat, morphology, migratory strategies (or lack thereof), and seasonal timing (i.e., phenology).

One general and unresolved paradox in avian speciation is how diversification is generated or maintained among such closely related forms within migratory lineages such as *Junco*, in particular when they have overlapping geographic distributions during at least some parts of the annual cycle (Winker 2010a). This is because greater migratory propensities—which are known to also correlate positively with dispersal abilities—are generally predicted to increase the potential for interbreed-

ing and gene flow among populations, leading to increased homogenization (Belliure et al. 2000; Winker 2010a). Below, we explore this problem
in the framework of Kevin Winker's (2010a) "heteropatric speciation"
model, as the junco system provides an ideal case study in which to consider the implications of variation in migratory behavior and seasonal
timing for population diversification (or homogenization).

Collectively, the goals of this chapter are (1) to lay before the reader
several categories, or "axes," of biogeographic variation that distinguish junco groups across their continent-wide range of habitats, including differences in morphology, migration, seasonality, and seasonal timing (i.e., phenology); (2) to consider in particular the role of variation
in migration and seasonal timing in facilitating or maintaining diversity
among *Junco* groups in the context of "heteropatry" (Winker 2010a) ; and
(3) to present suggestions for applying modern mensuration tools, analytical approaches, and emerging technologies to permit more extensive and
precise characterization of biogeographic variation among juncos.

With respect to the first goal, other sources have considered and summarized biogeographic variation among *Junco* groups. For example,
The Birds of North America: Life Histories for the 21st Century's "species accounts" summarize natural history information for dark-eyed
junco (*Junco hyemalis*; Nolan et al. 2002) and yellow-eyed junco (*Junco
phaeonotus*; Sullivan 1999) species, respectively. Atwell (2011) has previously provided an overview of intra-specific variation in migratory behavior among *Junco* forms, and Ferree (2013) recently examined biogeographic variation in body size and the amount of white in the tail feathers
in the context of local adaptation and sexual selection. In this volume,
Milá and colleagues (chapter 8) and Price and Hooper (chapter 9) also
consider primarily morphological and ecological diversity, respectively, in
the context of evaluating evolutionary relationships and processes related
to diversification. And, in arguably one of the most comprehensive surveys of intragenus avian biodiversity on record, Alden Miller (1941) evaluated over 11,774 *Junco* specimens from museums and field sites across
North America—describing, measuring, and grouping junco forms and
their respective habitats. However, despite its authoritative breadth and
depth of information about *Junco* biogeography, Miller's (1941) monograph is arguably so dense as to limit its utility for the modern reader.

Thus, our aim here is to provide a complement to these prior efforts
by presenting, in a single location, an easily digestible survey of several
important categories, or "axes," along which *Junco* groups are known to

vary biogeographically, emphasizing differences in habitat, morphology, migration, and seasonal timing. In this regard, we intend for this chapter to serve current and future students of evolutionary and integrative biology interested in the *Junco* system, perhaps stimulating new study questions, and to allow the reader of this volume to more fully appreciate the extent of phenotypic diversification evident across the *Junco*'s continent-wide range.

With respect to the second goal, that of addressing the evolutionary implications of variation in migration and seasonal timing among junco groups, we consider Kevin Winker's (2010a) concept of "heteropatric speciation," in which migration and reproductive timing can serve as drivers of divergence when populations are allopatric on breeding grounds but overlap in relative sympatry during winter. To date, others have focused on how plumage or morphometric diversification relate to evolutionary processes (e.g., Aleixandre et al. 2013; Ferree 2013; Milá et al. 2007; Miller 1941; Milá et al., chapter 8, this volume; Price and Hooper, chapter 9, this volume), but few have focused on the related importance of nonmorphological traits that also vary among populations. Biological timing, animal movements, and questions of when and where to reproduce are rapidly gaining recognition as critical to understanding population divergence.

With respect to the third goal, we aim to convey that—despite extensive historic and recent research on juncos—many basic pieces of information about natural history or phenotype remain entirely unknown, poorly documented, or unsummarized for most subspecies and populations, knowledge gaps that limit advances in organismal and evolutionary biology in many study systems, including juncos. Thus, throughout, we argue that future research should strive to more precisely and expansively characterize natural history and phenotype within and among *Junco* groups, and we provide suggestions for future work in this arena.

B. Axes of Biogeographic Variation among *Junco* Populations

B.1. Variation in Junco *Habitat*

A casual assessment of *Junco* habitat tolerances, for example from a bird field guide, suggests strong general similarity of habitat types that characterize all junco forms: juncos require temperate climate during summer, and are thus found breeding at either high latitudes, high altitudes, or coastal areas, and they utilize a variety of forests, woodlands, or parklands (with trees) during breeding and wintering (Nolan et al. 2002; Sullivan 1999). In describing the breeding habitat of the two most widely distributed forms, slate-colored and Oregon juncos, Miller (1941) indicates that "Bush and tree cover is essential with moist, somewhat open, plant-covered ground in which to forage" (320). These general breeding habitat requirements seem to apply to all *Junco* forms. They are not found breeding in entirely unshaded habitats, or when forests get too thick, or when ground cover gets too dry or very wet.

However, a closer look at the scope of variation in habitat types inhabited by different juncos during breeding or wintering sites across their range reveals a wide range of tolerances—including substantial environmental and climatic variation that may have led to ecological specialization among junco subspecies, races, and populations. Junco populations inhabit more than sixty degrees of latitude, from volcano juncos in Costa Rica to slate-colored juncos on the north slope of Alaska's Brooks Range, breeding or wintering across a vast array of local climatic conditions and habitat types (see color plate 2). The few exceptions where juncos are rarely found are those habitats that are either too dense, too warm, or too moist or dry, for example lowland tropical sites, marshes, or swamplands, or treeless deserts, steppes, and shrublands—habitat types where juncos neither breed nor winter (Miller 1941; Nolan et al. 2002; Sullivan 1999).

In the north, slate-colored juncos breed in coniferous (boreal) forests, but farther south they breed and winter in deciduous forests, oak savannahs, and parklands. In coastal areas, Oregon juncos inhabit both luxuriant rainforests (e.g., Pacific Northwest) and Mediterranean climates (e.g., coastal southern California), in contrast to the relatively arid and highly seasonal "sky island" mountain forests and meadows in the interior of the Oregon junco range. Throughout Mexico and parts of Central America, yellow-eyed juncos inhabit a variety of local mountain forest types where they breed and winter, including both coniferous and mixed pine-oak woodlands. And, farthest south, volcano juncos are endemic to the highest mountains of Costa Rica and Panama. Interesting endemic forms including Baja juncos and Guadalupe juncos inhabit highly localized habitats dominated by marine weather and unique vegetative assemblages.

With respect to elevation, juncos breed from sea level (e.g., coastal Oregon junco types) to over 3,500-meter elevation (e.g., pink-sided juncos in the Rocky Mountains and yellow-eyed juncos in southern Mexico and

Guatemala), providing the scope for underlying differences in metabolism or climatic tolerances associated with elevational clines. Although many montane-breeding *Junco* subspecies or races migrate along elevational gradients, other *Junco* taxa are sedentary and live year-round either at high elevation sites (i.e., on "sky islands," as with volcano juncos in Costa Rica or yellow-eyed juncos in the Guatemalan highlands) or at sea level coastal habitats (e.g., Oregon juncos that breed along the Pacific coast).

Such differences in climatic seasonality, elevation, and aridity are likely profound regulators of associated plant and animal communities, which form the basis for available foraging (seeds and insects) and nesting substrates for juncos, as well as the communities of predators or pathogens that impose selective pressures in any given habitat. Although we are unaware of any rigorous quantitative assessments of junco habitat types in the context of taxonomic specialization or local adaptation, Miller (1941) made extensive notes and conjectures about local adaptation to breeding and wintering habitats in his serial descriptions and delineations of the various junco groups. He even suggested for longer distance migrant forms that habitat preferences or tolerances hold true for both breeding and wintering ranges. For example, Miller (1941) wrote, in reference to certain groups of Oregon juncos that breed through the Pacific Northwest:

> The remarkable adherence of *oreganus* to humid forested areas in winter is highly suggestive of its preference and possible adaptation for comparable environments in summer. To a less degree this same tendency is seen in *shufeldti*. There is enough movement inland to indicate that it is not impossible, so far as migration routes and barriers for both these races are concerned, for them to spread to inland desert areas in winter. I believe it is preference for, or adjustment to, specific conditions in southerly latitudes which are most similar to those of their breeding habitat that largely determines the wintering range. (Miller 1941, 276)

Any such ecological habitat specialization would likely be linked to underlying phenotypic adaptations, for example physiological differences underlying thermal or aridity tolerances. Recent work in Andean sparrows (*Zonotrichia capensis*) completed by Zac Cheviron and colleagues, which examined the physiological and genetic basis of local adaptation to altitudinal clines, provides a model for this type of research (Cheviron et al. 2014). And more recent studies by the Cheviron group have begun to address genomic and physiological mechanisms underlying metabolic flexibility in response to temperature changes in slate-colored juncos (Cheviron et al. 2015), providing an exciting scope for future research on local (physiological) adaptations. Similarly, morphological or behavioral divergence linked to foraging, nesting, or predator avoidance strategies may be optimized among different habitats, although evidence for these types of local habitat adaptations remain generally unexplored in juncos.

Miller (1941) made mostly qualitative attempts to evaluate three classic biogeographic hypotheses relating junco morphology to habitat and climate: (1) Gloger's rule, which predicts that animals in warmer and more humid habitats are more darkly colored, in order to promote cryptic matching to the background habitat, hence reducing predation pressure (Gloger 1833); (2) Allen's rule, which predicts that appendages such as bills or limbs will be smaller or shorter in colder climates, reducing surface area to volume ratio and minimizing heat loss (Allen 1877); and (3) Bergmann's rule, which predicts generally larger body size (and lower surface area to volume ratio) under colder climates to minimize heat loss (Bergmann 1848). While Miller (1941) found limited support for these predictions across environmental gradients within certain junco groups, the pattern was lacking in others, and his results and interpretations were equivocal, with evidence "unsatisfactory or wanting" across the genus as whole (365). More recently a few evaluations of ecological specialization in response to local habitats have emerged from contemporary studies of the junco, and we discuss these in the sections on coloration and morphometrics, below.

B.2. Variation in Junco Morphology

Feather and eye coloration. Miller's (1941) detailed assessments of junco plumage coloration included two key components—qualitatively matching different feather patches with a known series of colored samples and a more detailed microscopic analysis of relative proportions of intracellular melanin pigments. While these two approaches were largely qualitative, Miller (1941) also scored the relative proportions of lighter (or white) versus darker (e.g., black, brown, gray) patches in both head (hood) and tail feathers, which he recorded as either continuous (tail) or ordinal (head) data. Nearly identical approaches to scoring "tail white" and "head black" are employed by more recent behavioral ecologists and evolutionary biologists interested in their roles as social signals (e.g., Atwell et al.

2014; Balph et al. 1979; Holberton et al. 1989; Ketterson 1979; McGlothlin et al. 2008).

We do not have access to Miller's collection of color standards, and they were not specifically referenced in his monograph (1941), but it is highly likely from his verbal descriptions of the junco plumage colors that they are based on Ridgway's (1912) reference guide intended to standardize colors and color names. Using a range of descriptive and finely graded color names, Miller worked carefully to compare and contrast both gross and fine differences in feather color in several anatomical regions, including differences in juncos' backs, heads, hoods, breasts, flanks, wings, wing coverts, rumps, underparts, lores, ocular regions, pileums, napes, throats, auricular regions, sides, and tails.

For example, as a particularly variable and diagnostic character, Miller included twenty-two different color descriptors of junco back color alone including Prout's brown, Brussels brown, raw umber, Sanford brown, Buffy citrine, dark mouse gray, dark olive gray, and slate, among others. Similarly, precise yet diverse color sets were applied to contrast junco flanks, sides, rumps, hoods, pileums, and napes (Miller 1941; Ridgway 1912), as well as the colors of unfeathered parts with variation in feet, legs, bill colors, and eyes (Miller 1941).

Contemporary researchers now apply photospectroscopic tools to quantify color variation in general (Hill and McGraw 2006), and this approach will soon be applied across *Junco* forms in particular (e.g., B. Milá, pers. comm.). These methods provide stark contrast to describing feather colors by matching feathers to chips as seen by the human eye. We now know that bird plumage reflects in the UV at wavelengths that birds can see, and we are far more knowledgeable in general about how to quantify color in relation to the avian visual system.

As current methods of color mensuration and statistical association analyses are applied across *Junco* taxa, it will be interesting to see the degree to which Miller's observations and ordinal classifications are borne forward by more precise and continuous quantitative assessments, including those that incorporate ultraviolet spectra that are visible only to the birds (Wilkie et al. 1998). Finally, one of the more exciting developments is the ability to co-consider the color of a bird's plumage or ornaments as seen by a rival or potential mate in contrast to the visual background in which the bird lives (Endler et al. 2005). Looking to the future, we have almost no idea about how detectable different plumage variants in the junco are against the light environments provided by the various habitat types listed in the section above. At present, we cannot say, for example, whether the paler plumage color set evidenced in the Baja junco's arid habitat is an example of crypsis or of Gloger's rule (which predicts paler forms in drier climates) (Huggett 2004), but modern mensuration and experimental approaches provide a framework for testing such predictions.

Feather color development is another area of research with huge potential. As a complement to his named colors, Miller (1941) also considered how cellular changes in melanin composition regulated the observed color differences, noting the proportions and hues of phaeomelanins versus eumelanins using ordinal categorizations. With respect to the exceptionally variable back colors in *Junco*, Miller characterized ten varieties or degrees of phaeomelanin saturation and eight categories of eumelanin saturation. For head color, Miller (1941) estimated head darkness across a ten-point scale. Today developmental biologists study melanocyte development in dark and white feathers of birds using an array of new techniques including immunohistocytochemistry (ICC) and genomics (Lin et al. 2013). Mikus Abolins-Abols and John Foley and colleagues are currently applying ICC to study melanocyte development in white and dark tail feathers in the junco (in progress); at the level of gene expression, Milá et al. are employing transcriptomics to compare pigment formation in growing feathers across junco subspecies (research currently in progress).

A question of long-standing interest is the relative contributions of genetic divergence and environmental variation to color patterns. Studies of captive juncos and hybrid offspring of mixed-race pairings provide strong evidence that most subspecies differences in feather plumage coloration among *Junco* forms are genetically based (McCarthy 2006; Stone 1893), although there is experimental evidence that environment (e.g., diet) also plays a role in shaping subtle differences in plumage elaboration (e.g., tail white) (McGlothlin et al. 2007). Importantly, although feather color variation provides the primary basis for the twenty-one distinct junco forms named by Miller (1941), intergradation between groups where the ranges overlap, as well as what he describes as (sometimes extensive) "nongeographical variation" (i.e., variation within groups, as with sex or age dimorphism or other polymorphisms) make it obvious that many of the among-group morphological distinctions are arbitrary. Although blending and intermediate forms with respect to color, shape, and size are the typical observation where different subspecies ranges meet (e.g., where Oregon and slate-colored juncos' ranges meet in British Columbia, Canada), in other cases, morphological delineations are relatively sharp (e.g., where gray-headed and red-backed juncos' ranges meet in the mountains of northern Arizona) (Miller 1941).

Finally, eye color is a trait that has long been used to distinguish the dark-eyed and yellow-eyed junco species (Nolan et al. 2002). However, we now know that eye color can be misleading because genetic analysis has revealed a dark-eyed form within the yellow-eyed clade (see chapter 8, this volume). Eye color also changes with age in the junco. Nestlings and juveniles have gray irises and the iris darkens during the autumn in the slate-colored junco (Nolan et al. 2002), whereas nestling and juvenile yellow-eyed juncos' irises transition from a dark umber/clay color to bright orange/yellow as they mature (Sullivan 1999). Learning how pigments are deposited in the iris as a bird matures or responds selectively to different solar regimes is a strong candidate for future research.

Morphometrics. Miller's monograph details a wide range of typical ornithological morphometric characters that vary among different junco populations, including linear measures of wing length (wing chord), tail length, bill length, bill depth, tarsal length, and body weight. Many of these traits are still of interest, and modern studies continue to report the same characters, along with bill width, and have shown them to differ among junco groups (Aleixandre et al. 2013; Bears et al. 2008; Ferree 2013). The range of variation in morphometric measures of body size is substantial, with linear morphometrics or body mass commonly ranging more than 20 percent among junco forms. As examples, junco males' average wing lengths range from 69 to 87 millimeters, tail lengths from 62 to 79 millimeters (Baja junco and white-winged junco, respectively), and tarsus lengths from 19.6 to 23.7 millimeters (Oregon junco to Guatemala junco, respectively) (Miller 1941). Common garden studies have indicated a genetic basis to variation in wing and tail lengths, although early developmental effects on such traits cannot be ruled out in these types of studies (Bears et al. 2008; Rasner et al. 2004).

Further, although speculative, observations from the field suggest potential ecological specialization in response to local habitats such as the Guadalupe Island junco's comparatively pointed and elongated bill, which may represent an adaptation to access seeds of the island's cypress tree forests (Aleixandre et al. 2013) (see color plate 5). Additionally, shorter wings have evolved in a recently established and now entirely sedentary population of Oregon juncos (*J. h. thurberi*), which could be linked to the cessation of migration (Rasner et al. 2004). Another study of geographic variation in wing length in the slate-colored junco asked whether there was support for Bergmann's rule among migratory populations during winter and found there was not (Nolan and Ketterson 1983). After correcting for variation in wing length attributable to sex and age, the study found that migrants that settled in the northern portion of the winter range did not differ in wing length from those that made longer migrations to the south. They did, however, vary geographically in seasonal fat deposition (see below).

While methods for morphometric assessments of shape and size of birds in the field or in museums have not altered much with time, change is on the way. For example, x-ray or ultrasound machines are now used to measure skeletal or tissue morphology, accompanied by multivariate analytical approaches to analyze correlated morphometric measures, and are expanding the study of anatomical evolution (Achterkirchen 2010). To our knowledge none of these more advanced morphometric techniques has been applied to the *Junco* system to date.

B.3. Variation in Junco *Migratory Behavior*

Seasonal migrations are most commonly understood as movement from breeding range to wintering range and back again each year (Dingle 1996). Species and populations differ in their migratory propensity, and migration involves changes in a suite of organismal systems as diverse as sociality, foraging strategies, lipid metabolism, muscle performance, and circadian activities (Piersma and van Gils 2010). Yet migration has been shown to evolve and be lost quite rapidly (Berthold et al. 2001; Zink 2011). Ornithologists, behavioral ecologists, and evolutionary biologists have long been fascinated by how migration has evolved in response to changing ecological conditions across geologic and contemporary time scales.

The extensive plumage color variation that evolved among *Junco* taxa in perhaps as few as the past 10,000 years (Milá et al. 2007; chapter 8, this volume) has provided a unique opportunity to study migratory variation. The color differences among closely related groups has allowed for general characterizations of migratory variation to be made from simple field observations of divergent morphs on wintering grounds (e.g., figure 3.1) and also from georeferenced museum specimens. As one example, although genetic data indicate virtually no detectable genetic structure in mtDNA between pink-sided juncos of the northern Rockies and the red-backed juncos of the southern Colorado Plateau (Mila et al. 2007), variation in feather color recorded in field and museum observations reveals that substantial proportions of pink-sided juncos make seasonal migrations spanning hundreds of miles, while red-backed juncos appear to be comparatively sedentary or strictly altitudinal migrants (Atwell et al. 2011; Miller 1941). Within more uniformly colored taxa (e.g., the chipping sparrow; Milá et al. 2006; Zink 2011), such dramatic variation in migratory behavior would be easily overlooked or much more difficult to summarize.

Migratory propensity. Although many avian lineages exhibit substantial intraspecific variation in migratory propensity (Alerstam et al. 2003; Winker 2010a; Zink 2011), the *Junco* system is unique in the degree to which a full spectrum of variation in migratory behavior has been documented among several apparently closely related *and* recently diverged groups of songbirds (Atwell et al. 2011). Juncos range from long-distance obligate migrants that apparently travel at minimum 1,200 kilometers between breeding and wintering habitats (e.g., consider *J. h. hyemalis* populations breeding in northern Canada or *J. h. hyemalis* wintering in Georgia) to entirely sedentary populations that remain on or near their territories year-round, including several forms from the Oregon group of dark-eyed juncos, most yellow-eyed forms that breed across Mexican and Central American highland habitats, as well as the isolated Baja and Guadalupe juncos. In between these extremes lies a full-spectrum of migratory propensities and strategies, including (not mutually exclusive) partial, altitudinal, facultative, and differential migrants within and among various populations and subspecies (Atwell et al. 2011; Peterson et al. 2013) (see table 3.1).

Rapid change in migratory propensity. Despite the array of physiological changes associated with shifts in seasonality and migration, recent studies clearly demonstrate that migratory and seasonal strategies *can* and *do* shift very rapidly in response to changing environmental conditions (Partecke and Gwinner 2007; Pulido and Berthold 2010). Some of the shifts are likely due to plasticity, but genetic change in migration can also be very rapid, as shown by the response of the blackcap (*Sylvia atricapilla*) to only a few generations of artificial selection (Berthold et al. 2001).

In the junco, two closely related "pairs of populations" demonstrate that variation in migratory behavior can emerge quickly in this lineage. In eastern North America, the slate-colored junco includes two "races": the northern junco (*J. h. hyemalis*), which breeds across the northern boreal

TABLE 3.1 **Variation in migratory strategies and average distances travelled among *Junco* populations, as inferred for populations sampled on breeding and wintering grounds and included in a study examining candidate genes associated with regulating migratory behavior**

Group	Subspecies	Sample Site (Latitude °N)	Range	Migratory behaviors	Migratory score
Slate-colored	*J. h. hyemalis*	Mississippi, USA	Wintering	LD2	6
Slate-colored	*J. h. hyemalis*	Indiana, USA	Wintering	LD1	5
Slate-colored	*J. h. hyemalis*	Michigan, USA	Wintering	R to LD	4
Carolina	*J. h. carolinensis*	Virginia, USA	Breeding	A, P, F	2
White-winged	*J. h. aikeni*	South Dakota, USA	Breeding	R to LD1; A, P, F	3
Oregon	*J. h. oreganus*	British Columbia, Canada	Breeding	R to LD1; P, F	3
Oregon	*J. h. thurberi*	Mt. Laguna, California, USA	Breeding	A, P, F	2
Oregon	*J. h. thurberi*	San Diego, California, USA	Year-round	S	1
Gray-headed	*J. h. caniceps*	Utah, USA	Year-round	R	3-5
Pink-sided	*J. h. mearnsi*	Wyoming, USA	Breeding	R to LD1	4-5
Guadalupe	*J. h. insularis*	Guadalupe Island, Mexico	Year-round	S	1
Yellow-eyed	*J. p. phaeonotus*	Durango, Mexico	Year-round	S	1
Yellow-eyed	*J. p. phaeonotus*	Mexico City, Mexico	Year-round	S	1
Guatemala	*J. p. alticola*	Huehuetenango, Guatemala	Year-round	S	1
Baja	*J. p. bairdi*	Baja California Sur, Mexico	Year-round	S	1

Notes: Group / subspecies designations follow Miller (1941), as summarized by Nolan et al. (2002) and Sullivan et al. (1999), with dark-eyed (*Junco hyemalis*) subspecies noted as "*J. h.*" species and yellow-eyed (*Junco phaeonotus*) subspecies as "*J. p.*" Migratory behavior codes indicate the range of migratory behaviors inferred from breeding and wintering distributions, primarily drawing from Miller (1941), and as follows: LD2 = long distance II, at minimum 1,600 km up to 5,600 km, depending on migratory connectivity; LD1 = long distance I, likely 400–700 km, possibly up to 5,000 km, depending on migratory connectivity; R = regional, typically greater than 200 km; A = altitudinal, typically less than 200 km; P = apparent partial migration documented (i.e., some individuals migrant while others do not); F = apparent facultative migration documented; S = sedentary (i.e., nonmigratory). Migratory scores are ordinal ranks of population migratory propensity, developed using methods similar to those found elsewhere (Mueller et al. 2011). Table adapted from Peterson et al. 2013.

forest and northern New England; and the Carolina junco (*J. h. carolinensis*), which breeds on higher elevation Appalachian mountaintops, intergrading northwards into the northern junco (Miller 1941). At Mountain Lake Biological Station, Virginia, in the southern Appalachians, Carolina juncos (*J. h. carolinensis*) breed in spring and summer and are apparently facultative altitudinal migrants during the winter months (Nolan, Jr. et al. 2002). At the same location, northern juncos (*J. h. hyemalis*) arrive in the autumn to spend the winter, having journeyed at minimum 500 kilometers, and possibly as far as 5,000 kilometers, from breeding ranges farther north. Evaluations of spatial memory in both species suggest migratory northern juncos (*J. h. hyemalis*) have better spatial memory when compared to sedentary Carolina juncos (*J. h. carolinensis*) and also display more densely packed hippocampal neurons (Cristol et al. 2003). Comparison of northern and Carolina slate-colored forms reveals how marked differences in behavior and physiology can emerge between very closely related, and "seasonally sympatric," groups. Recent studies of northern juncos' wintering distributions also reveal how climate change may be altering their migratory strategies (O'Neal and Ketterson, in review; see next section).

Similarly, in San Diego County, California, a recently established (ca. 1983) urban/coastal colonist population of *J. h. thurberi* has become sedentary year-round, while most members of the same subspecies are altitudinal migrants that flock in winter, facultatively travel to lower elevations, and abandon territorial defense in the winter (Atwell et al., in review; Unitt 2005; Yeh 2004) (see also chapter 10, this volume). Evidence from a common garden study of *Zugunruhe* (migratory restlessness behavior) suggests a genetic basis for this contemporary loss of migratory propensity, as the intensity of *Zugunruhe* was lower in birds originating from the sedentary colonist population than those from the higher elevation population (Atwell et al., in review). Future studies of free-living birds during the season of migration will be highly revealing with respect to how the hyperactivity of caged migrants maps onto their actual movement behavior in the wild.

Polymorphisms in migratory behavior. The term *partial migration* is used to refer to situations in which some populations within a given taxon (e.g., a species or subspecies) migrate and others do not (Chan 2001). More typically, however, "partial migration" refers to intrapopulation variation wherein some individuals within a population migrate and others do not (Pulido et al. 1996). A related phenomenon, differential migration, refers to within-population variation in distance migrated, often by individuals that differ in sex or age (Ketterson and Nolan, Jr. 1983). Regardless of usage, patterns of partial or differential migration may be difficult to detect for most groups of small birds with large and sometimes overlapping geographic ranges because populations that breed and winter in different locations exhibit relative uniformity of feather plumage coloration and morphometric characters. Until recently with the development of new tools for establishing connectivity among breeding and wintering populations (see below), knowing which birds originated where has presented a significant challenge and much remains to be learned.

In the 1970s and early 1980s, northern (slate-colored) juncos (*J. h. hyemalis*) were the subject of a series of studies documenting differential migration in the junco. By compiling information from museum skins and capturing wintering birds in the wild, these studies exhibited a gradient toward more female-biased sex ratios with decreasing latitude, indicating that females made longer migrations than males southwards from the boreal zone into the eastern US (Ketterson and Nolan, Jr. 1976). Further study showed that winter populations also differed in age structure, suggesting longer migrations by older than younger birds (Ketterson and Nolan, Jr. 1983). One study of juncos held in a common captive environment during autumn did not reveal a sex difference in the duration or intensity of nocturnal restlessness by sex despite differences in distance migrated (Ketterson and Nolan, Jr. 1985). These findings suggest that migratory distance may have an endogenous component but also be influenced by conditions, social or abiotic, encountered during migration.

One important outcome of these differential migration studies is the baseline they provide for further assessing the impact of environmental change on distances migrated. O'Neal and Ketterson (in review) compared the relative abundances of males and females (i.e., sex ratios) in present-day wintering flocks across a latitudinal gradient and compared these data to the historical male and female abundances at the same locations. They observed a more equal distribution of the sexes across the wintering range than previously, and these shifts correlated closely with changes in winter climate. They also found greater abundance in the populations wintering farther to the north—suggesting that climatic warming may be altering the timing and geography of junco migrations (O'Neal and Ketterson, in review).

However, even within this well-studied example of differential migration, it remains entirely unknown what specific biogeographic behavioral patterns generate the observed sex-ratio variation across the slate-colored junco's winter range. For example, are the females that winter farther north derived from more northern breeding populations or are they females from the breeding ranges that are closest to the winter range ("stratified" versus "leap-frog") (Atwell et al. 2011; Ketterson and Nolan 1983)? As with most migratory animals, particularly smaller-bodied taxa, specific information about geographic connectivity (or lack thereof) between breeding or wintering grounds remains virtually unknown within the *Junco* system, limiting the precision with which research may proceed (Bridge et al. 2011). Also lacking is quantitative information about the relative frequency of migration within partially migratory populations or among facultative versus obligate individuals.

Geographic connectivity among breeding and wintering populations. Current assessments of junco migratory variation, for example those generalized in Miller (1941) and presented here in table 3.1, are based chiefly on records of breeding versus wintering ranges of the most distinct junco morphotypes, taken from both field observations and georeferenced museum specimens, as a well as a handful of band recoveries. While this manner of characterizing migratory phenotype reveals the substantial variation within the genus, such observations lack quantitative specificity with respect to geographic connectivity, mean migratory distances travelled, frequency of obligate versus facultative migrants, or the extent of partial versus fully migratory populations.

The *oreganus* race of Oregon juncos (*J. h. oreganus*), which breeds across a fairly discrete coastal island and shoreline habitat of humid coastal forests ranging from Yakatut Bay south through the Queen Charlotte Islands (British Columbia, Canada), provides an example of this problem. Wintering *J. h. oreganus* are recorded as far south as New Mexico and Arizona (USA) but more commonly across western Washington, Oregon, and California, south through the Monterey Peninsula (Miller 1941). Based on the above records, one might conclude their migrations range from 500 to 2,500 kilometers; however, there are also winter records within the northern part of the breeding grounds, for example at Wrangell and Sitka, Alaska, suggesting that some proportion of *J. h. oreganus* individuals are nonmigratory (Miller 1941). Further, extensive study of slate-colored junco migration in the eastern US has demonstrated differential migration by sex and age cohorts (Ketterson and Nolan 1983) (see above), as well as the likelihood that juncos' migratory distances and distributions are changing in response to global climate change (Atwell et al. 2011; O'Neal and Ketterson, in review) (see above).

Such complexity and imprecision with respect to characterizing migratory variation is characteristic of nearly all junco groups except those completely sedentary or obligate, long-distance migratory extremes. Thus, although the full spectrum of migratory variation, coupled with recent diversification and plumage color differences, make the *Junco* system promising ground for future work, there is a huge gap in our quantitative understanding of migratory behavior among and within junco groups. Several emerging tools provide a scope to improve our understanding of junco migrations, most notably light-level geologgers that can provide information about geographic connectivity and phenology (Bridge et al. 2013; Bridge et al. 2011), as well as advancement in the use of isotopic and genomic data to infer geographic connectivity (Ruegg et al. 2014; Rundel et al. 2013).

B.4. Variation in Junco *Seasonal Timing*

Seasonally breeding animals are putatively adapted to time their reproduction to coincide with favorable ecological conditions such as temperature and food availability (Berthold et al. 2001). Similarly, classic life-history trade-offs such as investment in fecundity versus survival (i.e., current versus future reproduction) or mating versus parenting are predicted to vary with ecological conditions, for example across latitudinal or altitudinal gradients (Phillips et al. 2010; Réale et al. 2010). In addition to migratory propensities, junco taxa studied to date are known to vary substantially in the seasonal timing of their annual cycles, though detailed information on breeding and migratory phenology remains unknown because most groups have not received significant or standardized monitoring efforts.

Although migration can appropriately be understood as a component of the annual seasonal (reproductive) cycle, and the phenology of migration and breeding are inherently linked (Berthold et al. 2001), we treat migratory variation separately in the above section, in part arbitrarily for organizational purposes, in part because it represents such a dramatic behavioral phenomenon with related changes in physiology and morphology, and in part due to the unique geographic circumstances generated by migrations, which bear special consideration in the context of speciation and diversification. In this section, we focus specifically on differences in seasonal timing (phenology) of reproduction among junco groups, as well as potential variation in life-history parameters.

Duration of breeding season. With respect to the onset, duration, and termination of the breeding season, there is evidence for extensive biogeographic variation, even among very closely related *Junco* subspecies and populations. Five populations that have been the subject of systematic ecological and behavioral research provide informative examples to consider the range of seasonal variation present in the genus.

In southern California, two adjacent Oregon junco populations of the *thurberi* form exhibit strikingly divergent breeding phenologies, likely representing the extremes of the *Junco* phenological spectrum, despite their close geographic proximity and the fact that fewer than three decades have passed since the establishment of the urban population as separate (Atwell et al. 2014; Yeh and Price 2004) (see color plate 7 and chapter 10, this volume). In the milder coastal and urban environment, which also includes anthropogenic food and water sources, juncos begin breeding in February and continue into August (up to six months), with some pairs hatching up to four broods, and many pairs fledging at least two or more broods (Yeh and Price 2004). In contrast, in the nearby ancestral-range mountains to the east, more than a single breeding attempt is observed only rarely, and the climate and food availability are highly seasonal (i.e., cold winters and hot, dry summers), limiting the breeding season to late April through late June (two to three months) (Atwell et al. 2014). Interestingly, a common garden study to measure behavior and physiology under identical ambient aviary conditions suggests these differences in breeding phenology between the populations represent phenotypic plasticity, rather than genetic differences (Atwell et al. 2014).

At Mountain Lake Biological Station, Virginia, the length of the breeding season has been documented across multiple decades of long-term study by E. Ketterson and V. Nolan, Jr. and is intermediate to the situations in California. First eggs are observed during middle to late April, with breeding continuing through July and sporadically into August (three to four months), and second broods are relatively commonplace in pairs that evade nest predation (Nolan, Jr. et al. 2002). Kim Rosvall and colleagues report that this population is breeding earlier in recent years, in correlation with climatic warming trends (unpublished data, manuscript in prep.). In yellow-eyed junco populations in Arizona studied across multiple years by K. Sullivan (1999), the average breeding season begins at a similar time as in Virginia (mid-late April) but terminates a couple of weeks earlier (mid-late July), likely due to the warmer and more arid summer conditions. Yellow-eyed juncos in Arizona appear to most commonly be single brooded, though more experienced females can achieve more than one brood (Sullivan 1999). Gray-headed juncos under study by T. Martin over a twenty-year span were typically double-brooded and initiated clutches during the first two weeks of May. Importantly, in all five of these sample study populations, substantial annual (and presumably plastic) variation has been observed in the onset and termination of breeding in response to local climatic or ecological conditions (Nolan, Jr. et al. 2002; Sullivan 1999; Yeh and Price 2004).

Single-brooded *Junco* populations (i.e., short breeding seasons) may be common at higher elevations across the southern parts of the *Junco* range. For example, in higher elevation populations of red-backed juncos, yellow-eyed juncos, and Oregon juncos (e.g., *J. h. thurberi*), temperature extremes, aridity, or snowfall may more restrictively limit onset and continuation of the breeding season than in the double-brooded populations. Alternatively, in milder habitats, such as those in coastal areas along the Pacific coast inhabited by Oregon junco groups (e.g., *J. h. pinosus, J. o. oreganus*), longer breeding seasons with multiple broods may likely be the norm. Such generalizations are supported by the handful of observations of first eggs across the range (Nolan, Jr. et al. 2002), though it remains unclear what, if any, proportion of variation in breeding phenology among junco groups might have a genetic basis.

While the above observations from these intensively studied junco populations provide snapshots of phenology over several years, very few study populations have been observed longitudinally across years in order to characterize average breeding phenologies, and the vast majority of local populations and even entire subspecies remain uncharacterized with respect to their annual cycles.

For example, the Guadalupe junco and the Baja junco inhabit isolated island and/or high elevation habitats, respectively, and little is known about the breeding phenology of either group. In both of these habitats, local climatic conditions are apparently impacted heavily by ocean currents that drive ecological seasonality more strongly than simple latitudinal or regional ecotypic patterns. An observation of Guadalupe junco eggs on January 20 (Howell 1968) is complemented by observation of young juveniles and molting adults during a June field expedition (JWA, pers. obs.), suggesting a breeding season that begins unusually early (Jan) and is relatively long (five or more months), despite the seasonally intense aridity that characterizes much of Guadalupe Island's ecology. In contrast, an April (2011) field expedition to the Sierra de la Laguna, home of the Baja junco, observed scant territorial singing and no evidence for current or recent breeding (e.g., no cloacal protuberances, brood patches, nests, or juveniles) despite apparently favorable spring conditions (e.g., mild weather, abundant insects, blooming vegetation) (JWA, pers. obs.) which is most curious. Similarly, next to nothing is documented regarding the phenology of the volcano junco, *J. vulcani*.

Timing and duration of molt. Nutritionally costly and with potentially significant effects on flight, molt—the annual replacement of flight and body feathers—prepares individuals for migration and sometimes refreshes plumage used in breeding displays (Murphy 1996). To date, the general assumption is that definitive prebasic molt begins with the terminus of the breeding season and continues for several weeks or months and is completed before any subsequent autumnal migration. Where breeding season lengths (and hence termini) have been characterized (see above), the onset of molt has been observed to be accordingly variable. The only detailed summaries on timing and duration of molt available for juncos are from dark-eyed juncos breeding at Mountain Lake, Virginia (Nolan, Jr. et al. 2002) and yellow-eyed juncos breeding in Arizona (Sullivan 1999); however, these summaries do little to shed light on potential variation among populations or subspecies. Though undocumented as to whether there is significant geographical or phylogenetic variation in the timing or duration of molt among juncos, such information could be crucial in investigations of phenotypic responses to climate change or consideration of heteropatric models of speciation (see below).

Seasonal fat deposition. Fat deposits are also known to vary seasonally in association with preparing for energetically expensive migratory journeys and wintering in cold environments (Berthold 1995). Following a simple prediction that birds wintering farther north would have more seasonal fat, Rogers and colleagues (1993) measured northern (slate-colored) juncos across six sites on their wintering range from Michigan southwards to Alabama and Mississippi. They found that juncos at higher latitudes were indeed significantly fatter, but these differences did not persist after displacement to the field or in a common garden, indicating an environmental (plastic) basis for geographic variation in seasonal fattening within *J. h. hyemalis* (Rogers et al. 1993; see also Nolan, Jr. and Ketterson 1983)

In contrast to the findings summarized in Rogers et al. (1993), seasonal fat deposition across three seasons in a common garden was persistently lower in birds originating from a recently established sedentary *J. h. thur-*

beri population inhabiting coastal and urban San Diego, when compared to birds originating from the nearby montane native breeding range (Atwell et al., in review). Persistent differences of seasonal fat divergence in a common garden were also measured in association with contemporary divergence of European blackbirds (*Turdus merula*)—reinforcing the possibility that changes in the physiological mechanisms underlying seasonal fat deposition might evolve quickly in association with increased sedentariness and/or milder wintering habitats (Partecke and Gwinner 2007). To date, however, the timing or magnitude of seasonal fattening is unknown for most junco groups, limiting further consideration. Future efforts to characterize the phenology of breeding and migration should, where possible, include ordinal (e.g., scoring) or continuous (e.g., total) body electrical conductivity (e.g., Castro et al. 1990) assays of seasonal fat in the context of wintering and migration. Conveniently, some interpretations can be made based simply on seasonal variation in body mass within closely related populations or subspecies, as most circannual variation in body mass is attributable to seasonal fattening (Atwell et al., in review; Piersma and van Gils 2010).

C. Implications for Diversification under Heteropatry

One of our stated goals for this chapter was to demonstrate that variation in habitat, morphology, migration, and seasonality among junco groups is extensive and represents a "Gold Mine of Evolutionary Biology" (Winker 2010b) that should be considered when weighing hypotheses regarding the origins and maintenance of *Junco* diversification or, alternatively, homogenization. The fact that juncos have evolved into groups that are markedly differentiated phenotypically despite apparent genetic similarity (especially in the northern junco forms; see Milá et al. 2007) raises important questions about the extent to which reproductive isolating mechanisms exist and the degree to which they limit interbreeding between groups, now or in the past. Such mechanisms, if they exist, might be prezygotic (e.g., variation in timing of breeding or mating preferences) or postzygotic (e.g., genetic incompatibility, hybrid sterility, or reduced hybrid fitness).

In fact, the divergent groups of juncos exhibit substantial hybridization where ranges overlap, and the apparent viability of some hybrids suggests that at least limited gene flow *is* an omnipresent feature of *Junco* biogeography, particularly in the north (Milá et al. 2007; Miller 1941; Price 2008). What is currently not known is whether current patterns of divergence and interbreeding represent speciation in action or, conversely, homogenization in action, that is, the breakdown of previously effective barriers to interbreeding.

Like the vast majority of apparent speciation or diversification events, allopatry—the physical geographic separation of biological populations restricting or limiting gene flow—is generally assumed to be a central process underlying *Junco* diversity, with subsequent "hybridization" at zones of secondary contact (Coyne and Orr 2004; chapters 8 and 9, this volume). However, other models of speciation might also fit the junco system because its historic range cannot be known with certainty (chapter 9, this volume), and within its current range, suitable habitats are discontinuous (e.g., mountain ranges separated by lowland valleys, nonforested areas, etc.). For example peripatric speciation—in which a small population enters a new, geographically isolated niche that restricts gene flow from the main population—or parapatric speciation—in which populations occupy adjacent ranges—could both apply to the *Junco* (Coyne and Orr 2004). Such geographic restrictions or limitations to gene flow likely played a role in the diversification of color plumages that diagnose *Junco* forms, whether via local adaptation or genetic drift. Further, as noted above, additional morphological, behavioral, or physiological differences among junco populations could likely represent local ecological adaptations that could play a role in diversification. Chapter 9 (this volume) further considers "divergence under gene flow" in relation to drift, selection, hybridization, and introgression as they relate to the diversification of color plumage among junco forms.

Another explicitly defined pattern of distribution, "heteropatry," in which populations are sympatric in one season and allopatric in another, may apply to speciation or homogenization in general and to the junco in particular (Winker 2010a). Many dark-eyed junco forms exhibit marked migratory behavior (see above), and longstanding theories have argued that populations that are more migratory should be less likely to diverge (Montgomery 1896), because long-distance travel via migration or dispersal (which is thought to be on average farther in migrants) should promote gene flow and reduce divergence (Winker 2010a).

As was highlighted in figure 3.1, while divergent junco forms are allopatric when breeding, they share wintering grounds and may also overlap in their migratory routes and stopover sites. Winker (2010a) argues that such co-occurrence on wintering ranges and during migratory journeys, despite different breeding schedules, could actually provide the opportunity for substantial interbreeding *among* such populations via at least two possible mechanisms: (1) pair bonding during wintering or migration, coupled with flexibility to overcome breeding site fidelity; and (2) "gamete mobility," where males of many species exhibit sperm release during migration, female birds have some ability to store sperm, and, anecdotally at least, females copulate during migration (Winker 2010a).

Thus, Winker's theory of "heteropatric speciation" first concedes that such interbreeding—facilitated by overlap in both geographic distributions *and* reproductively compatible states—could promote gene flow, and hence homogenization, among populations. Waterfowl of the Anatidae family appear to provide an example of highly migratory and relatively homogenized taxa due to their frequent pair bonding on nonbreeding or wintering ranges (Winker 2010b). However, Winker also emphasizes that the existence of substantial divergence within several highly migratory lineages suggests that the same behavioral and physiological phenomena that provide the opportunity for interbreeding under heteropatry could alternatively *promote* diversification or reinforcement if immigrant or intermediate "hybrid" phenotypes fall between the adaptive peaks of either parent population (Winker 2010b).

Winker identifies key fitness-related traits linked to spatially and temporally variable resource peaks in migrants that are likely to undergo disruptive or reinforcement selection in the context of "heteropatry": "fat deposition (cyclic energetics), wing shape, timing and direction of migration, timing of reproduction, and other genetically controlled adaptations developed to exploit these different resource peaks" (Winker 2010b). Notably, as discussed above, distinct dark-eyed junco groups differ markedly in nearly all of these traits related to seasonality and migration across their continent-spanning range, making them a particularly compelling group within which to consider the potential evolutionary implications of heteropatric distributions.

To portray a hypothetical example of Winker's heteropatric speciation within the dark-eyed juncos, we can return to figure 3.1 and consider the Oregon juncos and gray-headed juncos that co-occur during early spring in the Colorado Rockies. At lower elevations, early-breeding resident gray-headed females could initiate reproduction in mid-April, during which time lingering Oregon juncos preparing to migrate back to British Columbia may produce viable sperm, providing the potential for interbreeding despite allopatry and asynchrony on either parent's breeding grounds. Further, pair bonding and flexibility in breeding site fidelity could result in recruitment of an Oregon female into the local gray-headed breeding population. However, such immigrants or "hybrids" would likely exhibit breeding phenology, migratory propensity, physiology, or morphology that was locally maladaptive. The response to such disruptive selection could favor prezygotic (e.g., mate choice) or even postzygotic (i.e., genetic) reproductive isolating mechanisms, though the latter may be less likely to develop over the modest time scales considered in the *Junco* system (Coyne and Orr 2004; Price 2008; Winker 2010a). Whether this hypothetical example occurs in nature remains to be seen.

Further testing the general predictions of Winker's heteropatric model for diversification in variably migratory lineages such as that of the *Junco* requires integrative approaches that combine behavioral ecological studies of migratory connectivity and mate choice, assays of neuroendocrine and physiological divergence, and transcriptomic and genomic data to address whether past or present evolutionary mechanisms associated with heteropatry are important in generating or maintaining divergence in variably migratory taxa. We stop short of prescribing step-by-step prescriptions for a research program addressing heteropatry in the *Junco*. Instead we encourage the reader to consider the unique mechanisms of ecological speciation under heteropatry when assimilating the range of biogeographic variation among junco groups and formulating new study questions.

D. Conclusions and the Future of *Junco* Natural History

To summarize, in this chapter we have described multiple axes of biogeographic variation among distinct subspecies, races, and local populations of juncos, including habitat, morphology, migration, and seasonal timing—in addition to the obvious plumage color differences that have long attracted evolutionary biologists to study the highly variable yet recently diversified *Junco* lineage. We also considered the evolutionary significance of variation in migration and seasonality among junco groups in the context of Winker's (2010a) model of "heteropatric speciation," which may play an important role in the origins and maintenance of diversification in the genus *Junco*. However, many additional data are needed to address the importance of heteropatry in the junco system, including more detailed characterizations of studies of migratory connectivity and timing and breeding phenology, alongside assays of neuroendocrine, behavioral, and genomic variation.

While one of our chief goals was to highlight the stunning scope of "intraspecies" variation and its importance to studies of juncos, perhaps more importantly, we have aimed to convey that detailed information about the phenotypic distributions and natural history of most junco populations remains either unquantified or entirely mysterious, particularly for behavioral, physiological, and life-history traits. This situation is hardly unique to the junco. For biologists, "natural history" is a field of science that, despite a plurality of definitions, includes observing, measuring, and documenting the traits of organisms—including their morphology, physiology, and behavior—as well as their individual and collective interactions with their biotic, abiotic, and social environments (Wilcove and Eisner 2000). Although descriptive natural history may be viewed by some as a fading enterprise as compared to hypothesis-driven, deductive, experimental, and increasingly lab-based science, several forces are contributing to a growing awareness that new insights can be gained by discovery of patterns in large volumes of data, including phenotypes in the wild (Wilcove and Eisner 2000). New life for old questions is occurring.

In recent years, the explosion of insights arising from genomics has led to a call for phenomics, the study of how genotypes map onto phenotypes (Houle 2010). This approach requires quantitative assessments of nature conducted at a large scale, and technology is generating new methods for measuring, monitoring, and analyzing environmental and phenotypic variation (Burleigh et al. 2013; Houle et al. 2010). Moving forward, we suggest a future emphasis on "modern natural history," incorporating emerging technologies and the power of citizen science to more accurately, precisely, and expansively characterize biogeographic variation within and among *Junco* populations, allowing for the integration of phenomic and environmental data under a framework of next-generation research on evolutionary and organismal systems.

Acknowledgments

This chapter was enhanced thanks to helpful comments and suggestions from three anonymous reviewers. We would also like to thank the natural historians and evolutionary biologists, past and present, who have spent countless hours characterizing variation among juncos across the continent—revealing an understated yet remarkable diversity of forms and behaviors. We are particularly grateful for informative and inspiring discussions with Trevor Price, Borja Milá, and Kevin Winker.

References

Achterkirchen, A. 2010. Portable CMOS x-ray system enables Harvard's lizard evolution research. Vision Systems Design. http://www.vision-systems.com.

Aleixandre, P., J. Hernández Montoya, and B. Milá. 2013. Speciation on oceanic islands: Rapid adaptive divergence vs. cryptic speciation in a Guadalupe Island songbird (Aves: *Junco*). *PLoS ONE* 8:e63242.

Alerstam, T., A. Hedenström, and S. Åkesson. 2003. Long-distance migration: Evolution and determinants. *Oikos* 103:247–60.

Allen, J. A. 1877. The influence of physical conditions in the genesis of species. *Radical Review* 1:107–40.

Atwell, J. W., G. C. Cardoso, D. J. Whittaker, T. D. Price, and E. D. Ketterson. 2014. Hormonal, behavioral, and life-history traits exhibit correlated shifts in relation to population establishment in a novel environment. *American Naturalist* 184:E147–E160.

Atwell, J. W., D. M. O'Neal, and E. D. Ketterson. 2011. Animal migration as a moving target for conservation: Intra-species variation and responses to environmental change, as illustrated in a sometimes migratory songbird. *Environmental Law* 41.

Atwell, J. W., R. J. Rice, and E. D. Ketterson. In review. Rapid loss of migratory behavior associated with recent colonization of an urban environment. Manuscript.

Balph, M. H., D. F. Balph, and H. C. Romesburg. 1979. Social status signaling in winter flocking birds: An examination of a current hypothesis. *Auk* 96:78–93.

Bears, H., M. C. Drever, and K. Martin. 2008. Comparative morphology of dark-eyed juncos *Junco hyemalis* breeding at two elevations: A common aviary experiment. *Journal of Avian Biology* 39:152–62.

Belliure, J., G. Sorci, A. P. Møller, and J. Clobert. 2000. Dispersal distances predict subspecies richness in birds. *Journal of Evolutionary Biology* 13:480–87.

Bergmann, C. 1848. *Über die Verhältnisse der Wärmeökonomie der Thiere zu ihrer Größe*. Göttingen: Vandenhoeck and Ruprecht.

Berthold, P. H. 1995. *Control of Bird Migration*. New York: Springer.

Berthold, P., H. G. Bauer, and V. Westhead. 2001. *Bird Migration: A General Survey*. New York: Oxford University Press.

Bridge, E. S., J. F. Kelly, A. Contina, R. M. Gabrielson, R. B. MacCurdy, and D. W. Winkler. 2013. Advances in tracking small migratory birds: A technical review of light-level geolocation. *Journal of Field Ornithology* 84:121–37.

Bridge, E. S., K. Thorup, M. S. Bowlin, P. B. Chilson, R. H. Diehl, R. W. Fleron, P. Hartl, et al. 2011. Technology on the move: Recent and forthcoming innovations for tracking migratory birds. *Bioscience* 61:689–98.

Burleigh, J., K. Alphonse, N. A. Alverso, H. Bik, C. Blank, A. Cirranello, H. Cui, et al. 2013. Next-generation phenomics for the Tree of Life. *PLOS Currents Tree of Life* 1.

Castro, G., B. A. Wunder, and F. L. Knopf. 1990. Total body electrical conductivity (TOBEC) to estimate total body fat of free-living birds. *Condor* 92:496–99.

Chan, K. 2001. Partial migration in Australian landbirds: A review. *Emu* 101: 281–92.

Cheviron, Z. A., C. Natarajan, J. Projecto-Garcia, D. K. Eddy, J. Jones, M. D. Carling, C. C. Witt, et al. 2014. Integrating evolutionary and functional tests of adaptive hypotheses: A case study of altitudinal differentiation in hemoglobin function in an Andean sparrow, *Zonotrichia capensis*. *Molecular Biology and Evolution* 31:2948–62.

Cheviron, Z. A., M. Stager, and D. Swanson. 2015. Regulatory mechanisms of metabolic flexibility in the dark-eyed junco (*Junco hyemalis*). *Journal of Experimental Biology* 218:767–77.

Coyne, J. A., and H. A. Orr. 2004. *Speciation*. Sunderland, MA: Sinauer Associates, Incorporated Publishers.

Cristol, D. A., E. B. Reynolds, J. E. Leclerc, A. H. Donner, C. S. Farabaugh, and C. W. S. Ziegenfus. 2003. Migratory dark-eyed juncos, *Junco hyemalis*, have better spatial memory and denser hippocampal neurons than nonmigratory conspecifics. *Animal Behaviour* 66:317–28.

Dingle, H. 1996. *Migration: The Biology of Life on the Move*. New York: Oxford University Press.

Endler, J. A., D. A. Westcott, J. R. Madden, and T. Robson. 2005. Animal visual systems and the evolution of color patterns: Sensory processing illuminates signal evolution. *Evolution* 59:1795–1818.

Ferree, E. D. 2013. Geographic variation in morphology of dark-eyed juncos and implications for population divergence. *Wilson Journal of Ornithology* 125: 454–70.

Gloger, C. W. L. 1833. *Das Abändern der Vögel durch Einfluss des Klima's*. Breslau: August Schulz & Co.

Hill, G. E., and K. J. McGraw. 2006. *Bird Coloration: Mechanisms and Measurements*. Cambridge: Harvard University Press.

Holberton, R. L., K. P. Able, and J. C. Wingfield. 1989. Status signaling in dark-eyed juncos, *Junco hyemalis*: Plumage manipulations and hormonal correlates of dominance. *Animal Behaviour* 37:681–89.

Houle, D., D. R. Govindaraju, and S. Omholt. 2010. Phenomics: The next challenge. *Nature Reviews Genetics* 11:855–66.

Huggett, R. J. 2004. Fundamentals of Biogeography. London: Routledge.

Ketterson, E. D. 1979. Status signaling in dark-eyed juncos. *Auk* 96:94–99.

Ketterson, E. D., and V. Nolan, Jr. 1976. Geographic variation and its climatic correlates in the sex ratio of eastern-wintering dark-eyed juncos (*Junco hyemalis hyemalis*). *Ecology* 57:679–93.

———. 1983. The evolution of differential bird migration. In *Current Ornithology*, edited by R. Johnston, 357–402. New York: Plenum Publishers.

———. 1985. Intraspecific variation in avian migration: evolutionary and regulatory aspects. *Migration: Mechanisms and Adaptive Significance* 27:553–79.

Lin, S. J., J. Foley, T. X. Jiang, C. Y. Yeh, P. Wu, A. Foley, C. M. Yen, et al. 2013. Topology of feather melanocyte progenitor niche allows complex pigment patterns to emerge. *Science* 340:1442–45.

McCarthy, E. M. 2006. *Handbook of Avian Hybrids of the World*. New York: Oxford University Press.

McGlothlin, J. W., D. L. Duffy, J. L. Henry-Freeman, and E. D. Ketterson. 2007. Diet quality affects an attractive white plumage pattern in dark-eyed juncos (*Junco hyemalis*). *Behavioral Ecology and Sociobiology* 61:1391–99.

McGlothlin, J. W., J. M. Jawor, T. J. Greives, J. M. Casto, J. L. Phillips, and E. D. Ketterson. 2008. Hormones and honest signals: Males with larger ornaments elevate testosterone more when challenged. *Journal of Evolutionary Biology* 21: 39–48.

Milá, B., J. E. McCormack, G. Castaneda, R. K. Wayne, and T. B. Smith. 2007. Recent postglacial range expansion drives the rapid diversification of a songbird lineage in the genus *Junco*. *Proceedings of the Royal Society B-Biological Sciences* 274:2653–60.

Milá, B., T. B. Smith, and R. K. Wayne. 2006. Postglacial population expansion drives the evolution of long-distance migration in a songbird. *Evolution* 60: 2403–9.

Miller, A. H. 1941. Speciation in the avian genus *Junco*. *University of California Publications in Zoology* 44:173–434.

Millien, V., S. Kathleen Lyons, L. Olson, F. A. Smith, A. B. Wilson, and Y. Yom-Tov. 2006. Ecotypic variation in the context of global climate change: Revisiting the rules. *Ecology Letters* 9:853–69.

Montgomery, T. H., Jr. 1896. Extensive migration in birds as a check upon the production of geographical varieties. *American Naturalist* 30:458–64.

Mueller, J. C., F. Pulido, and B. Kempenaers. 2011. Identification of a gene associated with avian migratory behaviour. *Proceedings of the Royal Society B-Biological Sciences* 278:2848–56.

Murphy, M. 1996. Energetics and nutrition of molt. In *Avian Energetics and Nutritional Ecology*, edited by C. Carey, 158–98. New York: Springer.

Nolan, Jr., V., and E. D. Ketterson. 1983. An analysis of body-mass, wing length, and visible fat deposits of dark-eyed juncos wintering at different latitudes. *Wilson Bulletin* 95:603–20.

Nolan, Jr., V., E. D. Ketterson, D. A. Cristol, C. M. Rogers, E. D. Clotfelter, R. C. Titus, S. J. Schoech, and E. Snajdr. 2002. Dark-eyed junco (*Junco hyemalis*). In *The Birds of North America*, edited by A. Poole and F. Gill, no. 716. Philadelphia: The Birds of North America, Inc.

O'Neal, D. M., and E. D. Ketterson. In review. Climate change and differential migration: Relaxation of sexual segregation in a songbird. Manuscript.

Partecke, J., and E. Gwinner. 2007. Increased sedentariness in European blackbirds following urbanization: A consequence of local adaptation? *Ecology* 88:882–90.

Peterson, M. P., M. Abolins-Abols, J. W. Atwell, R. J. Rice, B. Milá, and E. D. Ketterson. 2013. Variation in candidate genes CLOCK and ADCYAP1 does not consistently predict differences in migratory behavior in the songbird genus *Junco* [v1; ref status: indexed, http://f1000r.es/11p]. *F1000 Research* 2:115.

Phillips, B. L., G. P. Brown, and R. Shine. 2010. Life-history evolution in range-shifting populations. *Ecology* 91:1617–27.

Piersma, T., and J. A. van Gils. 2010. *The Flexible Phenotype: A Body-Centred Integration of Ecology, Physiology, and Behaviour*. Oxford: Oxford University Press.

Price, T. 2008. *Speciation in Birds*. Greenwood Village, CO: Roberts and Co.

Pulido, F., and P. Berthold. 2010. Current selection for lower migratory activity will drive the evolution of residency in a migratory bird population. *Proceedings of the National Academy of Sciences of the United States of America* 107:7341–46.

Pulido, F., P. Berthold, and A. J. van Noordwijk. 1996. Frequency of migrants and migratory activity are genetically correlated in a bird population: Evolutionary implications. *Proceedings of the National Academy of Sciences of the United States of America* 93:14642–47.

Rasner, C. A., P. Yeh, L. S. Eggert, K. E. Hunt, D. S. Woodruff, and T. D. Price. 2004. Genetic and morphological evolution following a founder event in the dark-eyed junco, *Junco hyemalis thurberi*. *Molecular Ecology* 13:671–81.

Réale, D., D. Garant, M. M. Humphries, P. Bergeron, V. Careau, and P.-O. Montiglio. 2010. Personality and the emergence of the pace-of-life syndrome concept at the population level. *Philosophical Transactions of the Royal Society B-Biological Sciences* 365:4051–63.

Ridgway, R. 1912. *Color Standards and Color Nomenclature*. Washington, DC: United States National Museum.

Rogers, C. M., V. Nolan, Jr., and E. D. Ketterson. 1993. Geographic-variation in winter fat of dark-eyed juncos—Displacement to a common garden. *Ecology* 74:1183–90.

Ruegg, K. C., E. Anderson, K. Paxton, V. Apkenas, S. Lao, R. Siegel, D. DeSante, et al. 2014. Mapping migration in a songbird using high-resolution genetic markers. *Molecular Ecology* 23:5726–39.

Rundel, C. W., M. B. Wunder, A. H. Alvarado, K. C. Ruegg, R. Harrigan, A. Schuh, J. F. Kelly, et al. 2013. Novel statistical methods for integrating genetic and stable isotope data to infer individual-level migratory connectivity. *Molecular Ecology* 22:4163–76.

Stone, W. 1893. A hybrid sparrow (*Zonotrichia albicollis* + *Junco hyemalis*). *Auk* 10:213–14.

Sullivan, K. A. 1999. Yellow-eyed junco (*Junco phaeonotus*). *The Birds of North America Online*, edited by A. Poole. Ithaca: Cornell Lab of Ornithology.

Unitt, P. 2005. *San Diego County Bird Atlas*. San Diego: San Diego Natural History Museum.

Wilcove, D. S., and T. Eisner. 2000. The impending extinction of natural history. *Chronicle of Higher Education*. http://www.chronicle.com.

Wilkie, S. E., P. M. Vissers, D. Das, W. J. Degrip, J. K. Bowmaker, and D. M. Hunt. 1998. The molecular basis for UV vision in birds: Spectral characteristics, cDNA sequence and retinal localization of the UV-sensitive visual pigment of the budgerigar (*Melopsittacus undulatus*). *Biochemical Journal* 330:541–47.

Winker, K. 2010a. On the origin of species through heteropatric differentiation: A review and a model of speciation in migratory animals. *Ornithological Monographs* 69:1–30.

———. 2010b. Subspecies represent geographically partitioned variation, a gold mine of evolutionary biology, and a challenge for conservation. *Ornithological Monographs* 67: 6–23.

Yeh, P. J. 2004. Rapid evolution of a sexually selected trait following population establishment in a novel habitat. *Evolution* 58:166–74.

Yeh, P. J., and T. D. Price. 2004. Adaptive phenotypic plasticity and the successful colonization of a novel environment. *American Naturalist* 164:531–42.

Zink, R. M. 2011. The evolution of avian migration. *Biological Journal of the Linnean Society* 104:237–50.

PART II

Hormones, Phenotypic Integration, and Life Histories

An Endocrine Approach

Part 2 addresses the relationships between hormones, phenotype, and fitness as revealed by studies of the dark-eyed junco at Mountain Lake Biological Station (MLBS) in Virginia, USA, beginning in 1983 and extending to the present.

To place this section in historical context, and to see how study of the evolution of behavior can lead to the study of mechanisms, consider that in the 1960s and 1970s many behavioral biologists and ornithologists were focused on explaining variability in avian mating systems. Inspired by David Lack, Gordon Orians, Steve Emlen, and Lew Oring, they were asking why some birds mate polygynously or polyandrously, when most mate monogamously[1]. The presumption was that male parental care was essential in monogamously mating birds, but under some ecological circumstances a

---

1. Emlen, S. T., and Oring, L, 1977, Ecology, sexual selection, and the evolution of mating systems, *Science* 197:215–23; Gordon, H. O., 1969, On the evolution of mating systems in birds and mammals, *American Naturalist* 103:589–603; Lack, D. L., 1968, *Ecological Adaptations for Breeding in Birds*, London: Methuen and Co. Ltd.

female might find it advantageous to choose territory quality over male help and elect to become the second mate of an already mated male.

We (Ketterson and Nolan) were curious about whether the presumption that male help was necessary in monogamous species was in fact true and began to study the junco at Mountain Lake. By capturing males at the nest, holding half captive for the summer, and comparing the reproductive success of females rearing young on their own with females that were receiving male help, we found that while the young were still in the nest, aided and unaided females did not vary detectably in reproductive success. While this experimental approach was satisfying in some ways, it led to a paradox. If females could rear young alone, why were they apparently monogamous? And if males did not benefit from caring for offspring, why did they do it?

While seeking a better method for asking why males help to care for their young, we read a paper by John Wingfield[2] reporting that male song sparrows treated with the hormone testosterone became polygynous when ordinarily they were monogamous, and another paper by Bob Hegner and Wingfield[3] that reported that male house sparrows treated with testosterone were less parental than untreated males. We decided to see whether we could "engineer" male juncos that would remain on their territories but be poor at caring for offspring. If yes, we could compare their fitness to untreated males and learn more about how mating systems evolve. We might even create males that were polygynous.

As is common, the more we learned the more complicated the simple question became. A key "aha" moment was when we realized that treating males with testosterone had many more effects than just altering parental behavior. For Ketterson the moment came while she was watching a nest in the woods to quantify male trips to the nest, and the male she was watching was indeed not feeding, but he was singing. In fact he was singing a lot! Clearly testosterone did not change just one thing about males.

Chapter 4 by Gerlach and Ketterson summarizes the research that followed from the awareness that experimental manipulation of hormones

2. Wingfield, J. C., 1984, Androgens and mating systems: Testosterone-induced polygyny in normally monogamous birds, *Auk* 101:665–71.

3. Hegner, R. E., and J. C. Wingfield, 1987, Effects of experimental manipulation of testosterone levels on parental investment and breeding success in male house sparrows, ibid. 104:462–69.

influenced many phenotypic traits concurrently in ways that relate to trade-offs in life histories. Phase one was to compare testosterone-treated males (T-males) and controls (C-males) for a whole array of phenotypes (home range, attractiveness to females, immune function, response to stressors, and more) and to relate those phenotypes to life span and reproductive success. Some results were again paradoxical (e.g., males treated with testosterone had higher fitness than controls). Why was that? Was there some constraint that prevented males from evolving towards a phenotype characterized by higher testosterone? Might a potential negative impact of elevated testosterone on female juncos influence the course of male evolution? These questions led to experimental alterations of testosterone in females as well and efforts to relate female T to phenotype and fitness. Chapter 4 reviews findings from both the male and the female experimental research conducted at Mountain Lake and highlights how long-term research can add to our understanding of complex phenotypes in nature.

Experimental approaches are ideal for creating variation between groups of animals that can be compared for phenotype and fitness, but the ultimate goal is to understand how phenotypes develop in natural environments and how they are maintained or altered by natural selection. Because hormones influence the expression of multiple traits, learning how individuals vary in hormonal pathways and how that variation relates to fitness was the goal of the research described in chapters 5, 6, and 7.

Chapter 5 by McGlothlin and Ketterson describes concepts and techniques from quantitative genetics and how they can be employed to relate hormonal pleiotropy (one hormone affecting many traits, an analog of genetic pleiotropy) to phenotypic matrices and correlational selection. Tight correlations among traits, if they are heritable, can facilitate or retard response to selection and thus have important implications for how animals have and will respond to environmental change.

Chapter 6 by Cain, Jawor, and McGlothlin returns to the field and empirical approaches to address the relationship of natural variation in hormone levels to phenotype and fitness in males and females. Correlations abound, and because they can be related to previous experimental results, the relationship between correlation and causation is made stronger. Selection acts on the total phenotype, and in males, hormone-related traits appear to trade off in their impact on survival, though not demonstrably on reproductive success. Males with the average testosterone phenotype were more likely to survive than either extreme. Females are more challenging to study, but natural variation in testosterone-related behavior also influences aspects of female fitness, some positively, some negatively. Many questions remain.

While field biologists were beginning to study hormones in natural populations, lab-based behavioral endocrinologists were moving from studying systemically circulating hormones to studying the action of hormones in the brain. New techniques from neuroanatomy, immunocytochemistry, pharmacology, and most recently transcriptomics led and continue to lead to a far deeper understanding of hormones and behavior. A key discovery that has yet to be fully integrated into evolutionary endocrinology is that the brain not only responds to circulating steroid hormones but also synthesizes steroids locally in the brain. This local production renders the brain more independent of the body than had previously been appreciated. How this independence relates to modules in the brain while still allowing the organism to respond as a whole to environmentally induced variation in hormone production is still being figured out.

Chapter 7 by Rosvall, Bergeon Burns, and Peterson addresses these issues by relating the sensitivity of various neural tissues to circulating or locally produced steroids and to behavior. Individuals exhibit variation in the abundance of steroid receptors and in patterns of gene expression mediated by hormones. This variation is presumably subject to selection and is sure to prove to be critical as we move toward building the map connecting hormones to life histories to population divergence over time.

Phenotypic Engineering

A Long-Term Study Using Hormones to Study Life History Trade-Offs and Sexual Conflict

Nicole M. Gerlach and Ellen D. Ketterson

... as Goethe expressed it, "in order to spend on one side, nature is forced to economise on the other side." I think this holds true to a certain extent with our domestic productions: if nourishment flows to one part or organ in excess, it rarely flows, at least in excess, to another part.... In our poultry, a large tuft of feathers on the head is generally accompanied by a diminished comb, and a large beard by diminished wattles.
—Darwin 1859, 147

A. Introduction

One of the major goals of organismal biology is to understand the evolution of life histories—how and why species, and individuals within species, differ in how they allocate their time and energy to the basic tasks of staying alive and producing offspring (Stearns 1989; Stearns 1992; Stearns 2000). Because the pool of resources available to any one individual is finite, the study of life history evolution is necessarily the study of trade-offs (Roff 2002; Stearns 1989; Stearns 1992). Time spent on one behavior (e.g., mate attraction) cannot be spent on other activities (e.g., being parental), and energy allocated to the demands of offspring production will be unavailable to satisfy other needs such as mounting an immune defense to fight an infection (Cody 1966; Ricklefs and Wikelski 2002). Selection would favor a hypothetical individual that could attract infinitely many mates *and* provide excellent parental care to each of its resulting offspring (Charnov 1993; Ricklefs 2000). However, as anyone who has attempted to balance a budget or a busy schedule knows, increasing allocation to one need requires that other needs receive less.

Each possible strategy of balance has its own costs and benefits, and the fact that many different aspects of an organism's phenotype are affected by these trade-offs can have significant consequences for the evolution of any single trait (Roff 2002; Stearns 1992).

Determining how trade-offs evolve requires that we learn how individuals vary in the way they balance allocation of their time and resources, in part by examining the mechanisms by which this variation arises (Ketterson and Nolan 1999; Ketterson et al. 1992; Ricklefs and Wikelski 2002; Williams 2012b). While all trade-offs ultimately arise from the finite nature of time and resources, the question remains: What organismal, evolutionary, and ecological factors determine where taxa or individuals fall along the major axes of possible patterns of allocation?

One set of mechanisms relates to the endocrine system, which includes the production of and response to circulating hormones that act as signaling molecules and have been shown to underlie suites of behavioral, immunological, and reproductive traits (Adkins-Regan 2005; Nelson 2005). As such, hormones are prime candidates for mechanisms by which life-history trade-offs may be mediated (Hau 2007; McGlothlin and Ketterson 2008; Nolan et al. 1992; Williams 2012a; Wingfield et al. 2001). This idea, that multiple traits may share a common endocrine basis, is frequently referred to as "hormonal pleiotropy," because just as two traits influenced by a single pleiotropic gene may evolve interdependently, so too can two traits affected by a single hormone influence each other's response to selection (Bartke 2011; Finch and Rose 1995; Flatt et al. 2005; Hau 2007; Ketterson and Nolan 1999).

Importantly, hormonal pleiotropy, like other mechanisms of phenotypic integration, can either delay or accelerate evolution depending on the circumstances (Flatt et al. 2005; Ketterson et al. 2009; Pigliucci and Preston 2004). For two traits mediated by a single endocrine mechanism, if selection favors a beneficial increase in one trait but that change is linked mechanistically to a detrimental decrease in the other trait, then their shared endocrine regulation may serve to constrain evolution, at least temporarily (Adkins-Regan 2008; Hau 2007; Ketterson and Nolan 1999; Lessells 2008; McGlothlin and Ketterson 2008). Conversely, hormonal pleiotropy can accelerate evolution in cases in which selection favors the coexpression of two independent traits. Without a shared regulatory mechanism, individuals that happen to express both traits must arise by chance and recombination before selection can favor the interaction between them. But when two traits are mediated by the same hormone, this common hormonal regulation can accelerate the evolution of the co-regulated traits above the rate that would be expected if the traits did not share a common mechanism: selection favoring one trait brings along the other (Agrawal and Stinchcombe 2009; Ketterson et al. 2009; Merilä and Björklund 2004). *By understanding how hormones are related to trade-offs between behaviors and other aspects of life history, we can better understand how selection might lead to integrated phenotypes.*

B. Hormones — Measurement and Manipulations

Hormones have been linked to a wide variety of physiological and behavioral phenotypes in a broad range of taxa (Eikenaar et al. 2011; Hau and Wingfield 2011). A classic way of demonstrating that a trait depends on a hormone for its expression is to remove the source of the hormone and show that the trait is altered, and then to add the hormone back and show that the trait returns (Nelson 2005). A classic example would be the cock's comb, which decreases in size and coloration if the cock is castrated but returns if the cock is treated with androgens (Champy and Kritch 1925, cited from Ludwig and Boas 1950; McGee et al. 1928, cited from Nelson 2005). Over time very sophisticated methods have been developed for knocking out and restoring hormonal effects (e.g., chemical agents that interfere with hormone synthesis, pharmacological compounds that block hormone receptors, or sequences of RNA that prevent normal gene transcription), but the principle remains the same (Nelson 2005).

Another way of establishing a relationship between a hormone and life histories is to employ a comparative approach by measuring concentrations of hormone in the circulation of a variety of species and correlating that measure with some aspect of the phenotype. The ability to measure hormone circulations in small samples of blood taken from free-living animals became available in the 1970s (Wingfield and Farner 1975). In a now classic study, Wingfield et al. (1987) compared circulating testosterone in bird species that differed in mating system and found that polygynous species that were more aggressive and less parental tended to maintain higher circulating levels of testosterone throughout the breeding season, while those that were more monogamous, less aggressive, and more parental elevated testosterone only briefly early in the breeding season. This pattern raised the possibility that the same hormone might have enhancing effects on aggression and suppressive effects on parental behavior. Wingfield et al. (1987) and a related study (1990) established a clear link between a hormone and a fundamental life history trade-off. More recent comparisons of this type have associated circulating levels of hormones more generally with geographic variation in life history traits and lifespan (Eikenaar et al. 2012; Goymann et al. 2004; Hau et al. 2010).

Hormone manipulations offer a valuable tool for studying life history trade-offs experimentally because they allow investigators to alter allocations to various aspects of life histories, such as the trade-offs between aggressive, sexual, and parental behavior, survival and reproduction, egg number and egg size, and timing of metamorphosis (Astheimer et al. 1992; Hegner and Wingfield 1987; Marler and Moore 1988; Runfeldt and Wingfield 1985; Silverin 1980; Silverin 1991; Sinervo and Licht 2006; Watson and Parr 1981; Wingfield 1984). By treating an individual with exogenous hormone, it is possible in effect to "push" that individual toward one end of the spectrum of potential allocations and measure the consequences for its physiology, behavior, and, ultimately, its relative fitness.

This technique, later called "phenotypic engineering" (Ketterson et al. 1996), can be compared to a classic approach to assessing the "adaptive value" or fitness consequences of single traits, such as gluing longer tail feathers on male widowbirds (*Euplectes progne*) to manipulate male morphology and mating success (Andersson 1982). The reasoning, drawn from game theory, is as follows: If a mutant phenotype were to arise with some unusual attribute, how would it fare in relation to other phenotypes, and could it "invade" the population? Importantly, hormone manipulations differ from alterations of single traits in that they alter multiple traits simultaneously and thus allow the study of effects of hormones on complex behavioral phenotypes and the relationship between those phenotypes and fitness. Although other hormones have been linked to life-history trade-offs, testosterone (T) has been the topic of much research in this area since Wingfield et al.'s classic paper (1987) relating seasonal profiles of circulating T to parental behavior, male-male aggression, and mating systems.

C. History of a Long-Term Study

About the time that Wingfield et al. (1987; 1990) and Hegner and Wingfield (1987) were conducting their research, our group had been studying male parental care in the junco at Mountain Lake Biological Station in Virginia (1983–1986). To quantify the impact of male parental behavior on fitness, we compared the behavior and reproductive success of females rearing young alone—because we had captured their mates at the time that the young hatched—to females that were rearing young with male help (Wolf et al. 1988; Wolf et al. 1990). Unaided females fed their offspring twice as often as did aided females, and they reared nearly as many young to nest-leaving and half as many young to independence compared to females that had male help (Wolf et al. 1988). One implication was that a male might achieve equal reproductive success whether he was mated to two females and provided no help or mated to just one female and helped. However, monogamy is the norm in this species, with polygyny rarely if ever observed in unmanipulated populations (Nolan et al. 2002), which suggested that some factor other than the marginal benefit of male care in terms of offspring survival must be limiting male juncos to monogamy. We developed a working hypothesis that females might be "enforcing" male parental behavior by being less faithful to unhelpful males, and if females were in short supply, then this enforcement might account for why males were generally helpful. To test this hypothesis, we needed a method to create males that were poor providers and then measure mate switching by females. Inspired by Hegner and Wingfield (1987), we began a program of treating some males with subcutaneous testosterone implants (see section D, below), with the expectation that the hormone would reduce male parental behavior and that females would be more faithful to control males.

As the study unfolded, however, we quickly realized that we had changed more than just one thing about the male juncos. Testosterone-treated males fed their offspring less often than control males, as predicted, but they also varied in other ways, including that they sang more (Ketterson et al. 1992). Thus, while our original intent was to use the hormone implants to create a class of males that provided less assistance to females in rearing young, we actually created males that might be expected to be *more* attractive to females because they sang more. We were fascinated by the result but needed to revisit the conceptual framework underlying the research. The question now became: How might selection act on males that differed in a whole suite of behavioral and physiological traits that affected allocation of time and energy to mating vs. offspring care?

Initially, we predicted that testosterone-treated males would shift their allocation away from offspring care and become polygynous. In dark-eyed juncos, as in other socially monogamous, temperate-zone songbird species (Goymann and Landys 2011; Wingfield et al. 1987; Wingfield et al. 1990), circulating testosterone levels peak during the early spring period

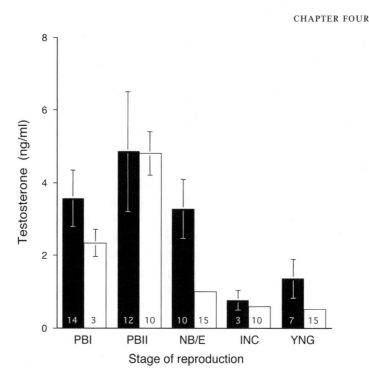

FIGURE 4.1. Plasma testosterone in the dark-eyed junco, compared by sex and stage of reproduction (solid bars males, open bars female) [ANOVA stage (P = .095), sex (P = .029), date
(P = .005), bleeding time (P = .043); ANOVA without date, stage (P = .000), sex (P = .008)].
Stages are prebreeding I (PBI), early April; prebreeding II (PBII), late April; nest building/
egg laying (NB/E); incubation (INC); and tending young (YNG). Testosterone varies with sex
and date or stage (figure and caption from Ketterson et al. 2005).

of territory establishment and pair bond formation (Ketterson et al. 1992)
and then decline as the nest cycle progresses, with levels during nestling
feeding being comparable to levels during the non-breeding season (see
figure 4.1). Hormone implants thus allowed us to alter the seasonal profile of testosterone in a monogamous bird to resemble the profile of a polygynous bird (as described by Wingfield et al. 1990) by maintaining peak
testosterone levels over the course of the breeding season, and we expected a corresponding shift in behavior from monogamy to polygyny. At
the time, the general consensus was that a species's apparent mating system reflected its genetic mating system such that females copulated only
with the male they to which they were socially mated and no others (Lack
1968; Wittenberger and Tilson 1980). But as it became apparent that extrapair paternity was a major factor contributing to the variance in re-

productive success in songbirds (Bray et al. 1975; Emlen and Oring 1977; Gowaty 1983; McKinney et al. 1984; Roberts and Kennelly 1980; Westneat 1987), it became possible and pressing to ask whether testosterone also affected allocation to extrapair mating effort.

Various publications have reviewed our findings for male juncos treated with testosterone (T) (Ketterson and Nolan 1999; Ketterson et al. 1996; Ketterson et al. 1992). Very briefly, treatment with T did not lead to polygyny in the classic sense of multiple females forming a pair bond with a single male, nor did testosterone-treated males have lower fitness than controls. Rather, males with elevated testosterone had higher fitness than control males, due to their increased numbers of extrapair offspring (Reed et al. 2006). This finding shaped the next decade of research: If males with experimentally elevated testosterone have increased fitness relative to males with unmanipulated testosterone, why were males with higher testosterone not more prevalent in the population? One possible explanation for the disparity between the apparent selective pressures that favor high testosterone and the absence of high-testosterone phenotypes in the population may be constraint by sexual conflict (Lande 1987; Lande and Arnold 1983). Specifically, if selection for high-T phenotypes in males were to result in females with elevated T, and if high-T females were at a selective disadvantage, then this fitness cost to females might constrain the evolution of testosterone in males. In order to investigate the phenotypic and fitness consequences of elevated testosterone in females, we again used phenotypic engineering to create a group of females that experienced continuous peak levels of testosterone. These findings were reported in a series of papers and reviewed in Ketterson et al. (2005).

In the following sections, we review the comparative effects of elevated testosterone on physiology, behavior, and life-history components in both male and female juncos and consider the net effects of T treatment on the fitness of each sex. We also raise caveats and make suggestions for future research projects.

D. Methods of Testosterone Manipulation in the Dark-Eyed Junco

Subcutaneous hormone implants provide a means by which hormone levels can be altered over a span of weeks or months in free-living individuals. Made of semiporous Silastic tubing (Dow Corning, i.d.=1.47 mm, o.d.=1.95 mm) packed with crystalline hormones and sealed at both ends, implants are placed beneath an individual's skin (Ketterson et al. 1991). The "dose" received can be adjusted by manipulating the length of the implant and thus the surface area across which the hormone can diffuse. Over time, the hormone will slowly diffuse across the tubing and enter the systemic circulation, raising the individual's circulating level of a hormone and maintaining that elevated level for as long as the crystalline hormone lasts. Males were given two 10-mm implants, while females were given one 5-mm implant; these implant sizes were sufficient to maintain testosterone at sex-typical peak levels throughout the breeding season (Clotfelter et al. 2004; Ketterson et al. 1992; Ketterson et al. 1991). Control individuals were given empty implants but were otherwise treated identically to individuals that received testosterone-filled implants. Treatment group was determined by coin flip after blocking by site of capture within the study area. In the earliest years of the study, male treatment was alternated between years. Beginning in 1990, individuals that were captured in multiple years received the same treatment of implant in each subsequent year. The exact dates during which implants were given varied but were typically early in the breeding season (mid-April to mid-May). At the end of the breeding season (late July to August), implanted individuals were recaptured and their implants were removed.

E. Effects of Testosterone Manipulation on Males and Females

E.1. Mating Effort

Testosterone is known from a variety of other taxa to have an activational effect on male-typical courtship behavior (Adkins-Regan 2005; Ketterson et al. 2005; Staub and De Beer 1997). In juncos, males with elevated T levels (T-males) sang long-range song more frequently relative to control males with empty implants (C-males), both in free-living conditions (Chandler et al. 1994; Ketterson et al. 1992) and while being observed in captivity (Enstrom et al. 1997). When presented with a female, T-males also performed courtship behaviors, such as singing short-range song, spreading the tail feathers, and carrying pieces of nest material, more frequently than did C-males (Enstrom et al. 1997). Perhaps as a result of this increased courtship intensity, females significantly preferred to spend time near and solicited more copulations from T-males as compared to their C-male counterparts during a two-choice mate test (Enstrom et al.

1997). Furthermore, T-males significantly increased their home range size relative to C-males, suggesting that T-males may be more actively seeking additional mating opportunities with nearby fertile females (Chandler et al. 1994). This effect occurred only during the incubation and nestling phases, not during the fertile period of the male's social mate (Chandler et al. 1997), suggesting that it is a flexible response to mating opportunities. This is supported by the fact that T-males did in fact sire more offspring in the nests of females other than their socially pair-bonded mate (extrapair offspring) (Raouf et al. 1997).

We have somewhat less evidence regarding testosterone's effect on female mating behavior. In other species in which females do not typically sing (or sing less than males), elevating female testosterone has been shown to induce females to sing (Ketterson et al. 2005; Staub and De Beer 1997), and this occurred in a small proportion of female juncos (Clotfelter et al. 2004). Courtship behaviors that are more female-typical, such as the precopulatory display, are typically thought to be stimulated by circulating estradiol rather than testosterone (Adkins-Regan 2005; Crews and Moore 1986). Nevertheless, circulating testosterone may play a role in female mate choice. In a two-choice mate test, similar to those described above, T-females were equally as likely as C-females to spend time near males but did not show a consistent preference for either T-treated or control males, suggesting that elevated testosterone may decrease a female's choosiness without decreasing her sexual motivation (McGlothlin et al. 2004). If this lack of discrimination between males is also present during natural courtship encounters, this may lead to T-females accepting copulations with extrapair males at a higher rate. We found no evidence that T-females produced higher rates of extrapair offspring than C-females (Gerlach and Ketterson 2013). However, because we did not directly measure copulation frequency or number of copulatory partners, we cannot say for certain whether elevated testosterone affects female mating behavior in the field.

E.2. Parental Behavior

The dark-eyed junco, like many other songbirds, divides parental care duties between the male and the female relatively equitably. In juncos, only the female incubates the eggs and broods the offspring, but both males and females contribute to feeding hatched offspring and defending the nest from predators throughout the mating cycle (Nolan et al. 2002). Testosterone significantly decreased the levels of parental care in both males and females. T-males decreased both their effort toward nest defense (Cawthorn et al. 1998) and toward offspring feeding during the nestling stage (Ketterson et al. 1992). In females, testosterone implants had no effect on rates of incubation or participation in nest defense during incubation (Clotfelter et al. 2004), nor did they affect nestling feeding (O'Neal et al. 2008). However, T-females spent less time brooding offspring than did C-females, and they also participated less in nest defense during this stage (O'Neal et al. 2008).

E.3. Offspring Production

Elevated testosterone played a role in offspring production in dark-eyed juncos, especially in females. Females with testosterone implants were significantly less likely to develop a brood patch (a featherless vascularized area on the abdomen, used in incubating eggs and developed as a precursor to reproduction) than were control females (Clotfelter et al. 2004). T-females were significantly less likely to build a nest than C-females (Gerlach and Ketterson 2013). They also delayed the laying of their first egg relative to C females, and while they did not differ in clutch size (Clotfelter et al. 2004), T-females laid fewer total eggs over the course of the breeding season than did C-females (Gerlach and Ketterson 2013). T-females as a group had fewer hatchlings and fewer successful fledglings than C-females, although this is likely due to accumulated differences during laying and incubation; once a female hatched offspring, T-females and C-females did not differ in their reproductive success (Gerlach and Ketterson 2013). If we are to consider the overall offspring output of individuals with elevated testosterone, however, on average T-females produced half the number of fledglings as C-females and thus would be at a substantial selective disadvantage (Gerlach and Ketterson 2013) (see figure 4.2A).

Because males can sire offspring with extrapair females as well as in their home nest, comparing testosterone's effect on reproductive success in males and females is more complicated. The first question is whether testosterone interferes with reproductive physiology in males in some way analogous to the reduced development of brood patches in females. When measuring sperm levels in wild-caught males, T-males were found to have smaller ejaculates than C-males, although when these same measurements were taken on captive males without access to females, sperm

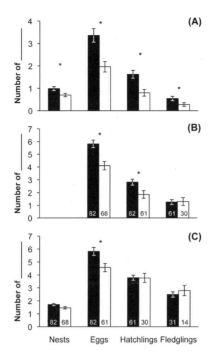

FIGURE 4.2. The impact of testosterone treatment on fitness of females. The top graph (A) indicates the reproductive success of all T-treated (white bars) and control females (black bars) at various reproductive stages. The middle graph (B) is the reproductive success only among females that had reached the previous stage (e.g., number of eggs laid by females that built at least one nest), and the bottom graph (C) is the success of females that had reached the current stage (e.g., number of eggs laid by females that laid at least one egg). Asterisks indicate significant differences between T-treated and control females (Gerlach and Ketterson 2013).

levels did not differ detectably based on hormone treatment (Kast et al. 1998). This suggests that elevated levels of T were not interfering with sperm production and that T-males' reduced sperm levels were likely a result of increased copulation number/frequency (see section E.1, "Mating Effort," above).

In males, T implants had no effects on the number of nests with which a male was associated or on mean clutch size produced by his socially pair-bonded partner (Raouf et al. 1997; Reed et al. 2006). Testosterone treatment also had no detectable effect on daily nest success, a measure of nest predation (Reed et al. 2006). In one report, C-males had higher apparent reproductive success (i.e., number of fledglings in the nest with his social mate regardless of sire) (Raouf et al. 1997), but a later analysis based on more years of study reported no difference in apparent success (Reed et al. 2006). With respect to extrapair paternity, we have reported that T implants led to losing fewer offspring in the home nest to extrapair paternity (Raouf et al. 1997) or had no effect (Reed et al. 2006). Both studies agreed that T-males were more likely than C-males to sire extrapair offspring (EPO) (Raouf et al. 1997; Reed et al. 2006), although they differed as to whether these EPO led to similar (Raouf et al. 1997) or greater total reproductive success (Reed et al. 2006) in T-males than C-males. The two studies differ primarily in the number of years of data available for analysis; the more inclusive study (Reed et al. 2006) suggests that T-treated males, on the basis of their increased reproductive success via extrapair fertilizations, were likely to be favored by selection over time as compared to C-males (see figure 4.3).

E.4. Survival

In order to understand the full extent of the life-history trade-offs mediated by testosterone, we also considered its effects on self-maintenance and survival. Experimentally elevated T tended to reduce investment in self-maintenance and survival-related traits in both sexes, but this resulted in reduced survival only in males. In early spring, experimentally elevated testosterone in males was shown to reduce body mass and fat stores, which may have made T-males more vulnerable to spring storms (Ketterson et al. 1991). Later during breeding, T-implants enhanced male activity levels but had no detectable effect on daily energy expenditure, indicating a shift in allocation away from rest and maintenance behaviors (Lynn et al. 2000). T was also shown to increase levels of the stress hormone corticosterone and to reduce immune function in both sexes (Casto et al. 2001; Schoech et al. 1999; Zysling et al. 2006) but not to reduce breeding season condition or body mass (Casto et al. 2001). Neither was T associated in males with the intestinal parasitic disease coccidiosis (Hudman et al. 2000); no comparable measure was made on females. Finally, T was also shown to delay or even halt the progression of postnuptial molt in both sexes (Clotfelter et al. 2004; Nolan et al. 1992). In the extreme case, some males did not molt at all and their plumage became exceedingly worn. More naturally, if selection favored prolonged elevation of testosterone in males such that molt was delayed, molt would have to proceed more rapidly, which can lead to plumage that is poorer in quality (Dawson et al. 2000). Despite the lack of a clear signal about which effect on T might have had the greatest effect on future reproductive

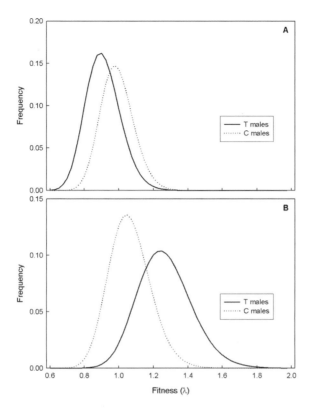

FIGURE 4.3. The impact of testosterone treatment on fitness of males. In the case of males, the method used was one from population ecology that predicts the relative rate of growth (lambda) of populations comprised of testosterone or control individuals when extrapair offspring are (A) ignored or (B) included. In an environment of ~50 percent controls and 50 percent experimentals, the method of analysis predicts that a population of T-treated males would have a higher growth rate and thus would gradually replace C-males over time. We note that this presumes that the benefit of traits exhibited by T-males would be independent of the relative frequency of T- and C-males in the environment (Reed et al. 2006).

opportunities, we found that testosterone implants led to a decrease in breeding-season survival in males (Reed et al. 2006) but had no effect on overwinter return rates in either sex (Gerlach and Ketterson 2013; Reed et al. 2006).

E.5. Extended Phenotypic Effects: Mates

Because we are interested in the effects of testosterone in a dynamic, free-living situation, we also considered that a hormone can affect not only the physiology and behavior of a single focal individual but also of other individuals that interact with the focal bird. Females, for example, often compensate for decreased feeding by their partners (Wolf et al. 1990), and in juncos, we observed more frequent feeding of offspring by females mated to T-males than by females mated to C-males, despite their not being directly subject to experimental manipulation (Ketterson et al. 1996). While this increased parental effort might have lowered these females' likelihood of survival to future breeding seasons (Wolf et al. 1991), this remains to be determined. Mates of T-males also had lower immune function (Ketterson et al. 2001, citing unpublished findings from Casto et al.), perhaps as a result of increased energy expenditure on parental care. The reciprocal question—what is the effect on males of being mated to a T-female?—has not been asked, although given the dynamic nature of a social pair bond and evidence in other species of a male's response to a mate implanted with estradiol, it is likely to be the case that the female's hormonal state can influence the behavior and/or physiology of the male (Runfeldt and Wingfield 1985; Silverin 1991).

E.6. Extended Phenotypic Effects: Offspring

The hormonal status of an individual is likely to affect the behavior and fitness of not only its pair-bonded partner but also its offspring. This is particularly true in females, where increased circulating levels of testosterone led to increased deposition of testosterone in the yolk of the eggs that they laid (Clotfelter et al. 2004). Yolk T can have a variety of effects on offspring development and fitness (Gil 2003; Groothuis et al. 2005; Müller et al. 2009; Schwabl et al. 2012). In the junco, we did not find any differences between offspring produced by T- vs. C-females in their mass at hatching (Gerlach and Ketterson 2013), which is unsurprising since T- and C-females also did not differ in their egg size or volume (Clotfelter et al. 2004). If increased yolk T did lead to increased begging vigor in nestlings, we might expect that offspring produced by T-females might grow faster or be heavier, particularly since we have no evidence that T-females decreased their feeding rates (O'Neal et al. 2008, see section E.2, "Parental Behavior," above); instead we found that there were no differences in mass between offspring of T- and C-females at any point during the nestling phase (Gerlach and Ketterson 2013). Offspring of T-females also did not differ in their immune response or in their degree of bilateral symmetry from offspring of C-females during the nestling stage (Gerlach,

Spevak, Ainsworth, and Ketterson, unpublished data). However, after nestlings leave the nest, there was some indication that offspring of T-females may have an advantage after they leave the nest. These offspring were somewhat more likely to survive to independence and to return to the population as adults (Gerlach and Ketterson 2013); understanding the developmental mechanisms underlying this difference remains an avenue of active research.

Experimentally elevated hormone levels in males are not directly transmissible to offspring in the same way as in females, but there are still indirect multigenerational effects of T implants in males. Social offspring of T-males grew more slowly and were smaller at fledging compared to offspring of C-males (Reed et al. 2006), perhaps due to decreased parental care by their fathers (see section E.2, "Parental Behavior," above). Because size at fledging is frequently a strong predictor of later survival and success (Galbraith 1988; Magrath 1991; Reed et al. 2006; Smith et al. 1989), this may represent a substantial fitness cost for the offspring of T-males. Offspring of T-males also had lower immune function as nestlings (Casto et al., unpublished data), although this did not appear to affect their likelihood of survival to independence or returning to the population as adults (Reed et al. 2006).

F. What Do Hormone Manipulations Tell Us about Trade-Offs and Selection?

One fact that has become increasingly evident from our work with experimental manipulations using hormone implants is that a wide variety of behavioral, physiological, and life-history traits are influenced by circulating testosterone and, furthermore, that several of these traits appear to be mediated in such a way that increased levels of one lead to decreased levels of the other. The most apparent of these is the trade-off between mating effort and parental effort in males; males with higher testosterone levels are more attractive and have increased rates of siring extrapair offspring but at the cost of parental care directed towards their own social offspring. There also may be a trade-off between survival and reproductive success; T-males have decreased immunity, self-maintenance behaviors, and survival, such that their increased annual reproductive success may come with a cost to future reproductive potential (Casto et al. 2001; Reed et al. 2006; Stearns 1992). The net effect of all of these trade-offs, however, is that the increased reproductive success that comes with elevated male T outweighs the costs (Reed et al. 2006), which suggests that if a mutant with a high T phenotype arose, we would expect that it should be favored by selection and should spread in the population, at least initially.

The evidence that T is mediating trade-offs in females is less clear. Elevated T in females appears to have primarily negative effects, interfering particularly with the early stages of reproduction (Gerlach and Ketterson 2013). It is possible that elevated T may be beneficial to females in terms of increased aggression and therefore increased success with territory or mate acquisition, as was shown in spotless starlings (*Sturnus unicolor*) (Veiga and Polo 2008). However, since these behaviors occur early in the spring (Nolan et al. 2002) before we placed implants in the majority of females, we cannot draw any direct conclusions. The two-fold cost of elevated T to reproductive success in females may be one factor constraining the evolution of T in males; we examine this and other possibilities in the next section.

G. How Is Selection on T in One Sex Likely to Affect the Other?

Because the sexes share the majority of their genome, selection acting on a trait in one sex can affect the evolutionary response of that trait in the other sex (Lande 1987; Lande and Arnold 1983). Just as selection on one T-mediated trait can cause a change in the expression of a second T-mediated trait, to the degree that testosterone levels are heritable and genetically correlated between the sexes, selection on testosterone levels in males can produce a correlated response in females, and vice versa. The evolutionary response in T levels will therefore be the net effect not only of selection on various T-mediated trade-offs within a sex but also the combination of selection on both sexes (see figure 4.4). If selection is acting in the same direction on both sexes, this correlated response should accelerate the rate of evolutionary change. However, if selection is acting in different directions in the two sexes, then the evolution of the trait in question may be constrained, resulting in an "optimum" that is optimal for neither sex (Cox and Calsbeek 2009; Ketterson et al. 2009; McGlothlin and Ketterson 2008).

We have documented the various selective pressures acting on testosterone levels in both male and female juncos. However, discussion of the potential evolutionary response presupposes two things: first, that testosterone levels are heritable within each sex (i.e., that males with high T will

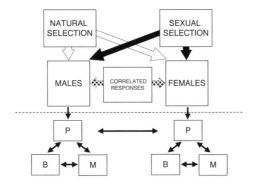

FIGURE 4.4. Creation of sex differences and resemblances by independent and correlated responses to natural and sexual selection. The boxes and arrows above the dotted line represent sexual (black arrows), natural (clear arrows), and correlated responses to selection (hatched double-headed arrow connecting male and female boxes) as they act on males and females. The boxes and arrows below the dashed line represent within-sex phenotypic variation in physiology, behavior, and morphology and the interactions among them. With respect to hormonal traits, in which the traits are correlated in their expression owing to hormonal pleiotropy, the phenotype can be constrained within each sex, owing to selection on one trait affecting the outcome of selection on others. Our emphasis here is on the potential for hormonal pleiotropy and correlated responses to selection to influence the coevolution of sex-typical phenotypes (Ketterson et al. 2005).

sire sons with high T), and second, that testosterone levels are genetically correlated between the sexes (i.e., that males with high T will sire daughters with high T). While we as yet have only limited evidence for either of these assertions in juncos (Atwell, J.W., pers. comm.), these are still reasonable assumptions based on several indirect lines of evidence. Testosterone has been found to be heritable in a variety of nonavian species (reviewed in Kempenaers et al. 2008; Pavitt et al. 2014), and maternal yolk T transfer is heritable in collared flycatchers (Tschirren et al. 2009). Furthermore, natural or artificial selection on testosterone or T-mediated traits in males have produced correlated responses in females in several species (Haley et al. 1990; McNeilly et al. 1988; Sandnabba 1996; Sinervo and Zamudio 2001), and circulating testosterone levels have been shown to be correlated between the sexes in birds, both across species and within species across the course of the breeding season (Ketterson et al. 2005, but see Garamszegi 2014; Goymann and Wingfield 2014; Ketterson 2014) (see figure 4.1). If this phenotypic correlation reflects an underlying genetic correlation, then selection on male T levels will likely affect their daughters, and selection on female T levels will have a corresponding response in their sons. That said, direct studies of genetic correlations between the sexes are still needed.

If there is a genetic correlation in testosterone levels between the sexes, then the results from implant studies may represent a case of sexual conflict. When considering males alone, our results suggest that if a mutation that produced an elevated testosterone profile arose, it would be favored over the current phenotype and would spread in the population. This raises the question of why males in our population currently have T profiles that appear to be sub-optimal. One possibility, stated simply, is that if these "mutant high-T males" also produced daughters with elevated testosterone, the negative effects on their daughters' fitness may counteract the fitness benefit that these males receive. Thus, selection against increased testosterone in females may constrain the evolution of testosterone in males, since males with high testosterone would produce daughters with a maladaptively high level of testosterone. It is not clear from our implant studies whether the converse is true: that is, whether selection on males may be constraining the evolution of testosterone levels in females and if females with lower T than the current population mean would be favored by selection. We therefore cannot say with certainty that the evolution of testosterone levels represents a true case of sexual conflict, although it may be the case that prevailing levels of testosterone represent a compromise between opposing selection on males and females.

An alternate possibility is that current mean testosterone levels in males represent a compromise not between selection on males and females but rather between selection on multiple male strategies. In much the same way that many vertebrate species display alternate male reproductive tactics, there may be multiple successful strategies with regard to testosterone level, with some males having a relatively higher level of testosterone and gaining fitness via extrapair copulations and others having a relatively lower level of testosterone and gaining fitness via increased effort to parental care and avoidance of the other costs of testosterone (Wikelski et al. 1999; Wingfield et al. 2001). We have seen that increasing male testosterone above the population mean also increases fitness, but the effects of lowering male testosterone on levels of parental care and fitness is unknown. These alternate physiological strategies need not be fixed but could vary with condition, physical or social environment, or age. Indeed, even within our implant studies, many of the effects of testosterone differed in magnitude between first-year and older males (Raouf and Ketterson 1998; Reed et al. 2006). Selection on male testosterone may also be frequency dependent, with the fitness benefits of elevated testosterone depending on the abundance of more parental, lower-T males. We must also consider that in our experimental studies, the increased testosterone was being provided from the exogenous implant; there may be costs to increasing endogenous production of testosterone (e.g., increased production of reactive oxygen, Zirkin and Chen 2000) that would affect the fitness of naturally high-T males.

H. The Value of Long-Term Studies

Another important factor to consider when studying the evolution of testosterone—or any phenotypic trait—is the role of temporal variation. Selection may vary not only between the sexes, between strategies, or between age classes but also over time. Because selective pressure is generated by the varying success of individuals interacting with their abiotic and biotic environments, a change in either or both of these environments can potentially lead to a change in the strength or even the direction of selection. We measure selection primarily during the breeding season, but variation in testosterone may also have effects during migration or the overwintering period that affect overall fitness. Selection may also vary between as well as within years: we have been studying the Virginia population of juncos for thirty years, and in that time we have seen annual variation in almost every feature we can measure. This includes long-term directional change, such as first egg dates being influenced by temperature variables (Friedline, Rosvall, Gerlach, and Ketterson, unpublished data); cyclical patterns, such as the level of nest predation, which is driven at least in part by acorn masting cycles (Clotfelter et al. 2007; Reed et al. 2006); and more stochastic variation, such as that of recruitment rates, adult sex ratio, and extrapair offspring production (Gerlach and Ketterson, unpublished data). We have also shown that there can be annual variation in selective gradients on mating success (Gerlach et al. 2012, supplemental materials), so it is reasonable to assume that there may be similar variation in selection on testosterone levels, particularly since T is so tightly linked to mating effort. Indeed, we can see the impact of this year effect by comparing treatment differences in net reproductive success of males in a study that consisted of four years of data (1990–1993, Raouf et al. 1997) vs. one that incorporated an additional seven years (1990–2000, Reed et al. 2006) (see section E.3, "Offspring Production," above). Thus, it is worth being aware that annual variation in a number of variables has the potential to strongly affect the selective pressures and evolutionary responses that we observe.

Annual variation in selection on testosterone levels may offer another potential explanation for the disparity between our experimental results regarding the increased fitness of high-T males and the mean T profile of the current population. Selection may favor elevated testosterone in some years (e.g., when the sex ratio is male-biased and males that invest most in mating effort are likely to have higher fitness) but not in others (e.g., when predation pressure is high and increased rates of parental behavior are necessary to achieve reproductive success). The previous examples are male specific, but the principle is equally applicable to females; elevated levels of testosterone may be generally maladaptive due to their effect on reproduction but may be favored in particular circumstances when competition for mates or other breeding resources is strong (Cain and Ketterson 2013). In both cases, if the variation between years is high, individuals with elevated T may be favored in one year, but their offspring may face a substantially different selective environment. The net effect of this type of varying selection is likely to be stabilizing selection, although selection in any shorter period may appear to be directional (or disruptive, if there are two "strategies" with equivalent fitness in the same environment, as discussed above). For this reason, it is critical when undertaking any kind of selection analysis that spans multiple years to acknowledge and examine the potentially large role of temporal variation.

I. The Subtleties of the Endocrine System, or, What Can't We Learn from Experimental Manipulations?

Experimental studies provide researchers with a valuable tool for understanding the relationships between hormones, physiology, behavior, and fitness. The experimental nature of the process and the group comparisons involved in the data analysis allow us to create a group of individuals at one extreme of the distribution of potential life-history trade-offs and to make statements about the causal nature of endocrine mechanisms. However, these same benefits have the effect of obscuring individual variation and its potential relationship with fitness. For example, implants provide manipulated animals with constant, high levels of testosterone, although we know that in unmanipulated animals, testosterone levels are extremely labile, in terms of both seasonal patterns and short-term changes related to social interactions. Because testosterone can mediate such a wide range of other traits, it is highly unlikely that any single constant level of circulating testosterone will be optimal. It is much more likely that selection will favor phenotypes that can vary their level of testosterone to be appropriate for the circumstances.

If selection is acting primarily on an individual's ability to produce these short-term bursts of testosterone, then measuring the fitness effects of long-term alterations in circulating levels may not provide a comprehensive picture of the likely evolution of hormonally mediated traits. In fact, we do see a clear relationship between an individual's ability to produce short-term rises in testosterone (as measured by a GnRH challenge) and a variety of fitness-related traits (see chapters 6 and 7, this volume). It may also be the case that variation in testosterone levels (whether baseline or short-term elevations) is less salient for determining behavioral phenotype than is sensitivity of the relevant neural area or other target tissue. For example, in tropical birds, individuals show year-round patterns of territorial aggression (typically a T-mediated behavior), despite very low circulating levels of testosterone (Dittami and Gwinner 2009; Hau et al. 2000; Levin and Wingfield 1992; Stutchbury and Morton 2001; Wikelski et al. 2003). If individuals vary in terms of their receptor density or sensitivity, or the amount of metabolic enzymes that convert testosterone to estradiol or other bioactive compounds, then the link between hormone production and phenotypic output is less straightforward (see chapter 7, this volume). Selection may therefore have many other aspects of the system on which to act, either instead of or in addition to circulating testosterone.

J. Summary

Studying evolution in an organismal context is difficult, particularly when studying those organisms in a free-living situation. Studying the evolution of the endocrine system is particularly complex, since it involves so many potential layers of variation: between individuals, between life-history stages, between time points, between social contexts, and so forth. We cannot study and understand the effects of all of this variation all at once, so by using manipulative experiments such as hormone implants, we can hold some aspects constant while looking at the effects on others. These types of studies give us valuable insights into the processes that are at work in unmanipulated birds across the full range of variation. However, it is important to bear in mind that even in most studies that do not involve manipulation, we are looking at only a small slice of time relative to the life of an individual, the history of the population, and the course of evolutionary change. By taking a long-term view, we can begin to see the effects of variation that occur over a much larger scale than a single breeding season.

Acknowledgments

First and foremost, we thank all of the researchers who have contributed their time, energy, and expertise to collecting field data on the junco over the thirty-plus years of this long-term project. There are far too many to list individually, but this research would not have been possible without their hard work and dedication. We would also like to thank Eric Snajdr, for coordinating many years of this research; Charles Ziegenfus, for contributing his time and expertise; James Murray, Henry Wilbur, Butch Brodie, and the Mountain Lake Biological Station for providing us with facilities and access to the juncos; and the Mountain Lake Hotel (now the Mountain Lake Lodge) and members of the Dolinger family, for allowing us access to their properties (and the juncos that live there). Thanks also to the anonymous reviewers of this book for providing helpful feedback on this chapter.

References

Adkins-Regan, E. 2005. *Hormones and Animal Social Behavior*. Monographs in Behavior and Ecology. Princeton: Princeton University Press.

———. 2008. Do hormonal control systems produce evolutionary inertia? *Philosophical Transactions of the Royal Society B-Biological Sciences* 363:1599–1609.

Agrawal, A. F., and J. R. Stinchcombe. 2009. How much do genetic covariances alter the rate of adaptation? *Proceedings of the Royal Society B-Biological Sciences* 276:1183–91.

Andersson, M. B. 1982. Female choice selects for extreme tail length in a widowbird. *Nature* 299:818–20.

Astheimer, L. B., W. A. Buttemer, and J. C. Wingfield. 1992. Interactions of corticosterone with feeding, activity, and metabolism in passerine birds. *Ornis Scandinavica* 23:355–65.

Bartke, A. 2011. Pleiotropic effects of growth hormone signaling in aging. *Trends in Endocrinology and Metabolism* 22:437–42.

Bray, O. E., J. J. Kennelly, and J. L. Guarino. 1975. Fertility of eggs produced on territories of vasectomized red-winged blackbirds. *Wilson Bulletin* 87:187–95.

Cain, K. E., and E. D. Ketterson. 2013. Costs and benefits of competitive traits in females: Aggression, maternal care, and reproductive success. *PLoS ONE* 8: e77816.

Casto, J. M., V. Nolan, Jr., and E. D. Ketterson. 2001. Steroid hormones and immune function: Experimental studies in wild and captive dark-eyed juncos (*Junco hyemalis*). *American Naturalist* 157:408–20.

Cawthorn, J. M., D. L. Morris, E. D. Ketterson, and V. Nolan, Jr. 1998. Influence of experimentally elevated testosterone on nest defence in dark-eyed juncos. *Animal Behaviour* 56:617–21.

Champy, C., and N. Kritch. 1925. Le tissu mucoélastique de la crête du coq, réactif de l'hormone sexuelle. *Comptes Rendus Hebdomadaires des Séances et Mémoires de la Société de Biologie et des ses Filiales* 92:683–85.

Chandler, C. R., E. D. Ketterson, and V. Nolan, Jr. 1997. Effects of testosterone on use of space by male dark-eyed juncos when their mates are fertile. *Animal Behaviour* 54:543–49.

Chandler, C. R., E. D. Ketterson, V. Nolan, Jr., and C. Ziegenfus. 1994. Effects of testosterone on spatial activity in free-ranging male dark-eyed juncos, *Junco hyemalis. Animal Behaviour* 47:1445–55.

Charnov, E. L. 1993. *Life History Invariants: Some Explorations of Symmetry in Evolutionary Ecology*. New York: Oxford University Press.

Clotfelter, E. D., D. M. O'Neal, J. M. Gaudioso, J. M. Casto, I. M. Parker-Renga, E. A. Snajdr, D. L. Duffy, V. Nolan, Jr., and E. D. Ketterson. 2004. Consequences of elevating plasma testosterone in females of a socially monogamous songbird: Evidence of constraints on male evolution? *Hormones and Behavior* 46: 171–78.

Clotfelter, E. D., A. B. Pedersen, J. A. Cranford, N. Ram, E. A. Snajdr, V. Nolan, Jr., and E. D. Ketterson. 2007. Acorn mast drives long-term dynamics of rodent and songbird populations. *Oecologia* 154:493–503.

Cody, M. L. 1966. A general theory of clutch size. *Evolution* 20:174–84.

Cox, R. M., and R. Calsbeek. 2009. Sexually antagonistic selection, sexual dimorphism, and the resolution of intralocus sexual conflict. *American Naturalist* 173: 176–87.

Crews, D., and M. C. Moore. 1986. Evolution of mechanisms controlling mating behavior. *Science* 231:121–25.

Dawson, A., S. A. Hinsley, P. N. Ferns, R. H. C. Bonser, and L. Eccleston. 2000. Rate of moult affects feather quality: A mechanism linking current reproductive effort to future survival. *Proceedings of the Royal Society B-Biological Sciences* 267:2093–98.

Dittami, J. P., and E. Gwinner. 2009. Annual cycles in the African stonechat *Saxicola torquata axillaris* and their relationship to environmental factors. *Journal of Zoology* 207:357–70.

Eikenaar, C., J. Husak, C. Escallón, and I. T. Moore. 2012. Variation in testosterone and corticosterone in amphibians and reptiles: Relationships with latitude, elevation, and breeding season length. *American Naturalist* 180:642–54.

Eikenaar, C., M. Whitham, J. Komdeur, M. Van Der Velde, and I. T. Moore. 2011. Endogenous testosterone is not associated with the trade-off between paternal and mating effort. *Behavioral Ecology* 22:601–8.

Emlen, S., and L. W. Oring. 1977. Ecology, sexual selection, and evolution of mating systems. *Science* 197:215–23.

Enstrom, D., E. D. Ketterson, and V. Nolan, Jr. 1997. Testosterone and mate choice in the dark-eyed junco. *Animal Behaviour* 54:1135–146.

Finch, C. E., and M. R. Rose. 1995. Hormones and the physiological architecture of life-history evolution. *Quarterly Review of Biology* 70:1–52.

Flatt, T., M.-P. Tu, and M. Tatar. 2005. Hormonal pleiotropy and the juvenile hormone regulation of *Drosophila* development and life history. *Bioessays* 27:999–1010.

Galbraith, H. 1988. Effects of egg size and composition on the size, quality, and survival of lapwing *Vanellus vanellus* chicks. *Journal of Zoology* 214:383–98.

Garamszegi, L. Z. 2014. Female peak testosterone levels in birds tell an evolutionary story. *Behavioral Ecology* 25:700–701.

Gerlach, N. M., and E. D. Ketterson. 2013. Experimental elevation of testosterone lowers fitness in female dark-eyed juncos. *Hormones and Behavior* 63:782–90.

Gerlach, N. M., J. W. McGlothlin, P. G. Parker, and E. D. Ketterson. 2012. Reinterpreting Bateman gradients: Multiple mating and selection in both sexes of a songbird species. *Behavioral Ecology* 23:1078–88.

Gil, D. 2003. Golden eggs: Maternal manipulation of offspring phenotype by egg androgen in birds. *Ardeola* 50:281–94.

Gowaty, P. A. 1983. Male parental care and apparent monogamy among eastern bluebirds (*Sialia sialis*). *American Naturalist* 121:149–57.

Goymann, W., and M. M. Landys. 2011. Testosterone and year-round territoriality in tropical and non-tropical songbirds. *Journal of Avian Biology* 42:485–89.

Goymann, W., I. T. Moore, A. Scheuerlein, K. Hirschenhauser, A. Grafen, and J. C. Wingfield. 2004. Testosterone in tropical birds: Effects of environmental and social factors. *American Naturalist* 164:327–34.

Goymann, W., and J. C. Wingfield. 2014. Male-to-female testosterone ratios, dimorphism, and life history—what does it really tell us? *Behavioral Ecology* 25:685–89.

Groothuis, T. G. G., W. Müller, N. Von Engelhardt, C. Carere, and C. M. Eising. 2005. Maternal hormones as a tool to adjust offspring phenotype in avian species. *Neuroscience and Biobehavioral Reviews* 29:329–52.

Haley, C. S., G. J. Lee, M. Ritchie, and R. B. Land. 1990. Direct responses in males and correlated responses for reproduction in females to selection for testicular size adjusted for body weight in young male lambs. *Journal of Reproduction and Fertility* 89:383–96.

Hau, M. 2007. Regulation of male traits by testosterone: Implications for the evolution of vertebrate life histories. *Bioessays* 29:133–44.

Hau, M., R. E. Ricklefs, M. Wikelski, K. A. Lee, and J. D. Brawn. 2010. Corticosterone, testosterone, and life-history strategies of birds. *Proceedings of the Royal Society B-Biological Sciences* 277:3203–12.

Hau, M., M. Wikelski, K. K. Soma, and J. C. Wingfield. 2000. Testosterone and year-round territorial aggression in a tropical bird. *General and Comparative Endocrinology* 117:20–33.

Hau, M., and J. C. Wingfield. 2011. Hormonally regulated trade-offs: Evolutionary variability and phenotypic plasticity in testosterone signaling pathways. In *Mechanisms of Life History Evolution: The Genetics and Physiology of Life History Traits and Trade-offs*, edited by T. Flatt and A. Heyland, Oxford Biology. Oxford: Oxford University Press.

Hegner, R. E., and J. C. Wingfield. 1987. Effects of experimental manipulation of testosterone levels on parental investment and breeding success in male house sparrows. *Auk* 104:462–69.

Hudman, S. P., E. D. Ketterson, and V. Nolan, Jr. 2000. Effects of time of sampling on oocyst detection and effects of age and experimentally elevated testosterone on prevalence of coccidia in male dark-eyed juncos. *Auk* 117:1048–51.

Kast, T. L., E. D. Ketterson, and V. Nolan, Jr. 1998. Variation in ejaculate quality in dark-eyed juncos according to season, stage of reproduction, and testosterone treatment. *The Auk* 115:684–93.

Kempenaers, B., A. Peters, and K. Foerster. 2008. Sources of individual variation in plasma testosterone levels. *Philosophical Transactions of the Royal Society B-Biological Sciences* 363:1711–23.

Ketterson, E. D. 2014. Male and female testosterone: Is one sex made in the image of the other? A comment on Goymann and Wingfield. *Behavioral Ecology* 25: 702.

Ketterson, E. D., J. W. Atwell, and J. W. McGlothlin. 2009. Phenotypic integration and independence: Hormones, performance, and response to environmental change. *Integrative and Comparative Biology* 49:365–79.

Ketterson, E. D., and V. Nolan, Jr. 1999. Adaptation, exaptation, and constraint: A hormonal perspective. *American Naturalist* 154:S4–S25.

Ketterson, E. D., V. Nolan, Jr., J. M. Casto, C. A. Buerkle, E. Clotfelter, J. L. Grindstaff, K. J. Jones, J. L. Lipar, F. M. A. Mcnabb, D. L. Neudorf, I. Parker-Renga, S. J. Schoech, and E. Snajdr. 2001. Testosterone, phenotype and fitness: A research program in evolutionary behavioral endocrinology. In *Avian Endocrinology*, edited by A. Dawson and C. M. Chaturvedi, 9–40. New Delhi: Narosa Publishing House.

Ketterson, E. D., V. Nolan, Jr., M. J. Cawthorn, P. G. Parker, and C. Ziegenfus. 1996. Phenotypic engineering: Using hormones to explore the mechanistic and functional bases of phenotypic variation in nature. *Ibis* 138:70–86.

Ketterson, E. D., V. Nolan, Jr., and M. Sandell. 2005. Testosterone in females: Mediator of adaptive traits, constraint on sexual dimorphism, or both? *American Naturalist* 166:S85–S98.

Ketterson, E. D., V. Nolan, Jr., L. Wolf, and C. Ziegenfus. 1992. Testosterone and avian life histories: Effects of experimentally elevated testosterone on behavior and correlates of fitness in the dark-eyed junco (*Junco hyemalis*). *American Naturalist* 140:980–99.

Ketterson, E. D., V. Nolan, Jr., L. Wolf, C. Ziegenfus, A. M. Dufty, Jr., G. F. Ball, and T. S. Johnsen. 1991. Testosterone and avian life histories: The effect of experimentally elevated testosterone on corticosterone and body mass in dark-eyed juncos. *Hormones and Behavior* 25:489–503.

Lack, D. 1968. *Ecological Adaptations for Breeding in Birds.* London: Methuen.

Lande, R. 1987. Genetic correlations between the sexes in the evolution of sexual dimorphism and mating preferences. In *Sexual Selection: Testing the Alternatives,* edited by J. W. Bradbury and M. B. Andersson, 83–95. Chichester: Wiley.

Lande, R., and S. J. Arnold. 1983. The measurement of selection on correlated characters. *Evolution* 37:1210–26.

Lessells, C. K. M. 2008. Neuroendocrine control of life histories: What do we need to know to understand the evolution of phenotypic plasticity? *Philosophical Transactions of the Royal Society B-Biological Sciences* 363:1589–98.

Levin, R. N., and J. C. Wingfield. 1992. The hormonal control of territorial aggression in tropical birds. *Ornis Scandinavica* 23:284–91.

Ludwig, A. W., and N. F. Boas. 1950. The effects of testosterone on the connective tissue of the comb of the cockerel. *Endocrinology* 46:291–98.

Lynn, S. E., A. M. Houtman, W. W. Weathers, E. D. Ketterson, and V. Nolan, Jr. 2000. Testosterone increases activity but not daily energy expenditure in captive male dark-eyed juncos, *Junco hyemalis. Animal Behaviour* 60:581–87.

Magrath, R. 1991. Nestling weight and juvenile survival in the blackbird, *Turdus merula. Journal of Animal Ecology* 60:335–51.

Marler, C. A., and M. C. Moore. 1988. Evolutionary costs of aggression revealed by testosterone manipulations in free-living male lizards. *Behavioral Ecology and Sociobiology* 23:21–26.

McGee, L. C., M. Juhn, and L. V. Domm. 1928. The development of secondary sex characters in capons by injections of extracts of bull testes. *American Journal of Physiology* 87:406–35.

McGlothlin, J. W., and E. D. Ketterson. 2008. Hormone-mediated suites as adaptations and evolutionary constraints. *Philosophical Transactions of the Royal Society B-Biological Sciences* 363:1611–20.

McGlothlin, J. W., D. L. H. Neudorf, J. M. Casto, V. Nolan, Jr., and E. D. Ketterson. 2004. Elevated testosterone reduces choosiness in female dark-eyed juncos (*Junco hyemalis*): Evidence for a hormonal constraint on sexual selection? *Proceedings of the Royal Society B-Biological Sciences* 271:1377–84.

McKinney, F., K. M. Cheng, and D. J. Bruggers. 1984. Sperm competition in apparently monogamous birds. In *Sperm Competition and the Evolution of Animal Mating Systems,* edited by R. L. Smith, 523–45. San Diego: Academic Press, Inc.

McNeilly, J. R., M. Fordyce, R. B. Land, G. B. Martin, A. J. Springbett, and R. Webb. 1988. Changes in the feedback control of gonadotrophin secretion in ewes from lines selected for testis size in the ram lamb. *Journal of Reproduction and Fertility* 84:213–21.

Merilä, J., and M. Björklund. 2004. Phenotypic integration as a constraint and adaptation. In *Phenotypic Integration: Studying the Ecology and Evolution of*

Complex Phenotypes, edited by M. Pigliucci and K. Preston, 107–29. Oxford: Oxford University Press.

Müller, W., J. Vergauwen, and M. Eens. 2009. Long-lasting consequences of elevated yolk testosterone levels on female reproduction. *Behavioral Ecology and Sociobiology* 63:809–16.

Nelson, R. J. 2005. *An Introduction to Behavioral Endocrinology*. Sunderland, MA: Sinauer Associates.

Nolan, V., Jr., E. D. Ketterson, D. A. Cristol, C. M. Rogers, E. D. Clotfelter, R. C. Titus, S. J. Schoech, and E. Snajdr. 2002. Dark-eyed junco (*Junco hyemalis*). In *The Birds of North America*, edited by A Poole and F. Gill, no. 716. Philadelphia: The Birds of North America, Inc.

Nolan, V., Jr., E. D. Ketterson, C. Ziegenfus, D. P. Cullen, and C. R. Chandler. 1992. Testosterone and avian life histories: Effects of experimentally elevated testosterone on prebasic molt and survival in male dark-eyed juncos. *Condor* 94: 364–70.

O'Neal, D. M., D. G. Reichard, K. Pavilis, and E. D. Ketterson. 2008. Experimentally-elevated testosterone, female parental care, and reproductive success in a songbird, the dark-eyed junco (*Junco hyemalis*). *Hormones and Behavior* 54:571–78.

Pavitt, A. T., C. A. Walling, J. M. Pemberton, and L. E. B. Kruuk. 2014. Heritability and cross-sex genetic correlations of early-life circulating testosterone levels in a wild mammal. *Biology Letters* 10:20140685.

Pigliucci, M., and K. A. Preston. 2004. *Phenotypic Integration: Studying the Ecology and Evolution of Complex Phenotypes*. Oxford: Oxford University Press.

Raouf, S. A., and E. D. Ketterson. 1998. Patterns of extra-pair fertilizations in dark-eyed juncos (*Junco hyemalis*): The effects of testosterone and age. *Ornithological Monographs* 49:81–101.

Raouf, S. A., P. G. Parker, E. D. Ketterson, V. Nolan, Jr., and C. Ziegenfus. 1997. Testosterone affects reproductive success by influencing extra-pair fertilizations in male dark-eyed juncos (Aves: *Junco hyemalis*). *Proceedings of the Royal Society B-Biological Sciences* 264:1599–1603.

Reed, W. L., M. E. Clark, P. G. Parker, S. A. Raouf, N. Arguedas, D. S. Monk, E. Snajdr, V. Nolan, Jr., and E. D. Ketterson. 2006. Physiological effects on demography: A long-term experimental study of testosterone's effects on fitness. *American Naturalist* 167:667–83.

Ricklefs, R. E. 2000. Density dependence, evolutionary optimization, and the diversification of avian life histories. *Condor* 102:9–22.

Ricklefs, R. E., and M. Wikelski. 2002. The physiology/life-history nexus. *Trends in Ecology and Evolution* 17:462–68.

Roberts, T. A., and J. J. Kennelly. 1980. Variation in promiscuity among red-winged blackbirds. *Wilson Bulletin* 92:110–12.

Roff, D. A. 2002. *Life History Evolution*. Sunderland, MA: Sinauer Associates.

Runfeldt, S., and J. C. Wingfield. 1985. Experimentally prolonged sexual activity in female sparrows delays termination of reproductive activity in their untreated mates. *Animal Behaviour* 33:403–10.

Sandnabba, N. K. 1996. Selective breeding for isolation-induced intermale aggression in mice: Associated responses and environmental influences. *Behavior Genetics* 26:477–88.

Schoech, S. J., E. D. Ketterson, and V. Nolan, Jr. 1999. Exogenous testosterone and the adrenocortical response in dark-eyed juncos. *Auk* 116:64–72.

Schwabl, H., D. Holmes, R. Strasser, and A. Scheuerlein. 2012. Embryonic exposure to maternal testosterone influences age-specific mortality patterns in a captive passerine bird. *Age* 34:87–94.

Silverin, B. 1980. Effects of long-acting testosterone treatment on free-living pied flycatchers, *Ficedula hypoleuca*, during the breeding period. *Animal Behaviour* 28:906–12.

———. 1991. Behavioral, hormonal, and morphological responses of free-living male pied flycatchers to estradiol treatment of their mates. *Hormones and Behavior* 25:38–56.

Sinervo, B., and P. Licht. 2006. Hormonal and physiological control of clutch size, egg size, and egg shape in side-blotched lizards (*Uta stansburiana*): Constraints on the evolution of lizard life histories. *Journal of Experimental Zoology* 257: 252–64.

Sinervo, B., and K. R. Zamudio. 2001. The evolution of alternative reproductive strategies: Fitness differential, heritability, and genetic correlation between the sexes. *Journal of Heredity* 92:198–205.

Smith, H. G., H. Källander, and J.-Å. Nilsson. 1989. The trade-off between offspring number and quality in the great tit *Parus major*. *Journal of Animal Ecology* 58:383–401.

Staub, N. L., and M. De Beer. 1997. The role of androgens in female vertebrates. *General and Comparative Endocrinology* 108:1–24.

Stearns, S. C. 1989. Trade-offs in life-history evolution. *Functional Ecology* 3: 259–68.

———. 1992. *The Evolution of Life Histories*. New York: Oxford University Press.

———. 2000. Life history evolution: Successes, limitations, and prospects. *Naturwissenschaften* 87:476–86.

Stutchbury, B. J. M., and E. S. Morton. 2001. *Behavioral Ecology of Tropical Birds*. San Diego: Elsevier Academic Press.

Tschirren, B., J. Sendecka, T. G. G. Groothuis, L. Gustafsson, and B. Doligez. 2009. Heritable variation in maternal yolk hormone transfer in a wild bird population. *American Naturalist* 174:557–64.

Veiga, J. P., and V. Polo. 2008. Fitness consequences of increased testosterone levels in female spotless starlings. *American Naturalist* 172:42–53.

Watson, A., and R. Parr. 1981. Hormone implants affecting territory size and aggressive and sexual behavior in red grouse. *Ornis Scandinavica* 12:55–61.

Westneat, D. F. 1987. Extra-pair fertilizations in a predominantly monogamous bird: Genetic evidence. *Animal Behaviour* 35:877–86.

Wikelski, M., M. Hau, W. D. Robinson, and J. C. Wingfield. 2003. Reproductive seasonality of seven neotropical passerine species. *Condor* 105:683–95.

Wikelski, M., S. Lynn, J. C. Breuner, J. C. Wingfield, and G. J. Kenagy. 1999. Energy metabolism, testosterone, and corticosterone in white-crowned sparrows. *Journal of Comparative Physiology* A 185:463–70.

Williams, T. D. 2012a. Hormones, life-history, and phenotypic variation: Opportunities in evolutionary avian endocrinology. *General and Comparative Endocrinology* 176:286–95.

———. 2012b. *Physiological Adaptations for Breeding in Birds*. Princeton: Princeton University Press.

Wingfield, J. C. 1984. Androgens and mating systems: Testosterone-induced polygyny in normally monogamous birds. *Auk* 101:665–71.

Wingfield, J. C., G. F. Ball, A. M. Dufty, Jr., R. E. Hegner, and M. Ramenofsky. 1987. Testosterone and aggression in birds. *American Scientist* 75:602–8.

Wingfield, J. C., and D. S. Farner. 1975. The determination of five steroids in avian plasma by radioimmunoassay and competitive protein-binding. *Steroids* 26: 311–21.

Wingfield, J. C., R. E. Hegner, A. M. Dufty, Jr., and G. F. Ball. 1990. The "challenge hypothesis": Theoretical implications for patterns of testosterone secretion, mating systems, and breeding strategies. *American Naturalist* 136:829–46.

Wingfield, J. C., S. E. Lynn, and K. K. Soma. 2001. Avoiding the "costs" of testosterone: Ecological bases of hormone-behavior interactions. *Brain Behavior and Evolution* 57:239–51.

Wittenberger, J. F., and R. L. Tilson. 1980. The evolution of monogamy: Hypotheses and evidence. *Annual Review of Ecology and Systematics* 11:197–232.

Wolf, L., E. D. Ketterson, and V. Nolan, Jr. 1988. Paternal influence on growth and survival of dark-eyed junco young: Do parental males benefit? *Animal Behaviour* 36:1601–18.

———. 1990. Behavioral response of female dark-eyed juncos to the experimental removal of their mates: Implications for the evolution of male parental care. *Animal Behavior* 39:125–34.

———. 1991. Female condition and delayed benefits to males that provide parental care: A removal study. *Auk* 108:371–80.

Zirkin, B. R., and H. Chen. 2000. Regulation of Leydig cell steroidogenic function during aging. *Biology of Reproduction* 63:977–81

Zysling, D. A., T. J. Greives, C. W. Breuner, J. M. Casto, G. E. Demas, and E. D. Ketterson. 2006. Behavioral and physiological responses to experimentally elevated testosterone in female dark-eyed juncos (*Junco hyemalis carolinensis*). *Hormones and Behavior* 50:200–207.

CHAPTER FIVE

Hormonal Pleiotropy and the Evolution of Correlated Traits[1]

Joel W. McGlothlin and Ellen D. Ketterson

... there are many unknown laws of correlation of growth, which, when one part of the organisation is modified through variation, and the modifications are accumulated by natural selection for the good of the being, will cause other modifications, often of the most unexpected nature.
—Charles Darwin, 1859, *Origin of Species*, chapter 4

A. The Evolution of Correlated Traits

In the *Origin of Species*, Darwin (1859) noted the generality of "correlations of growth" and speculated that when changes in one character are "accumulated through natural selection, other parts become modified" (143). Such correlated responses to selection are a common observation among animal breeders and horticulturists in their attempts to improve a trait of interest via artificial selection. For example, MacArthur (1948) imposed twenty-one generations of selection for body size in house mice (*Mus musculus*) and observed correlated responses in reproductive timing and scaling relationships of various body parts, among other traits. Genetic correlations are also ubiquitous in natural populations, and, depending on the traits in question, vary from weak to strong and from positive to negative (Roff 1996).

Genetic correlations are of interest to evolutionary biologists because they can bias a population's evolution along lines predicted by the genetic

1. This chapter incorporates portions of McGlothlin and Ketterson (2008) and Ketterson et al. (2009) with permission of the publishers.

correlation rather than the direction most favored by selection. In the 1970s and 1980s, Lande and colleagues (Lande 1979; Lande and Arnold 1983) formulated a general mathematical model for the evolution of continuous traits. This framework included statistical methods that allowed evolutionary biologists to estimate the strength and mode of selection acting on multiple quantitative characters. A central component of this theory was the genetic variance-covariance matrix, or **G**, which describes both the degree to which traits are heritable (genetic variance) and coinherited (genetic covariance). A population's response to natural selection on multiple traits can be predicted using Lande's multivariate breeder's equation, which states that the evolution of a set of correlated characters is equal to the **G** matrix multiplied by a vector of selection gradients (β), which measures the strength of directional phenotypic selection. Importantly, both these parameters are estimable in natural populations, and a wealth of data has been generated regarding the patterns of phenotypic evolution in natural populations as a result (Endler 1986; Kingsolver et al. 2001, 2012; Kruuk 2004; Wilson et al. 2010).

When genetic correlations among traits are strong, groups of traits should tend to respond to selection as a unit. Depending on the direction of selection, genetic correlations may either accelerate or impede response to selection. The accelerating effect has received much attention in the study of sexual selection, where genetic correlations between the sexes can lead to runaway evolution of extreme ornaments and extreme preferences (Lande 1981). However, genetic correlations among traits are more often seen as a constraint on rather than a facilitator of phenotypic evolution (Cheverud 1984; Maynard Smith et al. 1985; Arnold 1992; Blows and Hoffmann 2005). When we consider a two-trait example, a response to selection should be most rapid when the direction of selection is aligned in the direction of the genetic correlation (figure 5.1), also known as the "genetic line of least resistance" (Schluter 1996). With two positively correlated traits, this occurs when selection favors a simultaneous increase or decrease in both traits. Response to antagonistic selection (i.e., against the direction of the genetic correlation) should be more difficult. In other words, a positive genetic correlation may prevent a response to selection for a decrease in one trait and an increase in the other (figure 5.1). If genetic correlations are persistent, they may constrain the divergence of populations, with population differences tending to reflect underlying genetic architecture more so than selection (Schluter 1996).

From another perspective, genetic correlations among traits may be the products of natural selection and may thus be viewed as adaptations as well as evolutionary constraints (Merilä and Björklund 2004). Patterns of covariation may arise because selection favors the integration of traits performing a common function (Olson and Miller 1958; Cheverud 1982). Theory predicts that selection, particularly a form known as correlational selection, should be able to change both the magnitude and direction of genetic covariances, aligning the pattern of genetic correlations in a population with the adaptive landscape (Lande 1980; Lande and Arnold 1983; Phillips and Arnold 1989; Rice 2000; Sinervo and Svensson 2002; Jones et al. 2003; Phillips and McGuigan 2006). Correlational selection describes situations in which traits interact in their effects on fitness; that is, the selective advantage of one trait depends on the value of another. The classic example of correlational selection comes from an experiment on garter snakes (*Thamnophis ordinoides*) by Brodie (1992), who showed that spotted snakes were more likely to survive if they performed evasive behavior, whereas striped snakes survived better if they did not perform the behavior. In juncos, we have demonstrated correlational sexual selection on male body size and plumage; in particular, the effect of tail white on male mating success depends upon whether the plumage trait is borne by a large or small male (McGlothlin et al. 2005).

When correlational selection favors a certain combination of traits, the degree to which the traits share a common genetic basis is predicted to increase over time (Lande 1980; Phillips and Arnold 1989; Phillips and McGuigan 2006). Indeed, correlational selection has been shown to coincide with existing genetic correlations in a number of cases (Brodie 1989, 1992; Conner and Via 1993; Morgan and Conner 2001; McGlothlin et al. 2005; Roff and Fairbairn 2012). In these studies, selection was found to favor combinations of traits that were known to be genetically correlated, as one would predict if past selection accounted for the present existence of the genetic correlations.

B. Hormones as Mediators of Correlated Suites of Traits

Genetic correlations are merely abstractions that measure the statistical outcomes of a myriad of biological processes. We can achieve a richer understanding of how traits are interrelated by moving a step beyond statistical description and examining the mechanisms that underlie the correlations we observe. At the level of genes, genetic correlations are the result of both pleiotropy, the common effect of loci on multiple traits, and linkage, the physical or statistical association of loci affecting one trait to loci affecting a second trait (Lande 1980). Genes exert their effects on the phenotype largely by being translated into proteins, which are then participants in the complex biochemical network of the cell and the physiological and developmental processes of the multicellular organism. Although we cannot ever hope to fully understand the complexity of the translation from genotype to phenotype, targeting certain pathways or systems can provide insight into the mechanistic underpinnings of correlated trait evolution.

Because of their coordination of multiple aspects of physiology, morphology, and behavior, hormones and their associated endocrine pathways are uniquely positioned to inform our understanding of the causes and consequences of trait correlations. Due to their multifaceted phenotypic effects, hormones are often considered physiological analogs of genes with pleiotropic effects (Finch and Rose 1995; Ketterson and Nolan 1999; Flatt et al. 2005). Hormones can also serve as a link between an organism's environment and the coordinated expression of an appropriate set of phenotypes. For example, in many seasonally breeding birds, increasing day length influences gonadal growth and secretion of gonadal hormones, which then mediate expression of a suite of territorial and reproductive behaviors. The physiological details of the links among environment, hormones, and phenotype are usually complex; for simplicity, we refer to a set of such interconnections as a hormonal pathway. Many aspects of such pathways are likely to vary among individuals, generating phenotypic variation (chapters 6–7, this volume) (Hau 2007; Ball and Balthazart 2008). For example, individuals may differ in the rate of hormone synthesis, release, and degradation, leading to variation in circulating hormone levels. There is ample room for individual variation in how a hormonal pathway responds to environmental stimuli, as well as how hormones interact with each other or with target tissues to cause phenotypic effects.

Like genetic correlations, hormonal pleiotropy arising from common mechanistic mediation of multiple aspects of the phenotype may be seen both as an adaptive product of past selection and as a potential constraint on future evolution. One case where this dual nature is apparent is in the role of hormones as physiological mediators of life-history trade-offs (chapter 4, this volume) (Ketterson and Nolan 1992; Sinervo and Svensson 1998; Zera and Harshman 2001; Ricklefs and Wikelski 2002; Flatt and Heyland 2011). Hormone mediation may allow an organism to optimally allocate limited time or energy among competing functions such as growth, reproduction, or immune function. However, this hormone mediation also has the potential to impede a population's ability to respond to novel selection pressures encountered in a new environment, particularly if selection favors new relationships among hormone-mediated traits.

In the following sections, we address the potential ways that hormonal pleiotropy may affect a population's response to natural selection and, in turn, how selection may influence suites of hormone-mediated traits. We then recommend several empirical approaches for studying the potential evolutionary effects of hormones in natural populations. Applications of these approaches in studies of the dark-eyed junco will be discussed in chapters to follow (chapters 6, 7, and 10, this volume).

C. Natural Selection and Hormone-Mediated Suites

The extent to which a population's evolutionary response to selection will be enhanced or impeded by hormonal regulation of traits depends on the mode of selection (the shape of the relationship between phenotype and fitness) as well its strength and consistency (whether selection fluctuates or continues in the same direction for many generations). As discussed above, genetic correlations can drastically alter the response to directional selection (figure 5.1). Similarly, mediation by a common hormone may facilitate simultaneous evolutionary change in two traits in the direction of the hormone-mediated correlation. At a physiological level, this type of selection may lead to an evolutionary increase in the circulating hormone signal or a coordinated change in receptor expression across multiple tissues. In male juncos and other songbirds, traits related to mating and territorial aggression are enhanced by experimentally elevated testosterone levels, while parental behavior is diminished (chapter 4, this volume). If selection favored, for example, a simultaneous increase in territoriality and courtship behavior, both might respond rapidly, with a corresponding increase in testosterone levels or receptor expression, which may evolve as a consequence. Carere et al. (2003) observed such a response in a study that imposed artificial selection for coping strategy, a composite measure including aggression and exploratory behavior in great tits (*Parus major*). Birds that were bolder and more aggressive displayed a reduced corticosteroid stress response, while selection in the opposite direction led to an increased hormonal response to stress. At the other extreme, selection

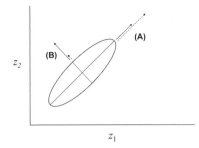

FIGURE 5.1 Schematic representation of the effect of genetic covariance on selection response. Here, traits z_1 and z_2 are positively correlated. Directional selection in two different directions is represented by the solid arrows, and response to selection is represented by the dashed arrows. Response to selection acting in the same direction on both traits (i.e., in the direction of the correlation [A]) is facilitated by genetic covariance, and the selection response is positive for both traits. Response to selection acting in opposite directions (i.e., against the direction of the correlation [B]) is slowed.

acting against the direction of the correlation should be constrained by hormonal mediation, at least in the short term. Returning to our previous example, if selection were to favor both increased territoriality and paternal behavior, the response may be slower because androgens tend to increase one trait while decreasing the other.

However, we predict that when selection to uncouple correlations is strong and consistent, as might be expected when a population colonizes a new environment, the power of hormonal mediation to act as a constraint may be overcome. This should occur because correlations among traits—even those mediated by hormones—should be able to evolve under the appropriate selective regime. To understand how this may occur, we must consider correlational selection. When traits are under correlational selection, that is, when traits interact in their effects on fitness, linkage disequilibrium (statistical associations among loci) is generated by differential survival or reproduction of individuals with favored combinations of alleles. At the same time, correlational selection should favor alleles that generate pleiotropy between the traits under selection, thus generating greater stability than linkage disequilibrium in the face of recombination. Correlational selection acting on sets of functionally related traits is likely to play a major role in shaping the composition of hormone-mediated suites of traits. When groups of traits are favorable when expressed together, selection should favor (or strengthen) their mediation by a common hormone. Conversely, when such relationships become unfavorable, the pattern of correlational selection should shift, leading to the loss of

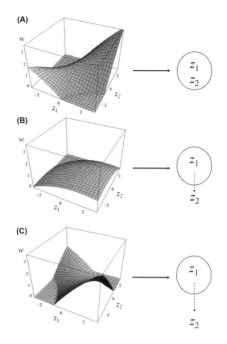

FIGURE 5.2. Hypothetical example of the effect of correlational selection on hormone-mediated trait suites. The plots at left are individual fitness surfaces, with two traits, z_1 and z_2, on the horizontal axes and relative fitness, w, on the vertical axis (Brodie et al. 1995). In (A), the two traits are regulated by a common hormone, as represented by the circle at the right. The fitness surface shows moderate directional selection on both traits as well as relatively strong correlational selection. This selective regime is predicted to maintain the correlation between the traits and, hence, their common hormonal basis. In (B), the direction of selection on z_2 is reversed, which should lead to a decrease in the genetic correlation and, potentially, the dissociation of z_2 from hormonal regulation. In (C), negative correlational selection also occurs, accelerating the dissociation of z_2.

traits from hormone-mediated suites. Several alternative scenarios of correlational selection are illustrated in figure 5.2. In these fitness surfaces, which are three-dimensional representations of selection in a population (Brodie et al. 1995), trait combinations that lead to high fitness (w, plotted on the vertical axis) are represented by peaks. When traits interact in their effects on fitness—that is, when there is correlational selection—the fitness surface shows greater curvature in directions at a diagonal to the two trait axes. Flatter surfaces indicate weaker correctional selection. In figure 5.2A, individuals with high values of two traits have higher fitness, and the traits' effects on fitness interact; that is, they are under both directional and correlational selection, creating a rising fitness ridge. Proximately, the correlation between the traits could be generated by a common hormonal mechanism, while ultimately this hormonal mechanism is maintained by correlational selection, which acts to strongly disfavor variants leading to a loss of hormonal regulation of one of the traits (e.g., by turning off the expression of the receptor in a specific tissue). When its direction corresponds to the direction of existing correlations among traits, correlational selection acts as the multivariate analogue of stabilizing selection, providing stability to correlated trait suites (Blows and Brooks 2003; Estes and Arnold 2007). Thus, when correlational selection matches existing hormone-mediated suites, their composition should be maintained over time.

Now imagine the population has colonized a new environment where one of the traits is disfavored. Figures 5.2B and 5.2C show that such a change in selection may lead to the decoupling of traits from a hormone-mediated suite. In Figure 5.2B, correlational selection is simply reduced, because the two traits are no longer favorable when expressed together. In this case, the genetic covariance between the traits should tend to decrease over time; mutations that decouple one of the traits from hormonal regulation may now invade, and the composition of the hormonal suite may slowly change. To the extent that new fitness interactions between traits occur in the new environment, the direction of correlational selection may also become reversed (figure 5.2C). Such negative correlation should tend to weaken hormonal mediation of traits that no longer work well with other hormone-mediated traits and may even be responsible for the eventual dissociation of traits from hormone-mediated suites such that hormonal mediation of a given trait is lost altogether.

When correlational selection favors an alteration of the composition of hormone-mediated suites, evolutionary change may occur in many different aspects of a hormonal pathway, due to the inherent complexity of such systems (Nijhout 2003; Adkins-Regan 2005; Hau 2007; Ball and Balthazart 2008). One likely mechanism for such change involves an alteration in the expression of hormone receptors. In general, hormones have no effect on a tissue unless the tissue expresses the appropriate receptor, and the expression of a receptor in a novel location or with novel timing could allow a trait to be co-opted into a hormone-mediated suite. Such co-option of existing physiological mechanisms is probably common because it requires fewer evolutionary steps than building a new pathway de novo (Nijhout 2003). Parental care behavior in male chestnut-collared longspurs (*Calcarius ornatus*, Lynn et al. 2002) might serve as an example of a trait that has become dissociated from a hormone-mediated suite. Although testosterone suppresses male parental behavior in many song-

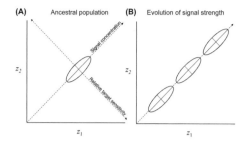

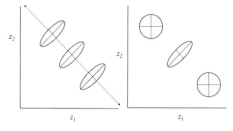

FIGURE 5.3. Evolutionary divergence of hormone-mediated suites. (A) In a hypothetical population, the correlation between z_1 and z_2 is mediated by variation in a hormone signal, which determines the direction of the correlation. Variation against the direction of the correlation may be determined by other factors, such as the relative sensitivity of the target tissues involved in the expression of the trait. In (B), (C), and (D), the ancestral population has given rise to two new populations where evolution of the hormonal suite has occurred. (B) Evolution via changes in strength of the hormone signal. (C) Evolution via changes in the relative sensitivity of targets. (D) One trait becomes dissociated from the hormonal suite altogether, perhaps by loss of target sensitivity to the signal as a result of negative correlational selection.

birds (chapter 4, this volume), longspur paternal care is insensitive to experimentally enhanced androgen levels. On an ultimate level, this change probably arises due to the extreme importance of male care in this species. Proximately, this loss of sensitivity may be related to altered location or abundance of testosterone receptors in neural or peripheral tissues.

The potential for hormonal pathways and the composition of hormone-mediated suites to evolve in response to selection suggests that hormones are not likely to act as rigid constraints on the divergence of populations over long time scales. However, it is likely that hormone-mediated suites (and their underlying physiological basis) will diverge in different ways depending on both the direction and mode of selection (e.g., the relative importance of directional vs. correlational selection) (figure 5.3). For example, if populations are under selection to diverge in directions delineated by existing hormonal pleiotropy, evolutionary change may be rapid and may occur by a change in the strength of the hormonal signal, coordinated change in target sensitivity, or both (figure 5.3B). Alternatively, if selection favors opposing changes in hormone-mediated traits, evolution might initially be quite slow, although some change may occur via evolutionary change in the relative sensitivities of target tissues (figure 5.3C). Yet another possibility is that negative correlational selection could favor an evolutionary loss of sensitivity in one target tissue, leading to a loss of one of the traits from the hormone-mediated suite (figure 5.3D). This would allow populations to diverge in any direction, as the two traits are no longer correlated by the hormonal signal. Of course, these evolutionary pathways are not mutually exclusive. We know of no examples, as yet, that reveal the relative frequency of these modes of divergence with respect to hormones, but research comparing populations of dark-eyed juncos has provided some initial insight (chapters 10–11, this volume).

D. Importance of Studies of Natural Variation in Hormone-Mediated Suites

Many studies examining potential effects of hormones on adaptation and constraint have used "phenotypic engineering" (chapter 4, this volume) (e.g., Ketterson and Nolan 1999; De Ridder, Pinxten and Eens 2000; Reed et al. 2006; Gerlach and Ketterson 2013). These studies, which typically examine the effects of experimentally enhanced hormone levels on behavior and components of fitness, allow inference of physiological mechanisms and experimental tests of adaptive hypotheses. However, they provide only limited insight into the evolutionary processes that shape hormone-mediated suites because experimental manipulation obscures individual variation, the raw material for natural selection. In order to explore the ways in which hormonal mediation influences evolution (and vice versa), phenotypic engineering should be accompanied by studies of natural variation in hormones and hormone-mediated traits. Below, we outline a few potentially rewarding avenues of research, some of which will be explored in juncos in later chapters.

One of the first steps is to quantify patterns of natural variation in hormonal pathways and their covariation with phenotypes of interest. Ideally, we would be able to estimate heritabilities and genetic correlations in order to predict the response of hormone-mediated traits to selection, but this is often quite difficult (though not impossible) to achieve in the field. However, estimates of phenotypic variance and phenotypic correlations are useful even in the absence of genetic data, as the presence of phenotypic variance is required in order for natural selection to act. Even phenotypic measures pose difficulties, however, because both behavior and hormones are notoriously variable within individuals. Perhaps for this reason, endocrinologists have generally remained skeptical about the predictive power of hormone signal-phenotype relationships among individuals (Adkins-Regan 2005; Hau 2007; Ball and Balthazart 2008; Fusani 2008; Williams 2008). The prevailing view among endocrinologists has been that plasma hormones are "permissive." In this view, once hormone concentration exceeds a threshold, further elevation of the hormone does not enhance expression of the trait (Adkins-Regan 2005; 2008; Ball and Balthazart 2008; Williams 2008). Using multiple measurements and standardized behavioral and physiological assays (such as simulated territorial intrusions, stress series, and hormonal challenges) may help distinguish within-individual flexibility from consistent differences among individuals (chapter 6, this volume) (Wingfield et al. 1997; Jawor et al. 2006; Goymann et al. 2007). Promisingly, studies measuring individual (co)variation for multiple behavioral and hormonal traits are becoming more common (chapters 6–7, this volume) (e.g., Bell 2007; Pinxten et al. 2007; Kempenaers et al. 2008; Williams 2008; Cockrem et al. 2009; Ouyang et al. 2011a).

Despite the difficulties of measuring genetic variation in hormones and hormone-mediated suites, recent studies have begun to tackle the quantitative genetics of hormones. Evidence from both artificial selection experiments and various quantitative genetic analyses indicates the presence of genetic variance in hormone levels (Carere et al. 2003; Øverli et al. 2005; Evans et al. 2006; Kempenaers et al. 2008; Mills et al. 2009; Schroderus et al. 2010; Coviello et al. 2011; Mokkonen et al. 2011, 2012; Mills et al. 2012; Travison et al. 2013; Jenkins et al. 2014). One study in bank voles (*Myodes glareolus*) has even provided evidence that testosterone levels are negatively genetically correlated with immune function, suggesting that hormonal suppression of immunity has a genetic basis in this species (Schroderus et al. 2010). More quantitative genetic studies of hormones and hormone-mediated traits are sorely needed. In particular, despite the prediction that hormonal pleiotropy should underlie genetic correlations among traits, there are very few data that can be used to test this idea. One way to test this idea would combine phenotypic engineering with quantitative genetics, measuring traits in pedigreed populations with experimentally elevated or depressed hormone levels and comparing these to genetic correlations in unaltered controls.

We also know relatively little about how selection acts on natural variation in hormones and hormone-mediated trait suites, and consequently, the idea that hormone-mediated suites are maintained or altered by selection remains mostly untested. Over a decade ago, Kingsolver et al. (2001) reported that there had been many fewer measurements of selection on physiology, behavior, and life history than of selection on morphology, presumably because of the difficulty of measuring such traits; this observation remains true today. In order to measure selection on hormone-mediated suites of traits, one needs a large sample of multiple traits and relative fitness measured on the same individuals. Because hormone-mediated traits are often behavioral or life-history traits, and hence often difficult to measure with accuracy, the sampling effort required to measure selection on a suite of such traits may be substantial.

In the absence of measurements of selection on entire suites of hormone-mediated traits, estimates of selection on individual variation in hormones may also provide insight. Blows and Brooks (2003) argue that by using multivariate statistical methods, correlational selection may be understood as quadratic (stabilizing or disruptive) selection on linear combinations of traits. This suggests that a measurement of quadratic selection on physiological measurements such as hormone levels, though not providing information about which traits contribute to fitness differences, may indicate the action of correlational selection on the multivariate suite. For example, corticosterone levels were found to be under stabilizing selection in cliff swallows (*Petrochelidon pyrrhonota*) (Brown et al. 2005), and our own work in juncos has demonstrated stabilizing selection on testosterone production via both survival and reproduction (chapter 6, this volume). Directional selection may be common as well; recent studies have found directional selection on corticosterone metrics in two species of songbird (Ouyang et al. 2011b, 2013) and opposing directional selection on levels of two yolk hormones in a third (Tschirren et al. 2014). Directional selection on hormone levels may indicate coordinated selection on several hormone-mediated traits in the direction they are affected by that hormone. However, this pattern may also indicate that both hormones and fitness are influenced by a third variable, such as condition (Schluter et al. 1991). More studies of selection on hormones and hormone-mediated traits are clearly needed, and we recommend that these should be conducted using a multivariate framework whenever possible to disentangle the complex interactions among hormones, traits, and fitness.

In addition to their role in demonstrating genetic variances and covariances, artificial selection studies can also provide useful insight into the ways that hormone-mediated suites might evolve in response to natural selection (Conner 2003; Fuller et al. 2005). First, artificial selection studies can probe whether observed genetic correlations among traits have the expected effect of constraining responses to selection. A number of studies have applied selection against the direction of genetic correlation for correlated traits not known to be hormone-mediated (e.g. Frankino et al. 2005; Conner et al. 2011). A general conclusion from these studies is that strong correlations slow, but do not prevent, the evolution of phenotypes in this direction (Conner 2012). Interestingly, however, the only study that attempted such selection on a negative correlation representing a trade-off between traits (Dorn and Mitchell-Olds 1991) showed no response to selection (see discussion in Conner 2012). Another possible approach is to apply artificial correlational selection to genetically correlated traits in an attempt to strengthen or weaken the correlation. One such study in a plant with separate sexes was able to show that between-sex genetic correlations decreased rapidly under negative correlational selection (Delph et al. 2011). Similar studies on hormone-mediated traits would be hugely beneficial for our understanding of the roles of hormonal pathways as constraints versus products of selection.

Comparative studies using closely related populations or species may also prove informative. Although comparing divergent groups does not necessarily illuminate evolutionary mechanisms, such work can provide insight into the evolutionary patterns of hormone-mediated suites and the mechanisms that have been altered in the process. At the most basic level, patterns of correlation across groups can be compared to hormone-mediated correlations within groups to determine the extent to which evolution has proceeded along hormonal lines of least resistance. If patterns of correlation differ strongly across populations or species, it would suggest that selection has favored the remodeling of the hormone-mediated suite. Of course, common-garden or reciprocal-transplant studies would be needed to rule out the possibility that population differences can be explained solely by environmental differences. Once patterns of population differentiation have been described, mechanistically focused studies can explore what components of hormonal pathways have diverged. Of primary interest is the extent to which evolution in hormone levels versus receptor expression explains population differentiation. These topics are addressed in subsequent chapters.

E. Conclusion

Studies of hormones have the potential to enrich our understanding of phenotypic evolution, owing to the ability of hormone-mediated suites of traits to both constrain evolutionary change along certain lines and be shaped by natural selection. Although there is a wealth of knowledge about variation in hormone profiles in natural populations and about the impact of experimentally altered hormones on suites of phenotypic characters, we are only just beginning to dissect the mechanisms responsible for evolutionary change in hormonal pathways and hormone-mediated suites. When combined with other approaches, including molecular and developmental genetics, the synthesis appears likely to provide important insights as we strive to understand how the inside world of organisms becomes adapted to the outside world. Although there is still much to learn, studies of natural variation in dark-eyed juncos have begun to shed some light on the evolution of hormone-mediated suites both within and among populations (chapters 6, 7, 10, this volume).

Acknowledgments

Discussions with colleagues over the years, including Elizabeth Adkins-Regan, Jonathan Atwell, Christy Bergeon Burns, Creagh Breuner, Edmund Brodie III, Robert Cox, Michaela Hau, Britt Heidinger, Jodie Jawor, Sharon Lynn, Ignacio Moore, Trevor Price, Kim Rosvall, Dale Sengelaub, and John Wingfield have greatly influenced our thinking on these topics.

References

Adkins-Regan, E. 2005. *Hormones and Animal Social Behavior*. Princeton: Princeton University Press.
———. 2008. Do hormonal control systems produce evolutionary inertia? *Philosophical Transactions of the Royal Society B-Biological Sciences* 363:1599–1609.

Arnold, S. J. 1992. Constraints on phenotypic evolution. *American Naturalist* 140: S85–S107.

Ball, G. F., and J. Balthazart. 2008. Individual variation and the endocrine regulation of behavior and physiology in birds and other vertebrates: A cellular/molecular perspective. *Philosophical Transactions of the Royal Society B-Biological Sciences* 363:1699–1710.

Bell, A. M. 2007. Future directions in behavioural syndromes research. *Proceedings of the Royal Society B-Biological Sciences* 274:755–61.

Blows, M. W., and R. Brooks. 2003. Measuring nonlinear selection. *American Naturalist* 162:815–20.

Blows, M. W., and A. A. Hoffmann. 2005. A reassessment of genetic limits to evolutionary change. *Ecology* 86:1371–84.

Brodie, E. D., III. 1989. Genetic correlations between morphology and antipredator behaviour in natural populations of the garter snakes *Thamnophis ordinoides*. *Nature* 342:542–43.

———. 1992. Correlational selection for color pattern and antipredator behavior in the garter snake *Thamnophis ordinoides*. *Evolution* 46:1284–98.

Brodie, E. D., III, A. J. Moore, and F. J. Janzen. 1995. Visualizing and quantifying natural selection. *Trends in Ecology & Evolution* 10:313–18.

Brown, C. R., M. B. Brown, S. A. Raouf, L. C. Smith, and J. C. Wingfield. 2005. Effects of endogenous steroid hormone levels on annual survival in cliff swallows. *Ecology* 86:1034–46.

Carere, C., T. G. G. Groothuis, E. Mostl, S. Daan, and J. M. Koolhaas. 2003. Fecal corticosteroids in a territorial bird selected for different personalities: Daily rhythm and the response to social stress. *Hormones and Behavior* 43:540–48.

Cheverud, J. M. 1982. Phenotypic, genetic, and environmental morphological integration in the cranium. *Evolution* 36:499–516.

———. 1984. Quantitative genetics and developmental constraints on evolution by selection. *Journal of Theoretical Biology* 110:155–71.

Cockrem, J. F., D. P. Barrett, E. J. Candy, and M. A. Potter. 2009. Corticosterone responses in birds: Individual variation and repeatability in Adelie penguins (*Pygoscelis adeliae*) and other species, and the use of power analysis to determine sample sizes. *General and Comparative Endocrinology* 163:158–68.

Conner, J. K. 2003. Artificial selection: A powerful tool for ecologists. *Ecology* 84: 1650–60.

———. 2012. Quantitative genetic approaches to evolutionary constraint: How useful? *Evolution* 66:3313–20.

Conner, J. K., K. Karoly, C. Stewart, V. A. Koelling, H. F. Sahli, and F. H. Shaw. 2011. Rapid independent trait evolution despite a strong pleiotropic genetic correlation. *American Naturalist* 178:429–41.

Conner, J. K., and S. Via. 1993. Patterns of phenotypic and genetic correlations among morphological and life-history traits in wild radish, *Raphanus raphanistrum*. *Evolution* 47:704–11.

Coviello, A. D., W. V. Zhuang, K. L. Lunetta, S. Bhasin, J. Ulloor, A. Zhang, D. Karasik, et al. 2011. Circulating testosterone and SHBG concentrations are heritable in women: The Framingham Heart Study. *Journal of Clinical Endocrinology & Metabolism* 96:E1491–E1495.

Darwin, C. 1859. *The Origin of Species*. London: John Murray.

Delph, L. F., J. C. Steven, I. A. Anderson, C. R. Herlihy, and E. D. Brodie, III. 2011. Elimination of a genetic correlation between the sexes via artificial correlational selection. *Evolution* 65:2872–80.

De Ridder, E., R. Pinxten, and M. Eens. 2000. Experimental evidence of a testosterone-induced shift from parental to mating behaviour in a facultatively polygynous songbird. *Behavioral Ecology and Sociobiology* 49:24–30.

Dorn, L. A., and T. Mitchell-Olds. 1991. Genetics of *Brassica campestris*. 1. Genetic constraints on evolution of life-history characters. *Evolution* 45:371–79.

Endler, J. A. 1986. *Natural Selection in the Wild*. Princeton: Princeton University Press.

Estes, S., and S. J. Arnold. 2007. Resolving the paradox of stasis: Models with stabilizing selection explain evolutionary divergence at all timescales. *American Naturalist* 169:227–44.

Evans, M. R., M. L. Roberts, K. L. Buchanan, and A. R. Goldsmith. 2006. Heritability of corticosterone response and changes in life history traits during selection in the zebra finch. *Journal of Evolutionary Biology* 19:343–52.

Finch, C. E., and M. R. Rose. 1995. Hormones and the physiological architecture of life-history evolution. *Quarterly Review of Biology* 70:1–52.

Flatt, T., and A. Heyland. 2011. *Mechanisms of Life History Evolution: The Genetics and Physiology of Life History Traits and Trade-offs*. Oxford: Oxford University Press.

Flatt, T., M. P. Tu, and M. Tatar. 2005. Hormonal pleiotropy and the juvenile hormone regulation of *Drosophila* development and life history. *BioEssays* 27: 999–1010.

Frankino, W. A., B. J. Zwaan, D. L. Stern, and P. M. Brakefield. 2005. Natural selection and developmental constraints in the evolution of allometries. *Science* 307: 718–20.

Fuller, R. C., C. F. Baer, and J. Travis. 2005. How and when selection experiments might actually be useful. *Integrative and Comparative Biology* 45:391–404.

Fusani, L. 2008. Endocrinology in field studies: Problems and solutions for the experimental design. *General and Comparative Endocrinology* 157:249–53.

Gerlach, N. M., and E. D. Ketterson. 2013. Experimental elevation of testosterone lowers fitness in female dark-eyed juncos. *Hormones and Behavior* 63:782–90.

Goymann, W., M. M. Landys, and J. C. Wingfield. 2007. Distinguishing seasonal androgen responses from male-male androgen responsiveness—revisiting the challenge hypothesis. *Hormones and Behavior* 51:463–76.

Hau, M. 2007. Regulation of male traits by testosterone: implications for the evolution of vertebrate life histories. *BioEssays* 29:133–44.

Jawor, J. M., J. W. McGlothlin, J. M. Casto, T. J. Greives, E. A. Snajdr, G. E. Bentley, and E. D. Ketterson. 2006. Seasonal and individual variation in response to GnRH challenge in male dark-eyed juncos (*Junco hyemalis*). *General and Comparative Endocrinology* 149:182–89.

Jenkins, B. R., M. N. Vitousek, J. K. Hubbard, and R. J. Safran. 2014. An experimental analysis of the heritability of variation in glucocorticoid concentrations in a wild avian population. *Proceedings of the Royal Society B-Biological Sciences* 281.

Jones, A. G., S. J. Arnold, and R. Borger. 2003. Stability of the **G**-matrix in a population experiencing pleiotropic mutation, stabilizing selection, and genetic drift. *Evolution* 57:1747–60.

Kempenaers, B., A. Peters, and K. Foerster. 2008. Sources of individual variation in plasma testosterone levels. *Philosophical Transactions of the Royal Society B-Biological Sciences* 363:1711–23.

Ketterson, E. D., J. W. Atwell, and J. W. McGlothlin. 2009. Phenotypic integration and independence: Hormones, performance, and response to environmental change. *Integrative and Comparative Biology* 49:365–79.

Ketterson, E. D., and V. Nolan, Jr. 1992. Hormones and life histories: An integrative approach. *American Naturalist* 140:S33–S62.

———. 1999. Adaptation, exaptation, and constraint: a hormonal perspective. *American Naturalist* 154:S4–S25.

Kingsolver, J. G., S. E. Diamond, A. M. Siepielski, and S. M. Carlson. 2012. Synthetic analyses of phenotypic selection in natural populations: Lessons, limitations and future directions. *Evolutionary Ecology* 26:1101–18.

Kingsolver, J. G., H. E. Hoekstra, J. M. Hoekstra, D. Berrigan, S. N. Vignieri, C. E. Hill, A. Hoang, et al. 2001. The strength of phenotypic selection in natural populations. *American Naturalist* 157:245–61.

Kruuk, L. E. B. 2004. Estimating genetic parameters in natural populations using the "animal model." *Philosophical Transactions of the Royal Society B-Biological Sciences* 359:873–90.

Lande, R. 1979. Quantitative genetic analysis of multivariate evolution, applied to brain:body size allometry. *Evolution* 33:402–16.

———. 1980. The genetic covariance between characters maintained by pleiotropic mutations. *Genetics* 94:203–15.

———. 1981. Models of speciation by sexual selection on polygenic traits. *Proceedings of the National Academy of Sciences of the United States of America* 78:3721–25.

Lande, R., and S. J. Arnold. 1983. The measurement of selection on correlated characters. *Evolution* 37:1210–26.

Lynn, S. E., L. S. Hayward, Z. M. Benowitz-Fredericks, and J. C. Wingfield. 2002. Behavioural insensitivity to supplementary testosterone during the parental phase in the chestnut-collared longspur, *Calcarius ornatus*. *Animal Behaviour* 63:795–803.

MacArthur, J. W. 1948. Selection for small and large body size in the house mouse. *Genetics* 34:194–209.

Maynard Smith, J., R. Burian, S. Kauffman, P. Alberch, J. Campbell, B. Goodwin, R. Lande, et al. 1985. Developmental constraints and evolution: A perspective from the Mountain Lake conference on development and evolution. *Quarterly Review of Biology* 60:265–87.

McGlothlin, J. W., and E. D. Ketterson. 2008. Hormone-mediated suites as adaptations and evolutionary constraints. *Philosophical Transactions of the Royal Society B-Biological Sciences* 363:1611–20.

McGlothlin, J. W., P. G. Parker, V. Nolan, Jr., and E. D. Ketterson. 2005. Correlational selection leads to genetic integration of body size and an attractive plumage trait in dark-eyed juncos. *Evolution* 59:658–71.

Merilä, J., and M. Björklund. 2004. Phenotypic integration as a constraint and adaptation. In *Phenotypic Integration: Studying the Ecology and Evolution of Complex Phenotypes*, edited by M. Pigliucci and K. Preston, 107–29. Oxford: Oxford University Press.

Mills, S. C., A. Grapputo, I. Jokinen, E. Koskela, T. Mappes, T. A. Oksanen, and T. Poikonen. 2009. Testosterone-mediated effects on fitness-related phenotypic traits and fitness. *American Naturalist* 173:475–87.

Mills, S. C., E. Koskela, and T. Mappes. 2012. Intralocus sexual conflict for fitness: Sexually antagonistic alleles for testosterone. *Proceedings of the Royal Society B-Biological Sciences* 279:1889–95.

Mokkonen, M., H. Kokko, E. Koskela, J. Lehtonen, T. Mappes, H. Martiskainen, and S. C. Mills. 2011. Negative frequency-dependent selection of sexually antagonistic alleles in *Myodes glareolus. Science* 334:972–74.

Mokkonen, M., E. Koskela, T. Mappes, and S. C. Mills. 2012. Sexual antagonism for testosterone maintains multiple mating behaviour. *Journal of Animal Ecology* 81:277–83.

Morgan, M. T., and J. K. Conner. 2001. Using genetic markers to directly estimate male selection gradients. *Evolution* 55:272–81.

Nijhout, H. F. 2003. Development and evolution of adaptive polyphenisms. *Evolution & Development* 5:9–18.

Olson, E. C., and R. L. Miller. 1958. *Morphological Integration*. Chicago: University of Chicago Press.

Ouyang, J. Q., M. Hau, and F. Bonier. 2011a. Within seasons and among years: When are corticosterone levels repeatable? *Hormones and Behavior* 60:559–64.

Ouyang, J. Q., P. Sharp, M. Quetting, and M. Hau. 2013. Endocrine phenotype, reproductive success, and survival in the great tit, *Parus major. Journal of Evolutionary Biology* 26:1988–98.

Ouyang, J. Q., P. J. Sharp, A. Dawson, M. Quetting, and M. Hau. 2011b. Hormone levels predict individual differences in reproductive success in a passerine bird. *Proceedings of the Royal Society of London* B 278:2537–45.

Øverli, Ø., S. Winberg, and T. G. Pottinger. 2005. Behavioral and neuroendocrine correlates of selection for stress responsiveness in rainbow trout—a review. *Integrative and Comparative Biology* 45:463–74.

Phillips, P. C., and S. J. Arnold. 1989. Visualizing multivariate selection. *Evolution* 43:1209–22.

Phillips, P. C., and K. L. McGuigan. 2006. Evolution of genetic variance-covariance structure. In *Evolutionary Genetics: Concepts and Case Studies*, edited by J. B. Wolf and C. W. Fox. Oxford: Oxford University Press.

Pinxten, R., E. de Ridder, L. Arckens, V. M. Darras, and M. Eens. 2007. Plasma testosterone levels of male European starlings (*Sturnus vulgaris*) during the breeding cycle and in relation to song and paternal care. *Behaviour* 144:393–410.

Reed, W. L., M. E. Clark, P. G. Parker, S. A. Raouf, N. Arguedas, D. S. Monk, E. Snajdr, et al. 2006. Physiological effects on demography: A long-term experimental study of testosterone's effects on fitness. *American Naturalist* 167:667–83.

Rice, S. H. 2000. The evolution of developmental interactions: Epistasis, canalization, and integration. In *Epistasis and the Evolutionary Process*, edited by J. B. Wolf, E. D. Brodie, III, and M. J. Wade, 82–98. New York: Oxford University Press.

Ricklefs, R. E., and M. Wikelski. 2002. The physiology/life-history nexus. *Trends in Ecology & Evolution* 17:462–67.

Roff, D. A. 1996. The evolution of genetic correlations: An analysis of patterns. *Evolution* 50:1392–1403.

Roff, D. A., and D. J. Fairbairn. 2012. A test of the hypothesis that correlational selection generates genetic correlations. *Evolution* 66:2953–60.

Schluter, D. 1996. Adaptive radiation along genetic lines of least resistance. *Evolution* 50:1766–74.

Schluter, D., T. D. Price, and L. Rowe. 1991. Conflicting selection pressures and life history trade-offs. *Proceedings of the Royal Society B-Biological Sciences* 246: 117–22.

Schroderus, E., I. Jokinen, M. Koivula, E. Koskela, T. Mappes, S. C. Mills, T. A. Oksanen, et al. 2010. Intra- and intersexual trade-offs between testosterone and immune system: implications for sexual and sexually antagonistic selection. *American Naturalist* 176:E90–E97.

Sinervo, B., and E. Svensson. 1998. Mechanistic and selective causes of life history trade-offs and plasticity. *Oikos* 83:432–42.

———. 2002. Correlational selection and the evolution of genomic architecture. *Heredity* 89:329–38.

Travison, T. G., W. V. Zhuang, K. L. Lunetta, D. Karasik, S. Bhasin, D. P. Kiel, A. D. Coviello, et al. 2013. The heritability of circulating testosterone, oestradiol, oestrone, and sex hormone binding globulin concentrations in men: The Framingham Heart Study. *Clinical Endocrinology* 80:277–82.

Tschirren, B., E. Postma, L. Gustafsson, T. G. G. Groothuis, and B. Doligez. 2014. Natural selection acts in opposite ways on correlated hormonal mediators of prenatal maternal effects in a wild bird population. *Ecology Letters* 17:1310–15.

Williams, T. D. 2008. Individual variation in endocrine systems: Moving beyond the "tyranny of the Golden Mean." *Philosophical Transactions of the Royal Society B-Biological Sciences* 363:1699–1710.

Wilson, A. J., D. Réale, M. N. Clements, M. M. Morrissey, E. Postma, C. A. Walling, L. E. B. Kruuk, et al. 2010. An ecologist's guide to the animal model. *Journal of Animal Ecology* 79:13–26.

Wingfield, J. C., K. Hunt, C. Breuner, K. Dunlap, G. S. Fowler, L. Freed, and J. Lepson. 1997. Environmental stress, field endocrinology, and conservation biology. In *Behavioral Approaches to Conservation in the Wild*, edited by J. R. Clemmons and R. Buchholz, 95–129. Cambridge: Cambridge University Press.

Zera, A. J., and L. G. Harshman. 2001. The physiology of life-history trade-offs in animals. *Annual Review of Ecology and Systematics* 32:95–126.

Individual Variation and Selection on Hormone-Mediated Phenotypes in Male and Female Dark-Eyed Juncos

Kristal Cain, Jodie M. Jawor, and Joel W. McGlothlin

It may be said that natural selection is daily and hourly scrutinising, throughout the world, every variation, even the slightest; rejecting that which is bad, preserving and adding up all that is good . . .
—Charles Darwin, 1859, *Origin of Species*, chapter 4

A. Introduction

B iologists studying the evolution of phenotypes have often focused on ultimate questions, such as how and why certain traits are favored or disfavored by natural selection (Endler 1986; Andersson 1994; Kingsolver et al. 2001). Although such studies are often conducted without considering the mechanisms that underlie phenotypes of interest, it is increasingly clear that integrating proximate and ultimate perspectives can lead to a greatly enriched understanding of phenotypic evolution (Finch and Rose 1995; Ketterson and Nolan 1999; Dufty et al. 2002; Adkins-Regan 2005; Ketterson et al. 2009). Endocrine pathways have been particularly important in such integrative studies (Adkins-Regan 2008; Williams 2012b). Hormones are important for regulating life-history trade-offs, mediating transitions between life-history stages, and mediating organismal responses to changing environments (chapters 4 and 5, this volume) (Ketterson and Nolan 1992; Wingfield 2003; Williams 2012b). Hormones are also intimately involved in the coordinated expression of suites of functionally related traits and thus are likely to be important targets for selection (Dufty et al. 2002; Zera et al. 2007; Mc-Glothlin and Ketterson 2008).

This regulation of suites of traits by hormones has been called "hormonal pleiotropy" by analogy to genetic pleiotropy, the phenomenon whereby a single gene influences multiple traits (chapter 5, this volume) (Finch and Rose 1995; Ketterson and Nolan 1999; Zera et al. 2007; McGlothlin and Ketterson 2008; Ketterson et al. 2009; Williams 2012a). Like genetic pleiotropy, hormonal pleiotropy may have profound influences on the evolutionary process. By favoring simple changes in the level or timing of circulating hormones, natural selection can alter numerous aspects of the phenotype, causing functionally related traits to be expressed and to potentially respond to selection in concert. While such coevolution may facilitate rapid adaptation to a new environment, hormonal pleiotropy may impede evolutionary change as well. For example, males and females often have different optimal phenotypes. In some cases, selection on male hormone levels may produce correlated responses in female hormone levels, resulting in a change in female expression of hormonally mediated traits that may or may not be beneficial (Ketterson et al. 2005; Cain and Ketterson 2012, 2013a; Garamszegi 2014; but see Goymann and Wingfield 2014a, 2014b)

In vertebrates, steroid hormones have received a great deal of attention in integrative studies of phenotypic evolution because they tend to play central roles in important biological functions and can be measured with relative ease. Steroids, which include estrogens, androgens, and glucocorticoids, regulate gene expression via changes in the timing or concentration of hormones in circulation. One of the best-studied vertebrate steroids is the androgen testosterone (T) (chapter 7, this volume). In both males and females, T production is regulated by the hypothalamic-pituitary-gonadal (HPG) axis (color plate 6, this volume). In response to appropriate stimuli, the hypothalamus releases gonadotropin-releasing hormone (GnRH), which leads to the release of luteinizing hormone (LH) from the pituitary, which in turn stimulates the gonads to produce and release of T. Testosterone is then transported to a variety of target tissues where it may trigger a myriad of phenotypic effects (Adkins-Regan 2005).

Testosterone has long been recognized as an important mediator of male traits. Manipulative experiments in a variety of vertebrates, including juncos (chapter 4, this volume), have shown that T is instrumental in the regulation of a wide variety of morphological, physiological, and behavioral traits, particularly those important in a reproductive context.

Such "phenotypic engineering" experiments are powerful because they reveal the causal links between hormones and phenotypes. However, manipulative studies obscure variation among individuals—the material upon which natural selection acts—which may limit evolutionary interpretations (Adkins-Regan 2008; McGlothlin and Ketterson 2008; Williams 2008; McGlothlin et al. 2010). Thus, a full understanding of the evolution of hormones and hormone-mediated traits requires that manipulative studies be complemented by studies of individual variation in hormonal profiles and of the relationship between hormonal phenotype and fitness (chapters 4, 5, and 10, this volume).

Studies of the causes and consequences of individual variation in hormones and related traits have been relatively uncommon (Williams 2008). One reason for this is that hormone titers are often highly labile, fluctuating rapidly in response to a variety of stimuli such as season, day, time of day, reproductive stage, and age (Wingfield et al. 1987; Wingfield et al. 1990; Jawor et al. 2006a; Williams 2008). Most temperate vertebrates show seasonal patterns of hormone levels; for example, in many birds, T levels rise sharply just before breeding and fall later in the breeding season (Wingfield et al. 1990). Hormones also show changes on much shorter scales, rising and falling throughout the day in a circadian pattern or in response to weather or prey availability (Wingfield et al. 2001). Further, individuals may also produce transitory increases in circulating T levels in response to social interactions with conspecifics, particularly potential competitors or mates (Wingfield 1985; Wingfield et al. 1987, 1990, 2001; Goymann 2009). However, recent research suggests that such social modulation may not be as common as once thought (Apfelbeck and Goymann 2011; DeVries et al. 2012; Villavicencio et al. 2013).

One strategy for dealing with the lability of hormone levels is to expose individuals to a standardized stimulus that "challenges" the individual's endocrine system. Much like an antigen injection is used to estimate variation in immune response, such challenges can provide informative estimates of natural variation in hormone production, allowing researchers to control for background noise due to environmental factors that influence circulating hormone levels (Jawor et al. 2006a). In our own work, we have primarily relied upon a physiological challenge, a standardized dose of GnRH, which stimulates an individual to increase its T level to a maximum. As we will review in this chapter, GnRH challenges have been extremely informative regarding the links between T, phenotype, and fitness in dark-eyed juncos.

B. GnRH Challenges

The GnRH challenge provides researchers with a standardized means to test the responsiveness of an individual's HPG axis in the face of the many factors that may create noise in simple measurements of general circulating hormone levels. Traditionally, GnRH challenges have been used as a tool to determine reproductive condition (e.g., Wingfield et al. 1979; Lacombe, Cyr, and Matton 1991; Schoech et al. 1996; Hirschenhauser et al. 2000; Millesi et al. 2002; Moore et al. 2002). Individuals able to produce LH or T in response to the GnRH injection were considered to be in breeding condition. In contrast, our own work has employed this method to assess individual variation in HPG responsiveness among reproductively active adults (Jawor et al. 2006a, 2007).

The GnRH-challenge method is simple. After capturing a bird, an initial blood sample is collected and a solution containing GnRH is injected into the pectoral muscle, where it is absorbed into the circulation. In response to GnRH, T levels tend to increase initially, then, after approximately 30 minutes, begin to decrease, and return to initial levels within two hours (Jawor et al. 2006a). At the 30 minute mark, a second blood sample is collected and then the bird is released (see color plate 6, this volume). The GnRH challenge results in three different measurements of plasma T concentration: *initial* or *prechallenge T*, *maximum* or *postchallenge T*, and *GnRH-induced increase*, which is simply the difference between the two measures. As we show below, each of these measurements has proved informative about phenotype or fitness in juncos, suggesting that GnRH challenges are a useful tool for future research.

GnRH challenges are advantageous for measuring circulating hormones in that they both minimize the effect of environmental noise and mimic a physiological phenomenon with potential behavioral relevance. Males in some species show transitory elevations of T in response to challenging social stimuli such as a potential mate or territorial intruder (Wingfield et al. 1987; Wingfield et al. 1990). These contextual elevations in T may allow males to avoid some of the costs of constitutively high levels of T, such as reduced immune function or paternal care, while still supporting appropriate behavior (Wingfield et al. 2001). Because the endocrine response to a GnRH challenge reflects maximum T production, it potentially represents T levels that are more relevant to behaviors commonly associated with T (such as aggression) than initial T levels.

In the following sections, we discuss some of the results of using GnRH challenges to investigate associations between T and fitness-relevant traits. Because T often plays different roles in the two sexes, we discuss findings from males and females separately but cross-reference where appropriate. We then contrast these findings with previous work that examined these relationships in animals with experimentally elevated T (via implants, reviewed in chapter 4, this volume). We conclude with a discussion of what new insights these findings might provide regarding the evolution of hormone-mediated phenotypes in particular and sexual differences more generally. We leave discussion of how tissue-level sensitivity to circulating hormone varies among individuals to chapter 7 and between populations to chapters 10 and 11 (this volume).

C. Responses to GnRH Challenges in Males

C.1. Individual and Seasonal Variation

Over the course of two breeding seasons, we repeatedly administered GnRH challenges to free-living males to assess seasonal and individual variation in the activity of the HPG axis (Jawor et al. 2006a). We also quantified the level of repeatability of various measurements of T (i.e., whether males with a relatively high response at one point of the season also predictably produced relatively high responses at later points in the season). To address these questions, GnRH challenges were performed at four time points. The first two of these were conducted during early breeding (late April-early May), at a time when females were just beginning to build nests and lay eggs on territories defended by males. The next two challenges were performed later, the first when males were feeding nestlings and the second near the conclusion of breeding in late July and early August.

We found that GnRH challenges led to a significant elevation of T at all breeding stages (figure 6.1A) (Jawor et al. 2006a). However, the magnitude of this response declined somewhat over the course of the breeding season. Specifically, postchallenge T was highest at the beginning of the early breeding stage, declined slightly during the end of the early breeding stage, and was lowest late in the breeding season. Initial T levels did not show such a marked difference across breeding stages, but there was a trend toward lower levels later in the breeding season. When controlling for seasonal patterns, we found that GnRH-induced T levels were re-

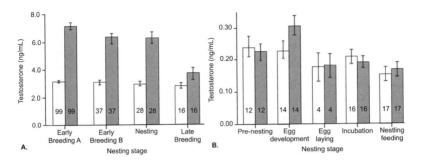

FIGURE 6.1. Male and female testosterone levels across nesting stage. (A) Male circulating T levels (± SE), before (white) and after (grey) a GnRH challenge, according to nesting stage (values adjusted for covariates). Early Breeding A samples were taken between mid-April and mid-May. Early Breeding B samples were from individuals challenged a second time 7–21 days later, but during the same time period. Late Breeding samples were taken at the end of the breeding season before molting began. (B) Female circulating T levels (± SE), before (white) and after (grey) a GnRH challenge, according to nesting stage (values adjusted for covariates). Prenesting samples were collected more than one week prior to the female's first egg of the season. Egg development samples were taken during the 7 days prior to the date of the first egg.

peatable. In other words, an individual with relatively low GnRH-induced T levels at one time tended to have a relatively low response at a later time point. Initial T levels were not similarly repeatable, suggesting that prechallenge measurements may be more susceptible to environmental noise.

C.2. Aggression

Because of the importance of competition for mates, territories, and other breeding resources, males in many species are more aggressive during breeding than in the nonbreeding season (Nelson 2005). In many species, T levels are highest at the beginning of breeding when competition and aggression are most intense (e.g., Wingfield and Goldsmith 1990). Experimental studies have shown that elevation of T tends to lead to increases in male aggression (reviewed in chapter 4, this volume). If so, individual variation in T production may be a good predictor of individual differences in aggression. To examine this relationship in juncos, we measured male ability to respond to GnRH challenges in the early prebreeding period and compared this to individual males' aggression level towards a simulated intruder (caged foreign male placed in the focal male's territory).

We found that the males capable of producing the most T in response to GnRH were also the most responsive to the simulated intruder (figure 6.2A) based on a composite measurement of aggression that included response latency, song, and time spent in the immediate area around the cage (McGlothlin et al. 2007). Initial T also showed a positive relationship with aggression, but it was weak and not statistically significant.

Because some songbird species are known to elevate T during aggressive encounters, we also examined the relationship between GnRH-induced T elevations and T levels produced during aggression (McGlothlin et al. 2008). Males were challenged with GnRH while they were feeding nestlings. As part of a separate experiment, the nestlings were removed, and two to four days later, these males received a simulated territorial intrusion and their T levels were measured. We found that T levels in response to GnRH were strongly correlated with T measured after aggression (figure 6.2B). More recent work has shown that testosterone does not always increase in response to territorial challenges in juncos, suggesting that the elevation in T seen in these simulated territorial intrusions may not be induced aggression per se but may instead reflect a physiological response to nest failure or mate fertility (Rosvall et al. 2012, 2014). In this view, the male HPG axis may be capable of upregulating in response to behavioral and/or social input only at certain stages of reproduction (e.g., after losing nestlings to a predator), which should facilitate courtship and the deterrence of neighboring males pursuing fertile mates.

C.3. Morphology

Along with increased investment in competitive behaviors, T is also often associated with the expression of other traits used in same-sex competition contexts. For example, in house sparrows (*Passer domesticus*), experimentally elevated T produces larger black bibs, a trait associated with social dominance (Evans et al. 2000; Gonzalez et al. 2001).

In Carolina juncos, plumage ornamentation consists mainly of a white patch on the outer tail feathers, or tail white (see color plate 8, this volume). Males tend to have larger white patches than females, and females prefer males with larger patches in mate choice trials (Hill et al. 1999). Males use tail white in both courtship and aggressive interactions, spreading their tails to increase the amount of visible white (Nolan et al. 2002). We found that the GnRH-induced increase in T was positively related to the size of a male's tail white patch (figure 6.2C), a relationship that was more pronounced in young males (McGlothlin et al. 2008). Because

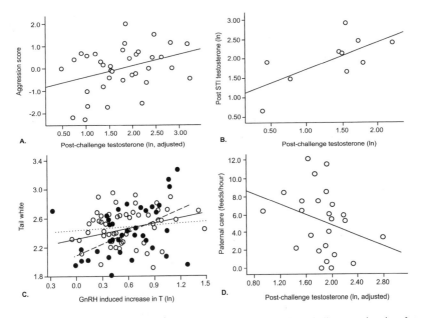

FIGURE 6.2. Relationships between individual response to GnRH challenge and traits of interest in males. (A) Aggression in response to a simulated intruder vs. postchallenge T levels (natural-log transformed, adjusted for covariates). (B) T after a social challenge (simulated territorial intrusion) vs. postchallenge T levels. (C) Age-specific relationship between ornamental plumage (tail white) and the GnRH-induced increase in T levels. First-year males are solid circles and the dotted line, older males are open circles and the dashed line, and the solid line is the overall best fit. (D) Provisioning (nestling feeding rate) and postchallenge T (natural-log transformed, adjusted for covariates).

GnRH response is also related to territorial aggression, this suggests that tail white may act as a reliable signal of male T-related behaviors, particularly in younger males. However, in contrast to the bibs of house sparrows, circulating T is unlikely to be directly responsible for mediating the expression of this plumage trait. T levels are very low during molt, and even very small doses of exogenous T can suppress feather growth (Nolan et al. 1992). Thus, if T does play a role in tail white expression, the variation likely lies at the level of receptors or cofactors in tissues responsible for feather development.

C.4. Paternal Care

Among the most important roles for T in free-living animals is mediating life-history trade-offs, which occur when there are multiple traits that might increase fitness but there is a limited amount of time, energy, or resources available to invest in these traits (Ketterson and Nolan 1992; Stearns 1992; Ketterson and Nolan 1999; Roff 2002). One such trade-off occurs when competition for mates conflicts with investment in offspring care. Individual males may resolve this trade-off between mating effort and parental effort differently according to quality or probability of obtaining additional mating success (i.e., high-quality males might invest more in mating effort, while low-quality males might be better served by investing in offspring care) (Magrath and Komdeur 2003). Because T is crucial in the regulation of competitive traits, it is also likely to be influential in mediating this parental care trade-off.

Numerous studies have shown that experimentally elevating T leads to a reduction in various forms of paternal care (chapter 4, this volume). However, in other species paternal care appears to be unrelated to circulating T levels (Lynn et al. 2005; Hau 2007; Lynn 2008). Investigating this relationship in juncos using the GnRH challenge, we found that males producing more T fed nestlings less frequently (figure 6.2D) (McGlothlin et al. 2007). In contrast, there was no detectable relationship between parental behavior and initial T levels. Taken together with the aggression and ornamentation results presented above, this result suggests that males capable of elevating T to higher levels invest more in traits and behaviors that improve mating success (e.g., territorial aggression and pursuing extrapair matings) at the expense of parental behavior.

C.5. Selection

The relationships between individual ability to respond to GnRH (post-challenge or GnRH-induced increase in T) and fitness-relevant traits suggest that individual variation in the responsiveness of the HPG axis is likely an important target for selection. However, few studies to date have examined the strength and direction of selection on individual hormonal profiles, and those that examined these relationships are inconsistent regarding the action of selection. A number of studies have reported a positive relationship between mating success and baseline circulating T levels (Borgia and Wingfield 1991; Alatalo et al. 1996; Mills et al. 2007), but others have found no relationship between baseline T and components of fitness (Brown et al. 2005; John-Alder et al. 2009). Given the stronger relationships between GnRH-induced increases in T and fitness-relevant behavioral traits, it is possible that such levels may be more informative about the relationships between T and fitness than are measures of initial, or baseline, T.

To address this question in the Virginia junco population, we examined the relationship between natural variation in T (initial and GnRH-induced increases) and fitness components (McGlothlin et al. 2010). Total annual selection was estimated over two breeding seasons by calculating annual survival (using recapture data) and annual offspring production (using DNA paternity analysis). We also partitioned total offspring production into success via social and extrapair mates. Previous work (discussed in chapter 4, this volume) has shown that males with experimentally elevated T suffered reduced annual survival but had greater offspring production, stemming chiefly from extrapair mating success (Raouf et al. 1997; Reed et al. 2006).

This examination of natural variation in T levels and fitness revealed some surprising patterns (McGlothlin et al. 2010). First, the magnitudes of GnRH-induced T increases, but not initial T levels, were associated with variation in fitness components. Second, the primary form of selection on GnRH-induced T production was stabilizing rather than directional. That is, males near the population mean tended to have higher fitness than males at either extreme. Third, selection on natural variation in T did not display the strong trade-off between survival and reproduction seen in implant studies. Rather, both survival selection and selection via reproductive success favored males that were slightly above average in ability to produce T in response to GnRH (figure 6.3). When reproductive success was separated into within-pair and extrapair components, there was some evidence of a reproductive trade-off. Whereas selection on within-pair success was directional, favoring males with higher T production, selection on extrapair success was stabilizing, favoring intermediate males. The conclusions that may be drawn from this work are somewhat limited, as selection was measured during only two breeding seasons and selection is often variable across time (Siepielski et al. 2009). Nevertheless, the differences between these results and those from implant studies are striking; we discuss some of the possible explanations for these differences later in the chapter (section E).

D. Responses to GnRH Challenges in Females

Although T has been well studied as a hormone influencing the physiology and behavior of male birds, its role in mediating female physiology and behavior is less well understood (Staub and De Beer 1997). Studies of T in females are important not only for understanding the evolution

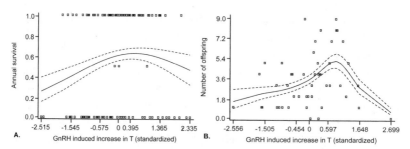

FIGURE 6.3 Relationships between GnRH-induced increase in T and components of fitness in males. Change in T was standardized and individual points represent average fitness of males for a given level of change in T. Solid lines are the predicted relationships as represented by a cubic spline; dashed lines indicate ± 1 standard error (calculated by bootstrapping). (A) Annual survival. (B) Annual reproductive success (total offspring production).

of female phenotypes but also because they can shed light on the evolution of sex differences and proximate mechanisms regulating life histories. Data from implant studies (chapter 4, this volume) have shown that experimental elevation of T can be beneficial for males but have negative consequences for females. This pattern is consistent with the possibility that sexually antagonistic effects of T may constrain each sex from reaching its evolutionary optimum (Ketterson et al. 2005; Bonduriansky and Chenoweth 2009). Understanding how selection is acting on naturally occurring variation in T-mediated phenotypes in females is crucial to understanding the potential role of T in such sexual conflict. Although the story is far from complete, we have made some progress toward addressing this question.

D.1. Seasonality and Reproductive State

Breeding female dark-eyed juncos have measurable levels of T, but throughout the majority of the breeding season, these levels are significantly lower than those in males (Ketterson et al. 2005). In an attempt to address individual variation in female T production, we performed GnRH challenges at multiple time points across two years (Jawor et al. 2007). Unlike males, which reliably responded to GnRH across the breeding season, females showed a measurable increase in T after a GnRH challenge under more limited circumstances. In one study, females displayed a significant increase in T only when they were challenged during egg development (figure 6.1B).

PLATE 1. Migratory junco populations with different breeding ranges often share wintering habitats with other migratory or with sedentary (nonmigratory) populations ("heteropatry," Winker 2010a; see chapter 3, this volume). Pictured here is a flock of wintering juncos in the foothills of the Rocky Mountains, Colorado, USA, that includes at least four subspecies including pink-sided, slate-colored, Oregon, and gray-headed forms. In spring, despite exposure to identical environmental stimuli, these different forms will follow divergent strategies with respect to the seasonal timing and the geography of their migratory journeys and breeding activities. (Photo: Peter Pereira).

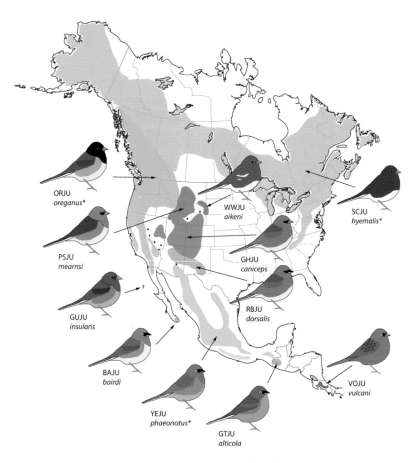

PLATE 2. Distribution of the junco species and morphs included in a recent evolutionary analysis by Milà and colleagues (chapter 8, this volume). Four-letter codes for vernacular names are as follows: VOJU, volcano junco; GTJU, Guatemala junco; YEJU, yellow-eyed junco; BAJU, Baja junco; GUJU, Guadalupe junco; PSJU, pink-sided junco; ORJU, Oregon junco; WWJU, white-winged junco; SCJU, slate-colored junco; GHJU, gray-headed junco; RBJU, red-backed junco. Asterisks indicate that other races exist that are not illustrated, including *palliatus* and *fulvescens* for YEJU; *carolinensis* for SCJU; and *thurberi, montanus, shufeldti, pinosus, townsendi,* and *pontilis* for ORJU. Black squares correspond to the approximate location of isolated hybrid populations. Individuals of mixed origin also exist where colored ranges come into contact.

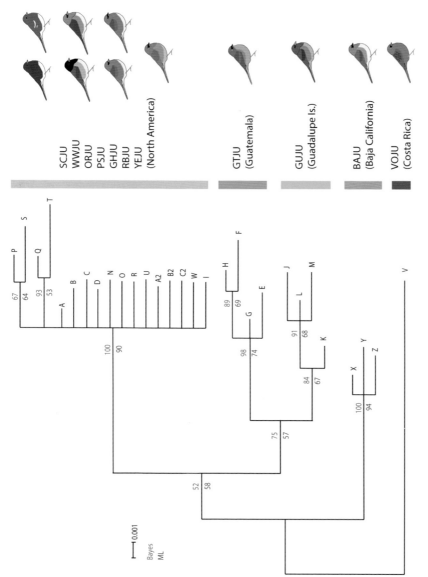

PLATE 3. Evolutionary relationships among the main forms within the genus *Junco* are illustrated in this preliminary phylogenetic tree, based on an analysis of mitochondrial DNA sequence data (cytochrome *c* oxidase I [COI], 690 base pairs). Branch support values correspond to a Bayesian analysis (above branches) and a maximum likelihood analysis (below branches). For details see chapter 8, this volume.

Slate-colored (Northern) junco
J. hyemalis hyemalis
(Ontario, Canada)

Slate-colored (Carolina) junco
J. hyemalis carolinensis
(Virginia, USA)

White-winged junco
J. hyemalis aikeni
(South Dakota, USA)

Pink-sided junco
J. hyemalis mearnsi
(Wyoming, USA)

Oregon junco
Junco hyemalis thurberi
(California, USA)

Gray-headed junco
J. hyemalis caniceps
(Utah, USA)

PLATE 4. In Alden Miller's definitive monograph *Speciation in the avian genus* Junco (1941, University of California Publications in Zoology), he delineated twenty-one distinct junco forms based on variation in morphology across their range, including striking differences in plumage and eye coloration, as well as quantitative differences in the size and shape of bills, tarsi, wings, tails, and feathers. Pictured here are examples of six distinct forms of juncos that breed in more northerly habitats (left page) and six forms that breed in more southerly habitats (right page). (Volcano junco photo: Boris Nikolov, www.fotobiota.com; all other photos: Borja Milá).

Red-backed junco
J. hyemalis dorsalis
(Arizona, USA)

Yellow-eyed junco
J. phaeonotus phaeonotus
(Distrito Federal, Mexico)

Guatemala junco
J. phaeonotus alticola
(Huehuetenango, Guatemala)

Guadalupe junco
J. insularis
(Guadalupe Island, Mexico)

Baird's (Baja) junco
J. bairdi
(Baja Sur, Mexico)

Volcano junco
J. vulcani
(Cartago, Costa Rica)

PLATE 5. In Alden Miller's definitive monograph *Speciation in the avian genus* Junco (1941, University of California Publications in Zoology), he delineated twenty-one distinct junco forms based on variation in morphology across their range, including striking differences in plumage and eye coloration, as well as quantitative differences in the size and shape of bills, tarsi, wings, tails, and feathers. Pictured here are examples of six distinct forms of juncos that breed in more northerly habitats (left page) and six forms that breed in more southerly habitats (right page). (Volcano junco photo: Boris Nikolov, www.fotobiota.com; all other photos: Borja Milá).

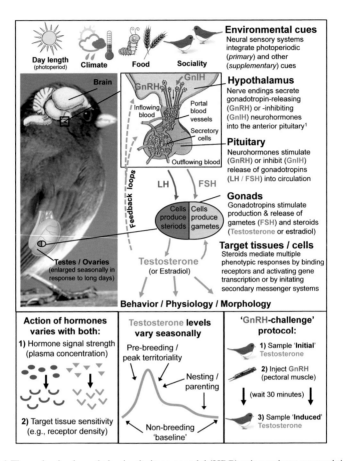

Environmental cues
Neural sensory systems integrate photoperiodic (*primary*) and other (*supplementary*) cues

Day length (photoperiod)   Climate   Food   Sociality

Brain   GnIH   GnRH

Inflowing blood   Portal blood vessels

Secretory cells

Outflowing blood

Feedback loops

LH   FSH

Cells produce steriods | Cells produce gametes

Testes / Ovaries (enlarged seasonally in response to long days)

Testosterone (or Estradiol)

Hypothalamus
Nerve endings secrete gonadotropin-releasing (GnRH) or -inhibiting (GnIH) neurohormones into the anterior pituitary[1]

Pituitary
Neurohormones stimulate (GnRH) or inhibit (GnIH) release of gonadotropins (LH / FSH) into circulation

Gonads
Gonadotropins stimulate production & release of gametes (FSH) and steroids (Testosterone or estradiol)

Target tissues / cells
Steroids mediate multiple phenotypic responses by binding receptors and activating gene transcription or by initating secondary messenger systems

Behavior / Physiology / Morphology

Action of hormones varies with both:	**Testosterone levels vary seasonally**	**'GnRH-challenge' protocol:**
1) Hormone signal strength (plasma concentration)	Pre-breeding / peak territoriality	**1)** Sample 'Initial' Testosterone
	Nesting / parenting	**2)** Inject GnRH (pectoral muscle)
2) Target tissue sensitivity (e.g., receptor density)	Non-breeding 'baseline'	(wait 30 minutes)
		3) Sample 'Induced' Testosterone

PLATE 6. The endocrine hypothalamic-pituitary-gonadal (HPG) axis regulates seasonal timing, reproductive development, and social behaviors in vertebrates. The HPG axis and the role of testosterone is a central feature of studies evaluating the organismal and evolutionary significance of endocrine systems in the junco (see chapters 4 through 7, 10, and 11). The main panel (*top*) provides an overview of the hormonal cascades that regulate testosterone in response to environmental cues. The lower panels illustrate (*left*) that the rates and magnitudes of hormonal effects are regulated by both the amount of hormone (signal) and the abundance and binding affinities of hormone-specific receptors, transport proteins, coactivators, or conversion factors; (*middle*) that testosterone expression varies seasonally in relation to life history across the annual cycle, in this example for a north-temperate breeding songbird; and (*right*) the "GnRH challenge" blood-sampling protocol used to evaluate an individual organism's ability to produce a short-term increase in plasma testosterone in response to an injection of gonadotropin-releasing hormone (GnRH) (see chapter 6 for details). Additional abbreviations as follows: GnIH, gonadotropin-inhibitory hormone; LH, luteinizing hormone; FSH, follicle-stimulating hormone. Studies to date have been biased towards males; we recommend a future emphasis on females.

[1]GnIH may act both locally in the brain to inhibit GnRH release and in the pituitary to inhibit release of gonadotropins; e.g., see Tsutsui et al. (2013, *Frontiers in Neuroscience* 7:60).

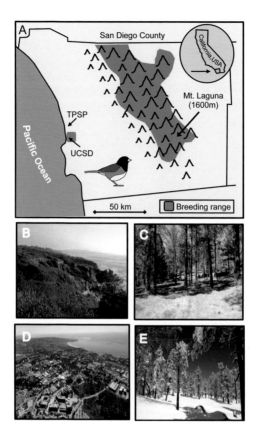

PLATE 7. The breeding range of dark-eyed juncos (*Junco hyemalis thurberi*) in San Diego County, CA, USA, is historically restricted to higher elevation (>1500m) forests and wooded canyons in the central and eastern parts of the county (panel A; adapted from P. Unitt's *San Diego County Bird Atlas*, 2005 edition), for example in habitats such as Mt. Laguna (panel C). Montane-breeding juncos are facultative altitudinal migrants, with most birds joining flocks and many migrating variable distances to lower elevations during winter, particularly during harsh climatic conditions including winter storms (panel E). In the early 1980s, a small isolated breeding population became established along the coast on the urbanized campus of the University of California, San Diego (UCSD, panel D), an area historically comprised of sage scrub and chaparral not suitable for junco breeding, as currently exhibited at the adjacent Torrey Pines State Park (TPSP, panel B). The UCSD juncos have nearly doubled the length of their breeding season and become entirely sedentary (i.e., nonmigratory), and they exhibit a host of other phenotypic differences in comparison with juncos from the nearby native range at Mt. Laguna. This study system is discussed in chapter 9 and reviewed comprehensively in chapter 10 (this volume).

PLATE 8. An Oregon junco photographed in flight, revealing the "tail white" feather plumage trait that has been studied extensively in juncos and discussed in nearly all chapters of this volume. Tail white is a social ornament used in dominance and courtship interactions, with a greater proportion of white preferred by females in mate choice trials, and conferring high social dominance in flocks. Tail white varies by age and sex (males and older birds have more white) and among subspecies and populations. The bottom panels show two ends of the spectrum of geographic variation in male tail white scores. Pictured are the right side tail feathers (six of twelve total retrices) from an Oregon junco from the University of California, San Diego (left panel, ~38 percent white, see chapter 10) and a white-winged junco from South Dakota (right panel, ~70 percent white, see chapter 11). Photos: Roy Hancliff, www .royhancliff.com (top); Mikus Abolins-Abols (left); Christine Bergeon Burns (right).

This result illustrates the importance of ovarian state (e.g., the presence or absence of currently yolking follicles) in female T production. The preovulatory follicle is very active in terms of hormone production and possesses numerous receptors for LH (Johnson 2000). Steroid hormones produced by preovulatory follicles may impact adult female behavior and be deposited into developing eggs (Whittingham and Schwabl 2002; Groothuis et al. 2005; von Engelhardt and Groothuis 2011). A further implication of this result is that linking individual difference in T profiles to variation in phenotype is likely to be substantially more difficult in females than in males. Because T tends to increase most robustly only when eggs are developing, GnRH challenges in females may often be uninformative. However, when care is taken to apply these challenges at the proper breeding stage, GnRH-induced increases in T may indeed provide insight into variation in female behavior and fitness, as we argue below.

D.2. Aggression

Females are aggressive toward same-sex conspecifics in a variety of contexts. Aggression over territory boundaries, wintering and breeding resources, mates, and offspring can all occur among females (Rosvall 2011, 2013). Previous research in a variety of species, including the junco, has found that experimental elevation of T with long-lasting implants can elevate aggression and facilitate the acquisition of resources or dominance status (chapter 4, this volume) (Veiga et al. 2004; Zysling et al. 2006; Rosvall 2013). These findings suggest that high levels of circulating T can be important for mediating female same-sex aggression. However, the relationships between endogenous T and aggressive behavior are not consistent across species and contexts. Female song sparrows (*Melospiza melodia*) respond aggressively to intruders but do not elevate T in response to the intrusion (Elekonich and Wingfield 2000). Jawor et al. (2006b) found that social dominance in captive female juncos did not covary with endogenous T levels; however, dominant females were more likely to initiate nesting than subordinate counterparts and to physically interfere with subordinates' nesting attempts, suggesting a reproductive benefit of dominance. In contrast, free-living female buff-breasted wrens (*Thryothorus leucotis*) showed a slight elevation of T in response to an intruding pair and a greater elevation in response to a single female intruder, but only in the prebreeding season (Gill et al. 2007).

These conflicting findings paint a complicated picture regarding the

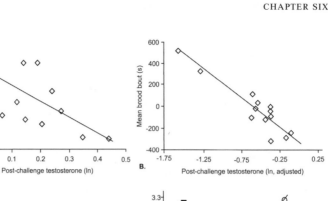

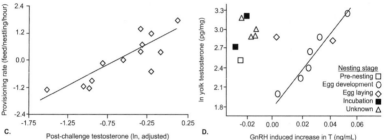

FIGURE 6.4. Relationships between individual response to GnRH challenge and traits of interest in females. (A) Aggression (latency to respond to a simulated intruder) vs. postchallenge T levels (natural-log transformed, adjusted for covariates). Shorter latency was interpreted as a more aggressive response, so the observed relationship with postchallenge T is actually positive. (B) Average amount of time spent brooding chicks and postchallenge T levels (natural-log transformed, adjusted for covariates). (C) Provisioning (nestling feeding rate) and postchallenge T levels (natural-log transformed, adjusted for covariates). (D) Yolk testosterone levels vs. GnRH-induced increase in T at different reproductive stages. Best-fit line uses only challenges administered during egg development.

role of T in mediating same-sex aggression in female birds. We used GnRH challenges to shed light on this relationship in free-living female juncos (Cain and Ketterson 2012). Challenges were administered during the prebreeding season just before egg laying was initiated (within the seven days leading up to egg laying). Later in the breeding season, female aggression towards a simulated female intruder was measured. Overall, female prebreeding T production was strongly related to female responsiveness to an intruder; females producing more T in response to GnRH challenge just before egg laying were quicker to attack a simulated intruder presented during incubation and were also generally more aggressive (figure 6.4A). These findings support the possibility that endogenous T is an important mediator of individual variation in aggression in female juncos.

D.3. Maternal Care

In the junco, as in many other avian species, females typically contribute more than males to parental care. Maternal care behavior includes incubation, nestling feeding, nestling defense, and fledgling care, and alteration in any one of these has the potential to reduce realized reproductive success. In bird species in which male care is essential, experimental elevation of T may have no effect on male care, a phenomenon referred to as behavioral insensitivity to T (Lynn et al. 2002; but see Van Roo 2004). Given the essential nature of maternal care, we might predict that females should also be insensitive to experimentally elevated T. However, research investigating these relationships has revealed a more complicated pattern. As predicted, T treatment has no effect on some measures of maternal care (Clotfelter et al. 2004; Ketterson et al. 2005; O'Neal et al. 2008), supporting the insensitivity hypothesis, but other studies have shown that experimentally increased T may reduce aspects of maternal behavior (chapter 4, this volume). As a result, there is currently no clear indication of whether and how T is important in mediating female life-history trade-offs.

Investigating the relationship between endogenous T and maternal behavior, we found that females that elevated T more strongly in response to GnRH spent less time brooding than females that produced less T (figure 6.4B) (Cain and Ketterson 2013b). However, in contrast to previous research on males, we found that more strongly responding females provisioned nestlings *more* frequently than more weakly responding females (figure 6.4C). These findings suggest that T is likely important in mediating maternal phenotypes and life-history trade-offs and, further, that the direction of the relationships between T and phenotype in females may differ from what is generally observed in males. Also worth noting is that these findings suggest that females may differentially invest in various forms of maternal care. Thus caution is warranted when examining only one form of care.

D.4. Maternal Effects

One of the most important consequences of T in females is the potential for females to influence the phenotype and fitness of offspring via maternal effects. Females deposit T in egg yolk, where it may affect embryonic development, which may have far-reaching consequences for adult phenotype and fitness (Schwabl 1993; Adkins-Regan 2005; Groothuis et al.

2005; von Engelhardt and Groothuis 2011). Elevated yolk steroids can generate fitness benefits or costs, depending on context. For example, elevated yolk T in eggs has been associated with greater development in hatchlings of the muscles that affect hatching and begging, and also with growth rates (Lipar and Ketterson 2000). However, experimental elevations in egg yolk T have also increased offspring mortality (Sockman and Schwabl 2000). Importantly, the vast majority of the data available on the effects of T on developing offspring stem from experimental manipulation studies, and little is known about the causes and consequences of natural variation in yolk T.

We investigated the relationship between GnRH-induced T production in female juncos and the amount of T they later deposited in the yolks of their eggs (Jawor et al. 2007; Cain and Ketterson 2013b). Females that responded more strongly to GnRH challenges also tended to place more T in the yolks of eggs they produced (figure 6.4D). As GnRH challenges are intended to mimic the endogenous release of GnRH, this result suggests that deposition of yolk T depends at least in part on the sensitivity of a female's HPG axis. It is likely that the greater sensitivity to GnRH during egg development reflects the temporary upregulation of the HPG axis that is a necessary functional step in the process of egg development and ovulation (Johnson 2000). More intriguingly, our results also suggest a simple mechanism by which a female's response to particular environments or experiences could influence early development in the next generation via differential deposition of yolk T.

D.5. Morphology

There is little research exploring potential relationships between female morphology and endogenous T. As noted above, male T in response to GnRH covaries with morphological and behavioral traits that are important for competition. Experimental manipulations have revealed similar causal relationships between T and competitive traits in females. In some species, elevated T in females leads to the development of morphological traits normally seen in males (Staub and De Beer 1997; Ketterson et al. 2005). For example, experimentally elevated T induced female budgerigars (*Melopsittacus undulatus*) to develop male-typical cere color (Nespor et al. 1996). We found that individual variation in ability to produce T in response to a GnRH challenge was positively related to overall body size in female juncos (as measured by tarsus, wing, and tail length) (Cain and Ketterson 2012) and to the size of a female's tail white patch (unpublished data). Juncos are mildly sexually dimorphic; males are on average larger and more ornamented. Thus, females producing more T in response to GnRH were closer to typical male morphology than females producing less T. In addition, larger females were more aggressive towards simulated intruders. Together, these findings suggest that, as in males, high levels of T production are associated with more competitive phenotypes in females.

D.6. Selection

The majority of studies examining the effects of experimentally elevated T (via implants) on female behavior and physiology have found that sustained high T levels are costly to females (Clotfelter et al. 2004; Garcia-Vigon et al. 2008; O'Neal et al. 2008; Veiga and Polo 2008; Gerlach and Ketterson 2013); however, T-implanted females can also experience some fitness advantages (Sandell 2007). As discussed in chapter 4 (this volume), T-implanted female juncos had lower reproductive success when measured during early stages of reproduction. However, they did not differ from control females in fledgling quality, extrapair offspring production, survival, or reproduction in the following year (O'Neal et al. 2008; Gerlach and Ketterson 2013) (chapter 4, this volume).

We have not yet been able to evaluate the strength and direction of selection on female ability to respond to GnRH as we have for males. As female response to GnRH outside of the egg-yolking phase is attenuated, acquiring data on a sufficient number of individuals to assess selection gradients has proved challenging. Our current results do allow us to draw some inferences about the relationship between T in response to GnRH and female fitness. As noted above, female aggression was positively related to GnRH response (Cain and Ketterson 2012), and more aggressive females produced smaller eggs and smaller newly hatched nestlings in some years (Cain and Ketterson 2013a). However, nestling size differences disappeared by nestling day three, suggesting some compensation mechanism. These data suggest that high T production may be somewhat costly to females but likely not as costly as the constitutively elevated T caused by implants (chapter 4, this volume). We also found fitness advantages associated with high T production. More aggressive females were more likely to have at least one successful nesting attempt in a single season than were less aggressive females, and these same aggressive females were also stronger responders to GnRH challenges (Cain and Ketterson 2012).

E. GnRH-Induced vs. Experimentally Elevated T

E.1. Phenotypes

The results reviewed above indicate that GnRH challenges provide a robust method for assessing natural variation in T in relation to phenotypes in males and females. Clearly, T has effects on the behavior and physiology of both sexes. When comparing the results of GnRH-challenge studies to those of implant studies, we conclude that the phenotypic correlates of being a naturally high T-producing individual are sometimes quite similar to the effects of having experimentally elevated T and sometimes quite different. Further, some of the patterns appear to be sex-specific and quite complex. Much remains to be learned about the mechanistic links that give rise to the correlations between T production ability and trait expression, and such examinations are likely to provide important insights (chapter 7, this volume).

In both sexes, the relationships between GnRH-induced T and behavior were similar to those seen with experimentally elevated T in implant studies (summarized in table 6.1; see also chapter 4, this volume). In males, both natural variation and experimentally elevated T were associated with a decrease in nestling feeding rate and an increase in same-sex competitive behaviors. Also in males, higher circulating (initial) T and experimentally elevated T were both associated with a reduction in immune function. In females, experimentally elevated T and T production ability following a GnRH challenge were both associated with a decrease in brooding and an increase in yolk T. In implant studies, T had no effect on female provisioning, and, unexpectedly, the relationship between natural T and provisioning rates in females was positive.

E.2. Selection

For males, the effects of experimentally elevated T on fitness components only partially predicted the patterns of covariation observed between T production and fitness (table 6.1). Experimentally elevated T enhanced reproductive success in males, primarily via enhanced extrapair mating success, which would predict positive directional selection on T produc-

TABLE 6.1 **Testosterone (T)-mediated traits in junco males and females in relation to experimentally elevated T versus endogenous T (either initial, postchallenge, or GnRH-induced increase)**

Trait	Sex	Effect of T implant Direction / measurement / source			Relationship with endogenous T measures Direction / measurement / source		
Territorial behavior	Male	+	song rates	1	+ (b)	simulated territorial intruder	8
	Female	+	resident-intruder (captive)	2	+ (b)	simulated intruder near nest	9
Immune function	Male	–	antibody production, cell-mediated immunity	3	– (a)	nonspecific immunoglobulin-G (IgG), and complement levels.	10
	Female	–	cell-mediated immunity	2	– (a)	cell-mediated immunity	2
Ornamentation	Male		unknown		+ (c)	tail-white patch size	11
	Female		unknown		+ (b)	tail-white patch size	16
Nestling provisioning	Male	–	feeding rate, days 4–10	1	– (c)	feeding rate, days 6 or 7	8
	Female	0	feeding rate, day 3	5	+ (b)	feeding rate, day 3	12
Brooding	Female	–	total time brooding, day 3	5	– (b)	average time brooding, day 3	12
Yolk T	Female	+	sample from yolk	6	+ (b)	entire yolk (homogenized)	13, 12
Survival	Male	+	annual survival rate	4	stabilizing (b)	annual survival	14
	Female	0	annual survival rate	7	unknown		
Reproductive success	Male	+	number of offspring (including extrapair)	4	stabilizing (b)	number of offspring (including extrapair)	14
	Female	–	number of offspring	5, 7	unknown	T-related traits + effect on nest success	9, 15

Notes: Letters indicate which measure of endogenous T: (a) Initial T; (b) Postchallenge T; (c) GnRH-induced change in T. Symbols indicate direction of relationships: (+) positive, (–) negative, and (0) no relationship detected. Numbers indicate references: (1) Ketterson et al. (1992); (2) Zysling et al. (2006); (3) Casto et al. (2001); (4) Reed et al. (2008); (5) O'Neal et al. (2008); (6) Clotfelter et al. (2004); (7) Gerlach and Ketterson (2013); (8) McGlothlin et al. (2007); (9) Cain and Ketterson (2012); (10) Greives et al. (2006); (11) McGlothlin et al. (2008); (12) Cain and Ketterson (2013b); (13) Jawor et al. (2007); (14) McGlothlin et al. (2010); (15) Cain and Ketterson (2013a); (16) Cain, unpublished data.

tion (see chapter 4, this volume, for details). However, the overall pattern of reproductive selection on T production following a GnRH challenge was stabilizing. That is, for annual reproductive success and extrapair reproductive success, both extremes were disfavored, and males near the population mean were the most successful.

Similarly, experimentally elevated T decreased survivorship, but viability selection on T production was also stabilizing rather than directional. Survival in strong responders to GnRH was reduced, which is arguably consistent with the pattern of reduced survival in T-implanted males as compared to controls. However, survival was also reduced in weak responders to GnRH. It is difficult to compare this result to experimental studies, as they did not include a treatment that reduced T. In sum, the effect of experimentally elevated T on fitness was an uneven predictor of the relationships between natural ability to elevate T and fitness in males.

There are several potential explanations for these differences. First, natural variation in T in response to GnRH may reflect variation in individual quality, whereas in implant experiments, individuals are assigned to implant or control groups randomly with respect to quality. This is a general problem in life-history evolution: allocation trade-offs may be obscured when individuals vary in quality (or "acquisition"), which means that individual allocation decisions may not be apparent at a population level (van Noordwijk and de Jong 1986; Roff and Fairbairn 2007). In the case of juncos, high quality males may be capable of greater allocation to both reproduction *and* survival, leading to a similar pattern of selection via both fitness components. Variation in quality may also be responsible for the decreased fitness of males with extremely low T responses to GnRH; low-quality males may be both unable to upregulate their HPG axis to the same extent as higher-quality males and also be ill equipped to survive and reproduce. Decreased fitness at the other extreme, however, may reflect fitness costs of high testosterone. As in implant experiments, males with extremely high GnRH responses showed reduced survival, which may result from increased risk taking, predation, reduced immune function, or other T-mediated costs (Wingfield et al. 2001). Unlike experimental studies, however, males with high responses to GnRH do not appear to be compensated by increased extrapair mating success.

Another potential explanation for different fitness consequences for high responders to GnRH and T-implanted males is the duration of elevated T involved. Implants sustain T at high levels over time and prevent levels from fluctuating as they do naturally by overriding negative feedback. Although GnRH-induced variation and experimentally elevated T are associated with similar effects on behavior, implants cause increases in T-mediated behavior over the full course of the breeding season. In contrast, the GnRH challenge is not an experimental treatment designed to alter behavior. Rather, it is a probe that acts briefly to reveal variation among individuals in their natural ability to elevate T in different contexts and over a very short period of time. Despite the repeatability of GnRH-induced T levels, T levels decrease immediately after the challenge, whereas T-implanted males are experimentally fixed at high T levels. Integrated over this long timespan, it may not be surprising that the fitness costs and benefits of T-mediated behaviors differ between T-implanted males and naturally high responders. Nevertheless, if a mutant phenotype with constitutively elevated T were to arise, results from implant studies of males suggest it would have the potential to spread.

In females, experimentally elevated T generally decreased fitness (chapter 4, this volume), but unlike in males, the cost was in fecundity rather than survival. Because we have not yet been able to measure selection on female GnRH response, we cannot make direct comparisons to males. Preliminary results suggest that while there are substantial fitness costs to higher T, there are also important benefits (Cain and Ketterson 2012, 2013a). Aggression was positively associated with individual T production and was strongly related to reproductive success in two of three years, with no relationship in the third, suggesting that higher T-production ability is likely to be less costly than the sustained high levels that result from implants. The positive association between individual T production and maternal provisioning further supports this possibility. Together, these results suggest that the fitness consequences of female T-production ability and T-mediated phenotypes in females are likely complex and dependent on the female's environment (Cain and Rosvall 2014).

F. Conclusions

Our combined results demonstrate natural variation, in both males and females, in the production of T by the HPG axis and its association with a suite of morphological, physiological, and behavioral traits. In general, high T production is related to greater investment in traits important for competition, but this increase may come at the expense of other fitness-relevant traits such as parental care. The studies described here are among the first to link individual ability to produce T with ecologically relevant traits. The collective findings point to the utility of physiological challenges as tools to reveal important relationships between individual variation in hormone and phenotype that would be difficult if not impossible to observe in relation to unperturbed (baseline) circulating hormone levels alone. Further, our work has shown that natural selection can and does act on natural variation in hormones and hormone-mediated traits, supporting the conclusion that hormones may represent a potent physiological source of evolutionarily relevant variation. Future studies like those described here can provide continued insights into how selection acts on hormone-mediated phenotypes in free-living animals, adding enormously to our understanding of the role of proximate mechanisms in shaping evolutionary responses.

Our research was conducted on a well-established population of juncos, and for this reason, it is perhaps unsurprising that the suite of T-mediated traits we studied was under stabilizing selection. However, we also expect that hormones often facilitate both plastic responses and rapid adaptation to changing and newly colonized environments where we might expect to detect directional selection (see chapters 8 and 11, this volume, for more discussion) (Ketterson et al. 2009). Understanding the ecological and evolutionary role of natural variation in hormones in response to environmental change is one of the grand challenges for future work in integrative evolutionary biology.

Acknowledgments

The research described in this chapter condenses over a decade of field-work on hormonal, behavioral, and morphological measures of hundreds of birds. We were able to complete this research only with the assistance of a cadre of hardworking and dedicated postdocs, graduate students, field assistants, undergraduates, and volunteers. We are particularly indebted to our coauthors on this work: George Bentley, Christy Bergeon Burns, Joe Casto, Nicki Gerlach, Tim Greives, Ellen Ketterson, Jenny Phillips, Sara Schrock, Eric Snajdr, and Danielle Whittaker. We would also like to thank Henry Wilbur and Butch Brodie (directors of Mountain Lake Biological Station), the Mountain Lake Lodge, and the Dolinger family for supporting our research efforts by allowing us to work on their properties. Finally, we thank the National Science Foundation, which provided crucial funding for this research via a series of research grants and fellowships.

References

Adkins-Regan, E. 2005. *Hormones and Animal Social Behavior*. Princeton: Princeton University Press.

———. 2008. Do hormonal control systems produce evolutionary inertia? *Philosophical Transactions of the Royal Society B-Biological Sciences* 363:1599–1609.

Alatalo, R. V., J. Höglund, A. Lundberg, P. T. Rintamäki, and B. Silverin. 1996. Testosterone and male mating success on the black grouse leks. *Proceedings of the Royal Society B-Biological Sciences* 263:1697–1702.

Andersson, M. 1994. *Sexual Selection*. Princeton: Princeton University Press.

Apfelbeck, B., and W. Goymann. 2011. Ignoring the challenge? Male black redstarts (*Phoenicurus ochruros*) do not increase testosterone levels during territorial conflicts but they do so in response to gonadotropin-releasing hormone. *Proceedings of the Royal Society B-Biological Sciences* 278:3233–42.

Bonduriansky, R., and S. F. Chenoweth. 2009. Intralocus sexual conflict. *Trends in Ecology & Evolution* 24:280–88.

Borgia, G., and J. C. Wingfield. 1991. Hormonal correlates of bower decoration and sexual display in the satin bowerbird (*Ptilonorhynchus violaceus*). *Condor* 93: 935–42.

Brown, C. R., M. B. Brown, S. A. Raouf, L. C. Smith, and J. C. Wingfield. 2005. Effects of endogenous steroid hormone levels on annual survival in cliff swallows. *Ecology* 86:1034–46.

Cain, K. E., and E. D. Ketterson. 2012. Competitive females are successful females: Phenotype, mechanism, and selection in a common songbird. *Behavioral Ecology and Sociobiology* 66:241–52.

———. 2013a. Costs and benefits of competitive traits in females: Aggression, maternal care, and reproductive success. *PLOS One* 8:e77816.

———. 2013b. Individual variation in testosterone and parental care in a female songbird: The dark-eyed junco (*Junco hyemalis*). *Hormones and Behavior* 64: 685–92.

Cain, K. E., and K. A. Rosvall. 2014. Next steps for understanding the selective relevance of female-female competition. *Frontiers in Ecology and Evolution* 2:1–3.

Casto, J. M., V. Nolan, Jr., and E. D. Ketterson. 2001. Steroid hormones and immune function: Experimental studies in wild and captive dark-eyed juncos (*Junco hyemalis*). *American Naturalist* 157:408–20.

Clotfelter, E. D., D. M. O'Neal, J. M. Gaudioso, J. M. Casto, I. M. Parker-Renga, E. A. Snajdr, D. L. Duffy, et al. 2004. Consequences of elevating plasma testosterone in females of a socially monogamous songbird: Evidence of constraints on male evolution? *Hormones and Behavior* 46:171–78.

DeVries, M. S., C. P. Winters, and J. M. Jawor. 2012. Testosterone elevation and response to gonadotropin-releasing hormone challenge by male northern cardinals (*Cardinalis cardinalis*) following aggressive behavior. *Hormones and Behavior* 62:99–105.

Dufty, A. M., Jr., J. Clobert, and A. P. Møller. 2002. Hormones, developmental plasticity, and evolution. *Trends in Ecology & Evolution* 17:190–96.

Elekonich, M. M., and J. C. Wingfield. 2000. Seasonality and hormonal control of territorial aggression in female song sparrows (Passeriformes: Emberizidae: *Melospiza melodia*). *Ethology* 106:493–510.

Endler, J. A. 1986. *Natural Selection in the Wild*. Princeton: Princeton University Press.

Evans, M. R., A. R. Goldsmith, and S. R. A. Norris. 2000. The effects of testosterone on antibody production and plumage coloration in male house sparrows (*Passer domesticus*). *Behavioral Ecology and Sociobiology* 47:156–63.

Finch, C. E., and M. R. Rose. 1995. Hormones and the physiological architecture of life-history evolution. *Quarterly Review of Biology* 70:1–52.

Garamszegi, L. Z. 2014. Female peak testosterone levels in birds tell an evolutionary story: A comment on Goyman and Wingfield. *Behavioral Ecology* 25:700–701.

Garcia-Vigon, E., P. J. Cordero, and J. P. Veiga. 2008. Exogenous testosterone in female spotless starlings reduces their rate of extrapair offspring. *Animal Behaviour* 76:345–53.

Gerlach, N. M., and E. D. Ketterson. 2013. Experimental elevation of testosterone lowers fitness in female dark-eyed juncos. *Hormones and Behavior* 63:782–90.

Gill, S. A., E. D. Alfson, and M. Hau. 2007. Context matters: Female aggression and testosterone in a year-round territorial neotropical songbird (*Thryothorus leucotis*). *Proceedings of the Royal Society B-Biological Sciences* 274:2187–94.

Gonzalez, G., G. Sorci, L. C. Smith, and F. de Lope. 2001. Testosterone and sexual signalling in male house sparrows (*Passer domesticus*). *Behavioral Ecology and Sociobiology* 50:557–62.

Goymann, W. 2009. Social modulation of androgens in male birds. *General and Comparative Endocrinology* 163:149–57.

Goymann, W., and J. C. Wingfield. 2014a. Correlated evolution of female and male testosterone—internal constraints or external determinants? A response to comments on Goymann and Wingfield. *Behavioral Ecology* 25:704–5.

———. 2014b. Male-to-female testosterone ratios, dimorphism, and life history—what does it really tell us? *Behavioral Ecology* 25:685–99.

Greives, T. J., J. W. McGlothlin, J. M. Jawor, G. E. Demas, and E. D. Ketterson. 2006. Testosterone and innate immune function inversely covary in a wild population of breeding dark-eyed juncos (*Junco hyemalis*). *Functional Ecology* 20:812–18.

Groothuis, T. G. G., W. Müller, N. von Engelhardt, C. Carere, and C. Eising. 2005.

Maternal hormones as a tool to adjust offspring phenotype in avian species. *Neuroscience and Biobehavioral Reviews* 29:329–52.

Hau, M. 2007. Regulation of male traits by testosterone: implications for the evolution of vertebrate life histories. *BioEssays* 29:133–44.

Hill, J. A., D. A. Enstrom, E. D. Ketterson, V. Nolan, Jr., and C. Ziegenfus. 1999. Mate choice based on static versus dynamic secondary sexual traits in the dark-eyed junco. *Behavioral Ecology* 10:91–96.

Hirschenhauser, K., E. Möstl, P. Péczely, B. Wallner, J. Dittami, and K. Kotrschal. 2000. Seasonal relationships between plasma and fecal testosterone in response to GnRH in domestic ganders. *General and Comparative Endocrinology* 118: 262–72.

Jawor, J. M., J. W. McGlothlin, J. M. Casto, T. J. Greives, E. A. Snajdr, G. E. Bentley, and E. D. Ketterson. 2006a. Seasonal and individual variation in response to GnRH challenge in male dark-eyed juncos (*Junco hyemalis*). *General and Comparative Endocrinology* 149:182–89.

―――. 2007. Testosterone response to GnRH in a female songbird varies with stage of reproduction: Implications for adult behaviour and maternal effects. *Functional Ecology* 21:767–75.

Jawor, J. M., R. Young, and E. D. Ketterson. 2006b. Females competing to reproduce: dominance matters but testosterone does not. *Hormones and Behavior* 49:362–68.

John-Alder, H. B., R. M. Cox, G. J. Haenel, and L. C. Smith. 2009. Hormones, performance and fitness: Natural history and endocrine experiments on a lizard (*Sceloporus undulatus*). *Integrative and Comparative Biology* 49:393–407.

Johnson, A. L. 2000. Reproduction in the female. In *Sturkie's Avian Physiology*, edited by G. C. Whittow, 569–96. San Diego: Academic Press.

Ketterson, E. D., J. W. Atwell, and J. W. McGlothlin. 2009. Phenotypic integration and independence: Hormones, performance, and response to environmental change. *Integrative and Comparative Biology* 49:365–79.

Ketterson, E. D., and V. Nolan, Jr. 1992. Hormones and life histories: an integrative approach. *American Naturalist* 140:S33–S62.

―――. 1999. Adaptation, exaptation, and constraint: A hormonal perspective. *American Naturalist* 154:S4–S25.

Ketterson, E. D., V. Nolan, Jr., and M. Sandell. 2005. Testosterone in females: Mediator of adaptive traits, constraint on sexual dimorphism, or both? *American Naturalist* 166:S85–S98.

Ketterson, E. D., V. Nolan, Jr., L. Wolf, and C. Ziegenfus. 1992. Testosterone and avian life histories: Effects of experimentally elevated testosterone on behavior and correlates of fitness in the dark-eyed junco (*Junco hyemalis*). *American Naturalist* 140:980–99.

Kingsolver, J. G., H. E. Hoekstra, J. M. Hoekstra, D. Berrigan, S. N. Vignieri, C. E. Hill, A. Hoang, P. Gilbert, and P. Beerli. 2001. The strength of phenotypic selection in natural populations. *American Naturalist* 157:245–61.

Lacombe, D., A. Cyr, and P. Matton. 1991. Plasma LH and androgen levels in the red-winged blackbird (*Agelaius phoeniceus*) treated with a potent GnRH analog. *Comparative Biochemistry and Physiology A-Physiology* 99:603–7.

Lipar, J. L., and E. D. Ketterson. 2000. Maternally derived yolk testosterone enhances the development of the hatching muscle in the red-winged blackbird *Agelaius phoeniceus*. *Proceedings of the Royal Society B-Biological Sciences* 267:2005–10.

Lynn, S. E. 2008. Behavioral insensitivity to testosterone: Why and how does testosterone alter paternal and aggressive behavior in some avian species but not others? *General and Comparative Endocrinology* 157:233–40.

Lynn, S. E., L. S. Hayward, Z. M. Benowitz-Fredericks, and J. C. Wingfield. 2002. Behavioural insensitivity to supplementary testosterone during the parental phase in the chestnut-collared longspur, *Calcarius ornatus*. *Animal Behaviour* 63:795–803.

Lynn, S. E., B. G. Walker, and J. C. Wingfield. 2005. A phylogenetically controlled test of hypotheses for behavioral insensitivity to testosterone in birds. *Hormones and Behavior* 47:170–77.

Magrath, M. J. L., and J. Komdeur. 2003. Is male care compromised by additional mating opportunity? *Trends in Ecology & Evolution* 18:424–30.

McGlothlin, J. W., J. M. Jawor, T. J. Greives, J. M. Casto, J. L. Phillips, and E. D. Ketterson. 2008. Hormones and honest signals: Males with larger ornaments elevate testosterone more when challenged. *Journal of Evolutionary Biology* 21: 39–48.

McGlothlin, J. W., J. M. Jawor, and E. D. Ketterson. 2007. Natural variation in a testosterone-mediated trade-off between mating effort and parental effort. *American Naturalist* 170:864–75.

McGlothlin, J. W., and E. D. Ketterson. 2008. Hormone-mediated suites as adaptations and evolutionary constraints. *Philosophical Transactions of the Royal Society B-Biological Sciences* 363:1611–20.

McGlothlin, J. W., D. J. Whittaker, S. E. Schrock, N. M. Gerlach, J. M. Jawor, E. A. Snajdr, and E. D. Ketterson. 2010. Natural selection on testosterone production in a wild songbird population. *American Naturalist* 175:678–701.

Millesi, E., I. E. Hoffmann, S. Steurer, M. Metwaly, and J. P. Dittami. 2002. Vernal changes in the behavioral and endocrine responses to GnRH application in male European ground squirrels. *Hormones and Behavior* 41:51–58.

Mills, S. C., A. Grapputo, E. Koskela, and T. Mappes. 2007. Quantitative measure of sexual selection with respect to the operational sex ratio: A comparison of selection indices. *Proceedings of the Royal Society B-Biological Sciences* 274: 143–50.

Moore, I. T., N. Perfito, H. Wada, T. S. Sperry, and J. C. Wingfield. 2002. Latitudinal variation in plasma testosterone levels in birds of the genus *Zonotrichia*. *General and Comparative Endocrinology* 129:13–19.

Nelson, R. J. 2005. *An Introduction to Behavioral Endocrinology*. Sunderland, MA: Sinauer.

Nespor, A. A., M. J. Lukazewicz, R. J. Dooling, and G. F. Ball. 1996. Testosterone induction of male-like vocalizations in female budgerigars (*Melopsittacus undulatus*). *Hormones and Behavior* 30:162–69.

Nolan, V., Jr., E. D. Ketterson, D. A. Cristol, C. M. Rogers, E. D. Clotfelter, R. C. Titus, S. J. Schoech, and E. Snajdr. 2002. Dark-eyed junco (*Junco hyemalis*). In *The Birds of North America*, edited by A. Poole and F. Gill, no. 716. Philadelphia: The Birds of North America, Inc.

Nolan, V., Jr., E. D. Ketterson, C. Ziegenfus, D. P. Cullen, and C. R. Chandler. 1992. Testosterone and avian life histories: Effects of experimentally elevated testosterone on prebasic molt and survival in male dark-eyed juncos. *Condor* 94: 364–70.

O'Neal, D. M., D. G. Reichard, K. Pavlis, and E. D. Ketterson. 2008. Experimentally-elevated testosterone, female parental care, and reproductive success in a songbird, the dark-eyed junco (*Junco hyemalis*). *Hormones and Behavior* 54:571–78.

Raouf, S. A., P. G. Parker, E. D. Ketterson, V. Nolan, Jr., and C. Ziegenfus. 1997. Testosterone affects reproductive success by influencing extrapair fertilizations in male dark-eyed juncos (Aves: *Junco hyemalis*). *Proceedings of the Royal Society B-Biological Sciences* 264:1599–1603.

Reed, W. L., M. E. Clark, P. G. Parker, S. A. Raouf, N. Arguedas, D. S. Monk, E. Snajdr, V. Nolan, Jr., and E. D. Ketterson. 2006. Physiological effects on demography: A long-term experimental study of testosterone's effects on fitness. *American Naturalist* 167:667–83.

Roff, D. A. 2002. *Life History Evolution*. Sunderland, MA: Sinauer Associates.

Roff, D. A., and D. J. Fairbairn. 2007. The evolution of trade-offs: Where are we? *Journal of Evolutionary Biology* 20:433–47.

Rosvall, K. A. 2011. Intrasexual competition in females: Evidence for sexual selection? *Behavioral Ecology* 22:1131–140.

———. 2013. Life history trade-offs and behavioral sensitivity to testosterone: An experimental test when female aggression and maternal care co-occur. *PLOS One* 8:e54120.

Rosvall, K. A., M. P. Peterson, D. G. Reichard, and E. D. Ketterson. 2014. Highly context-specific activation of the HPG axis in the dark-eyed junco and implications for the challenge hypothesis. *General and Comparative Endocrinology* 201:65–73.

Rosvall, K. A., D. G. Reichard, S. M. Ferguson, D. J. Whittaker, and E. D. Ketterson. 2012. Robust behavioral effects of song playback in the absence of testosterone or corticosterone release. *Hormones and Behavior* 62:418–25.

Sandell, M. I. 2007. Exogenous testosterone increases female aggression in the European starling (*Sturnus vulgaris*). *Behavioral Ecology and Sociobiology* 62: 255–62.

Schoech, S. J., R. L. Mumme, and J. C. Wingfield. 1996. Delayed breeding in the co-operatively breeding Florida scrub-jay (*Aphelocoma coerulescens*): Inhibition or the absence of stimulation? *Behavioral Ecology and Sociobiology* 39:77–90.

Schwabl, H. 1993. Yolk is a source of maternal testosterone for developing birds. *Proceedings of the National Academy of Sciences of the United States of America* 90:11446–50.

Siepielski, A. M., J. D. DiBattista, and S. M. Carlson. 2009. It's about time: The temporal dynamics of phenotypic selection in the wild. *Ecology Letters* 12:1261–76.

Sockman, K. W., and H. Schwabl. 2000. Yolk androgens reduce offspring survival. *Proceedings of the Royal Society B-Biological Sciences* 267:1451–56.

Staub, N. L., and M. De Beer. 1997. The role of androgens in female vertebrates. *General and Comparative Endocrinology* 108:1–24.

Stearns, S. C. 1992. *The Evolution of Life Histories*. New York: Oxford University Press.

van Noordwijk, A. J., and G. de Jong. 1986. Acquisition and allocation of resources: Their influence on variation in life history tactics. *American Naturalist* 128:137–42.

Van Roo, B. L. 2004. Exogenous testosterone inhibits several forms of male parental behavior and stimulates song in a monogamous songbird: The blue-headed vireo (*Vireo solitarius*). *Hormones and Behavior* 46:678–83.

Veiga, J. P., and V. Polo. 2008. Fitness consequences of increased testosterone levels in female spotless starlings. *The American Naturalist* 172:42–53.

Veiga, J. P., J. Vinuela, P. J. Cordero, J. M. Aparicio, and V. Polo. 2004. Experimentally increased testosterone affects social rank and primary sex ratio in the spotless starling. *Hormones and Behavior* 46:47–53.

Villavicencio, C. P., B. Apfelbeck, and W. Goymann. 2013. Experimental induction of social instability during early breeding does not alter testosterone levels in male black redstarts, a socially monogamous songbird. *Hormones and Behavior* 64:461–67.

von Engelhardt, N., and T. G. G. Groothuis. 2011. Maternal hormones in avian eggs. *Birds*, vol. 4 of *Hormones and Reproduction of Vertebrates*, 91–127. San Diego: Academic Press.

Whittingham, L. A., and H. Schwabl. 2002. Maternal testosterone in tree swallow eggs varies with female aggression. *Animal Behaviour* 63:63–67.

Williams, T. D. 2008. Individual variation in endocrine systems: Moving beyond the "tyranny of the Golden Mean." *Philosophical Transactions of the Royal Society B-Biological Sciences* 363:1699–1710.

———. 2012a. Hormones, life-history, and phenotypic variation: Opportunities in evolutionary avian endocrinology. *General and Comparative Endocrinology* 176:286–95.

———. 2012b. *Physiological Adaptations for Breeding in Birds*. Princeton: Princeton University Press.

Wingfield, J. C. 1985. Short-term changes in plasma levels of hormones during establishment and defense of a breeding territory in male song sparrows, *Melospiza melodia*. *Hormones and Behavior* 19:174–87.

———. 2003. Control of behavioural strategies for capricious environments. *Animal Behaviour* 66:807–15.

Wingfield, J. C., G. F. Ball, A. M. Dufty, Jr., R. E. Hegner, and M. Ramenofsky. 1987. Testosterone and aggression in birds. *American Scientist* 75:602–8.

Wingfield, J. C., J. W. Crim, P. W. Mattocks, and D. S. Farner. 1979. Responses of photosensitive and photorefractory male white-crowned sparrows (*Zonotrichia leucophrys gambelii*) to synthetic mammalian luteinizing-hormone releasing hormone (syn-LHRH). *Biology of Reproduction* 21:801–6.

Wingfield, J. C., and A. R. Goldsmith. 1990. Plasma levels of prolactin and gonadal steroids in relation to multiple brooding and renesting in free-living populations of the song sparrow, *Melospiza melodia*. *Hormones and Behavior* 24: 89–103.

Wingfield, J. C., R. E. Hegner, A. M. Dufty, Jr., and G. F. Ball. 1990. The "challenge hypothesis": Theoretical implications for patterns of testosterone secretion, mating systems, and breeding systems. *American Naturalist* 136:829–46.

Wingfield, J. C., S. E. Lynn, and K. K. Soma. 2001. Avoiding the "costs" of testosterone: Ecological bases of hormone-behavior interactions. *Brain, Behavior, and Evolution* 57:239–51.

Zera, A. J., L. G. Harshman, and T. D. Williams. 2007. Evolutionary endocrinology: The developing synthesis between endocrinology and evolutionary genetics. *Annual Review of Ecology Evolution and Systematics* 38:793–817.

Zysling, D. A., T. J. Greives, C. W. Breuner, J. M. Casto, G. E. Demas, and E. D. Ketterson. 2006. Behavioral and physiological responses to experimentally elevated testosterone in female dark-eyed juncos (*Junco hyemalis carolinensis*). *Hormones and Behavior* 50:200–207.

CHAPTER SEVEN

Diving Deeper into Mechanism

Individual and Sex Differences in Testosterone Production, Sensitivity, and Genomic Responses

Kimberly A. Rosvall, Christine M. Bergeon Burns, and Mark P. Peterson

The fastest snake has a burst speed ten times that of the slowest ... Assuming for a moment that these individual differences are real, these observations immediately suggest two sorts of questions. First, what is the functional basis of these individual performance differences? Which physiological or morphological factors make a fast snake fast? ... Second, what are the ecological and evolutionary consequences of these differences? ... These questions ... are obscured if one concentrates only on central tendency. This is the tyranny of the Golden Mean.
—Bennett (1987), "Interindividual variability: An underutilized resource," in *New Directions in Ecological Physiology*

A. Introduction

Over thirty years ago, Bennett highlighted the tyranny of the golden mean and what we as biologists miss if we ignore variation around the mean. From an evolutionary perspective, individual variation is of paramount importance. Individual variation is the raw material of evolution (Darwin 1859), and the strength of selection is proportional to the variance among individuals (Lande 1979). Bennett reminds us that if we focus solely on the average differences among two or more groups of organisms, then two types of questions are essentially lost. First, we would fail to identify the proximate mechanisms that account for natural individual differences in phenotype, and second, by not asking whether (or why) some individuals have an advantage over others in a particular evolutionary and ecological context, we would fail to ascertain the proximate mechanisms by which phenotypes might evolve.

As evolutionary endocrinologists, our research attempts to understand how hormone-mediated traits evolve, and the integration of Bennett's two individual-based questions is essential to this process. In short, we need to identify the causes and consequences of individual variation in hormones and hormone-mediated traits. Historically, endocrinologists have not typically focused on individual variation, in part because individual differences were assumed to be noise attributable to technical variance (e.g., imprecise assays or high inter-run variability in autoradiography) (Bennett 1987). Modern techniques that measure hormone levels and protein or transcript abundance with improved precision have mitigated these concerns (Williams 2008) (e.g., quantitative PCR, in situ hybridization, or commercial-grade enzyme immunoassays, to name a few). Nevertheless, traditional approaches to the study of how hormones affect phenotypes have yet to incorporate a full appreciation of interindividual variability. Specifically, hormones are often thought to be permissive in their control of behavior: above a certain threshold of circulating hormone, a phenotype is likely to be expressed, but continuous variation in hormone levels or in the abundance of hormone receptors is often thought to bear little quantitative meaning for the expression of phenotype (Adkins-Regan 2005, 2008; Ball and Balthazart 2008; Hews and Moore 1997, but see chapter 6, this volume).

A related issue that has pervaded evolutionary endocrinology relates to the degree to which phenotypic differences are a function of variation in hormone "signal" (e.g., concentration of hormones circulating in the blood) or variation in target tissue "sensitivity" (e.g., abundance of hormone receptors, cofactors, or metabolizing enzymes in the brain or other tissues where hormones directly act) (Adkins-Regan 2008; Hau 2007; Ketterson et al. 2009; Wingfield 2012). While this debate has historically focused on comparing group means, such as castrate vs. intact, or receptor-antagonist vs. saline control, this core issue (signal vs. sensitivity) is also central to our quest to understand the mechanisms underlying *individual* variation in hormones and hormone-mediated traits.

One way to unravel the mechanisms underlying this variation is to separately examine each of the constituent parts of the endocrine system, asking, for example, how varying the amount of hormones, receptors, or transport molecules independently affects phenotypic variation. The complement to this more reductionist approach to biological complexity is a systems biological or whole-organism integrative approach, one that also examines covariation and interconnectedness of the many different components of the endocrine system and how they collectively contribute to phenotypic variation. After all, like traits themselves, the mechanisms underlying trait variation also evolve, and we need a clearer understanding of how different components of the endocrine system covary or possibly coevolve in ways that affect phenotypic evolution (see chapter 5, this volume, for a more detailed theoretical discussion). It is our view that a comprehensive understanding of phenotypic evolution will require both broad and focused perspectives.

Our goal in this chapter is to use our research on testosterone and phenotype in the dark-eyed junco to begin to unpack the mechanistic black box of hormonal phenotypes—to understand how different components of testosterone production, sensitivity, and downstream genomic effects work together or vary independently to yield a particular phenotypic output. Previous chapters have largely focused on testosterone levels themselves and their relation to phenotype and fitness, showing that, at least in the junco, continuous variation in circulating testosterone seems to be functionally important—it relates to aggression, parenting, immune function, and reproductive success in the wild (chapter 6, this volume). We have also learned that testosterone titers may reflect a compromise between males and females as a result of sexual conflict over optimal testosterone levels (chapters 4 and 6, this volume). If we are to understand how hormones and hormone-mediated traits evolve, we need to understand the proximate mechanisms underlying these individual differences in hormones and their effect on physiology, behavior, and, ultimately, fitness. We also need to understand how the relationships among endocrine components and phenotype vary between the sexes. Like Bennett, our perspective in this chapter emphasizes variation within natural populations, though interpopulation comparisons of this sort clearly also bear on these same questions (see especially chapter 11, this volume).

B. Diving Deeper into the Mechanisms Underlying Hormonally Mediated Phenotypes

The endocrine system is complex, and there are many components that can vary in ways that ultimately lead to individual or sex differences in T or T-mediated phenotypes (Hau and Wingfield 2011; Wingfield 2012).

The signaling axis that is primarily responsible for T production and secretion is the hypothalamo-pituitary-gonadal (HPG) axis (see color plate 6, this volume). Typically, the HPG axis is activated in response to an external stimulus—this could be, for example, a threat from a rival or extended day length. These stimuli are relayed to the hypothalamus, which secretes gonadotropin releasing hormone (GnRH), a neuropeptide that acts on the pituitary gland to secrete peptides including luteinizing hormone (LH). LH enters the circulation and acts on the gonad, which produces and secretes sex steroids, including testosterone. Recent work has addressed variation at the level of the brain and especially in GnRH-producing neurons (including variation in the neuropeptide GnIH) as important sources of variation in T release (Tsutsui et al. 2010). Considerably less attention has been paid to downstream components of the HPG axis, particularly with respect to individual variation (Adkins-Regan 2005, 2008).

Gonadal steroids bring about physiological and phenotypic effects by moving from the circulation into target tissues in the brain and body, where hormones bind to hormone-processing molecules. Binding globulins and other cofactors assist with transport in the circulation, and the percent of free versus bound hormone may affect the local salience of a steroid (Malisch and Breuner 2010). Within target tissues, additional cellular and molecular features also influence the degree to which hormones affect phenotype. For example, testosterone can bind nuclear androgen receptors (ARs), initiating a suite of changes in gene expression, often referred to as genomic effects because of the hormone-receptor interaction with genomic DNA. Testosterone can also be metabolized into inactive forms (e.g., androstenedione, via 17β-reductase) or other active hormones (e.g., dihydrotestosterone, via 5α-reductase). Many of the effects of T on the brain occur after the enzyme aromatase (AROM) converts T to estradiol (E_2), another potent steroid hormone with massive genomic effects that are mediated via nuclear estrogen receptors (ERs). Estradiol can influence phenotype via rapid, nongenomic effects, in which hormones interact directly with membrane receptors to alter cellular physiology without changing gene expression (Vasudevan and Pfaff 2008). The specific physiological or genomic changes induced by steroid hormones can vary by tissue (Van Nas et al. 2009) and may be affected by developmental history or epigenetics (Crews et al. 2012). Furthermore, several additional components of the endocrine system add further complication, including feedback at various levels of the

HPG axis (Adkins-Regan 2008; Wingfield 2012), interactions with other endocrine axes such as the hypothalamo-pituitary-adrenal axis (Viau 2002), and local steroid biosynthesis in the brain or other tissues (London et al. 2009; Schmidt et al. 2008). In sum, there are dozens and dozens of interacting components that may underlie the variation in hormones and hormone-mediated phenotypes seen among individuals and between the sexes.

So, how can we begin to address Bennett's challenge? Why do some individuals have the ability to produce a lot of T while some produce far less? Are "high T" individuals the most sensitive to LH in the gonad or the least sensitive to negative feedback in the hypothalamus? And, when we see associations between T and phenotype, are these linkages the consequence of variation in hormone signal, sensitivity, or signal transduction? How interconnected are the different components of the endocrine system, and how easily can they be broken apart by evolution?

C. Endocrine Mechanisms and Evolution: Integration and Independence

An important first step towards predicting how hormone-mediated traits might evolve is to understand the degree to which the different components of the endocrine system can change independently. Just as multiple morphological or behavioral traits mediated by a common hormone may be tightly integrated (and therefore likely to coevolve) or be more loosely connected (and therefore free to evolve independently), different aspects of the endocrine system that mediate these traits may be linked (integrated) or decoupled (independent) in their expression (Adkins-Regan 2005; Hau 2007; Ketterson et al. 2009; Ketterson and Nolan, Jr. 1999; McGlothlin and Ketterson 2008). (For more detail on the continuum between integration and independence, see chapter 5, this volume.)

A few examples illustrate the point: If phenotypic variation is primarily a consequence of variation in hormone signal, then evolution could theoretically bring about correlated changes in many traits at once (integration). If phenotypic variation instead (or also) lies in tissue-specific sensitivity to hormones or in the downstream effects of hormones, then there could be multiple pathways for evolutionary modification. Mechanistically, this could allow for selection to independently alter one trait without affecting other aspects of organismal physiology or behavior (independence). Understanding the mechanistic underpinnings responsible for phenotypic integration versus independence requires knowing whether hormone signal, hormone sensitivity, and genomic responses to hormones covary in natural systems. This detailed knowledge of existing variation is necessary if we are to make accurate predictions about the evolutionary processes by which changes in the endocrine system can bring about phenotypic evolution (see chapter 11, this volume).

The degree to which males and females vary in tandem in different components of the endocrine system represents yet another layer of integration or independence with important evolutionary implications. Chapters 4 and 6 in this volume provide an in-depth account of the similarities and differences in the effects of T in the two sexes in juncos and in other species. To summarize briefly, we know that T does more than just facilitate male reproductive behaviors: females also produce endogenous testosterone, and this hormone mediates many physiologically important processes, regardless of sex (Staub and De Beer 1997). However, optimal T levels are often different for males and female, leading to sexually antagonistic selection that yields intermediate and potentially suboptimal T levels in each sex (Ketterson et al. 2005; Lande 1980; Møller et al. 2005; Van Doorn 2009). It is also possible for the effects of T to become decoupled between the sexes by modifying, for example, the density of androgen receptors in a particular brain area in one sex and not the other or by changing the particular genes that are expressed in response to T in each sex. Detailed knowledge of the patterns of covariation between phenotype, signal, sensitivity, and genomic responses—in both sexes—is a necessary step in understanding the mechanisms by which sexual conflict is resolved (Peterson et al. 2013; Rosvall 2013).

D. Variation within and among Tiers of the HPG Axis

With respect to the HPG axis, much recent research in behavioral endocrinology has focused on the top tier, examining GnRH secretion from the hypothalamus as the source of variation in T (Tsutsui et al. 2010). In contrast, relatively little research has examined sources of variation along multiple levels of the HPG axis, particularly within the same study or the same individuals. As a consequence, we know that the tiers of the axis are functionally connected (i.e., one tier stimulates the next, etc.), but it is not clear whether the constituent parts of the HPG axis work together or separately to produce a high- versus low-T phenotype (Adkins-Regan 2008). One of a handful of important exceptions comes from a comparison of the two morphs of the white-throated sparrow (*Zonotrichia albicollis*). Males of the white-striped (more aggressive) and tan-striped (more parental) morphs vary in T production, though the morphs do not differ in LH production (Spinney et al. 2006). The morphs also differ in expression of ERα in behaviorally relevant areas of the brain (Horton et al. 2014). These findings demonstrate morph-related variation at multiple levels of the endocrine system, but, because the morphs differ in gonadal response to LH but not in pituitary response to GnRH, the pituitary and gonad appear to vary somewhat independently. Research on the physiological changes that occur during ascent to social dominance in the African cichlid *Astatotilapia burtoni* instead suggests greater integration of different components of the HPG axis because males that are rising exhibit changes in the hypothalamus, pituitary, and gonads at largely the same time (Huffman et al. 2012; Maruska and Fernald 2011; Maruska et al. 2011). While these comparisons of genetic and environmentally induced morphs provide important insights into the mechanisms underlying phenotypic diversity, the relative lability or interconnectedness of various endocrine components under selection remains poorly understood, as these results do not address individual differences, which are key to understanding whether components of the HPG axis might evolve in concert or independently.

D.1. Variation among Individuals

Past work in the junco has shown repeatable and functional variation in the degree of T elevation following an injection of GnRH (or "GnRH challenge"). Furthermore, this individual variation in T covaries with several phenotypic measures (aggression, parenting, etc.), and it predicts reproductive success in the wild (Jawor et al. 2006; McGlothlin et al. 2010) (see chapter 6, this volume). These results from studies using standardized GnRH challenges indicate that at least some of the meaningful individual variation in ability to elevate T must reside along the HPG axis *below* the level of GnRH secretion. In other words, selection on T titers and associated phenotypes could operate via changes at the level of the pituitary (e.g., in the abundance of GnRH receptors, in the amount of LH released), at the level of the gonad (e.g., in the abundance of LH receptors, in the amount of T released), or any combination of these options (see

Bergeon Burns et al. 2014 and Rosvall et al. 2013 for a more thorough discussion of these options and alternatives).

We explored individual variation along several parts of the HPG axis in juncos, asking how different components of the axis relate to an individual male's ability to elevate T (Bergeon Burns et al. 2014). We measured the amount of T produced in response to a GnRH challenge, and we then dissected the classic GnRH challenge into its constituent parts by also measuring the amount of LH produced in response to a GnRH injection and the amount of T produced in response to an LH injection. We also used quantitative PCR to precisely measure transcript abundance for gonadal LH-R, as a measure of the gonad's sensitivity to LH. Finally, we measured transcript abundance for AR and AROM in the rostral hypothalamus, an area of the brain that (among other things) is a prime target for negative feedback in the regulation of the HPG axis. We have used these data to investigate integration and independence along the HPG axis, and our results show considerable independence between different tiers of the axis, despite some integration within a tissue (Bergeon Burns et al. 2014).

Beginning first with the hypothalamus, we did not find any relationship between a male's ability to elevate T and the abundance of AR or AROM mRNA in the rostral hypothalamus. To the extent that transcript abundance is a repeatable attribute of an individual and predicts protein abundance, this finding suggests that variation in T production among individuals is unlikely to be related to sources of steroid-mediated feedback at the top of the HPG axis. Males with the highest T production are not simply those that are less sensitive to negative feedback (or vice versa). This lack of covariation between T and neural sensitivity to sex steroids also shows that different components of the endocrine cascade are not necessarily correlated, suggesting that signal and target could theoretically change independently of one another. On the other hand, we found significant positive correlations between AR and AROM mRNA abundance within the rostral hypothalamus ($R^2 = 0.38$, n = 44 males, p < 0.0001), suggesting tight integration of endocrine components within a tissue, despite independence of signal and target, an idea that we will expand upon below.

The results of our different hormone injections reveal that the pituitary is also not a major source of interindividual variation in T production. We found that individual variation in LH produced in response to GnRH was not related to individual variation in T produced in response to LH (figure 7.1A). However, the amount of T produced by an individual

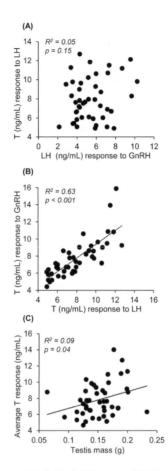

FIGURE 7.1. Gonad as source of individual differences. (A) The amount of LH a male produces following a GnRH injection is not correlated with the amount of T produced in response to an LH injection, but (B) the amount of T produced following GnRH and LH injections are highly correlated. (C) Testis mass predicts a male's average T level following LH or GnRH challenges. Each point represents one individual male, and statistics reflect the bivariate relationships shown here.

was very similar whether the testes were stimulated via a standardized peripheral LH injection or whether the pituitary was stimulated via a peripheral GnRH injection that stimulates endogenous LH production (figure 7.1B). These findings suggest that the primary source of among-individual variation in T production resides at the level of the gonad, not the pituitary, and that pituitary and gonadal responsiveness are not correlated. The observation that repeatable individual variation in T production was not reflective of upstream variation in LH is consistent with hypotheses that LH may operate more like a step function than a graded response. Thus, we suggest that variation in LH is critical to the functioning of the system (i.e., LH must be released for gonadal T production), but variation in LH production does not contribute to quantitative, continuous individual variation in T production.

Because of the tight correlations between T production in response to LH and GnRH challenges, it appears that *something* about the gonad is likely to account for repeatable variation in T production. We did not find relationships between T and LH-R abundance, but we did find that testicular mass predicted significant individual variation in T production (Bergeon Burns et al. 2014). That is, males with testes of greater mass produced more T (figure 7.1C). Collectively, these data provide support for the hypothesis that different components of the HPG axis can vary independently and that variation in characteristics of the gonad, but not LH signal or LH sensitivity, are likely to be responsible for repeatable individual variation in T titers.

These results lend themselves to the hypothesis that the evolution of T-mediated phenotypes in this species may relate more to changes in the gonad, rather than correlated changes along the HPG axis as a whole. It is not yet clear why testicular mass and T output are correlated. One hypothesis is that inherent individual variation in T secretion affects the rate of spermatogenesis, which is known to influence gonadal mass. It could also be that inherent differences in gonadal size lead to variation in the cellular machinery that produces T. Quantification of the expression of enzymes involved in gonadal steroidogenesis may further elucidate the mechanisms underlying meaningful variation in T production, particularly since environmentally induced changes in T levels are known to relate to changes in the abundance of steroidogenic acute regulatory protein or 3-β-hydroxysteroid-dehydrogenase (Huffman et al. 2012; Pradhan et al. 2010). It is currently unknown, however, whether there are functional individual differences in the abundance of these enzymes in the gonad.

D.2. Variation between Males and Females

As a first step towards understanding the degree to which the sexes have coevolved endocrine mechanisms, we compared the above-described results for captive male juncos to parallel research on captive female juncos (Rosvall et al. 2013). If mechanisms of T production in females are largely coevolved with males, then we predicted that the sexes would show similar patterns of variation (i.e., that female T production would also largely be a consequence of gonadal variation, instead of correlated variation along the HPG axis as a whole). If, however, the sexes have some ability to evolve independently from one another, then we might anticipate different sources of interindividual differences in each sex.

When we compared captive male and female juncos that were in early breeding condition (i.e., all individuals had recrudesced gonads), the sex differences we observed were consistent with other studies (Cornil et al. 2011; Dawson et al. 1985). Females had less hypothalamic AROM mRNA than males, and females elevated T less than males did in response to LH or GnRH challenges. Females also had less gonadal LH-R mRNA than males, perhaps because females were not yet yolking eggs, but GnRH challenges led to a greater LH surge in females than in males, possibly related to the central role of LH surges in ovulation (Farner and Wingfield 1980). Similar to males, our investigation of sources of variation in female T production revealed that neither LH sensitivity (LH-R mRNA), LH signal (in response to GnRH), nor neural sex steroid sensitivity (AR and AROM mRNA) was related to female T production, despite repeatable T production in response to either GnRH or LH challenges (Rosvall et al. 2013). Thus, our among-individual results in females again point to remarkable independence among different tiers of the axis because female T production was apparently primarily a consequence of variation at the level of the gonad.

Collectively, these data point to repeatable individual variation in T production that arises from gonadal variation in both sexes, and we do not see covariation between the different tiers of the axis among individuals within either sex. An individual that can produce a large amount of T does not necessarily also have relatively greater gonadotropin production or sensitivity (LH, LH-receptor), nor is s/he least (or most) sensitive to T-induced negative feedback (hypothalamic AR or AROM). The observation of similar patterns in the two sexes suggests that sex-specific differences in the degree of integration or independence along the HPG axis are not an evolved response to sexual conflict over T levels.

E. Sensitivity to Testosterone in Neural Target Tissues

Once T is produced and released into circulation, hormone sensitivity in target tissues provides a gate through which circulating hormone can affect phenotype. While the data are too numerous to fully review here, there is very strong evidence that neural androgen receptors, estrogen receptors, and the enzyme aromatase are functionally linked with the expression of steroid-mediated behavior, including aggression, parenting, and mating behavior. Pharmacological and molecular manipulations have demonstrated causal links between AR, ERα, and AROM and various T-mediated behaviors (oftentimes, sexually dimorphic behavior) (Forlano et al. 2006; Matsumoto et al. 2003; Ogawa et al. 1997; Sperry et al. 2010). Comparative neuroendocrinology confirms these linkages as well: seasonal, sex, strain, or species comparisons of the abundance of AR, ERα, and AROM often map onto these "group" differences in behavior (Canoine et al. 2007; Goncalves et al. 2010; Goodson et al. 2012; Riters et al. 2001; Voigt and Goymann 2007).

Thus, the evolution of T-mediated trait expression can occur via changes in hormone sensitivity, potentially circumventing constraints posed by the multiple effects of T on organismal physiology and behavior. However, the ease with which tissue-specific changes in hormone sensitivity actually do occur is not clear, and an individual-based approach to this question can shed light onto the relative likelihood of independent changes in behavior, hormone signal, and sensitivity in a given population or lineage. For example, might selection favor a male that has low testosterone in circulation but high T sensitivity in certain tissues, so that this male can reap the benefits of T, while minimizing the costs? Upregulation of T sensitivity in, for example, areas of the brain controlling aggression or downregulating T sensitivity in parental care centers would be an adaptive route to express the beneficial sides of T, while minimizing the costs (Canoine et al. 2007; Voigt and Goymann 2007; Wingfield et al. 2001).

For evolution to shape T-mediated traits via changes in sensitivity to T in target tissues, individual variation must be present in natural populations. It has long been hypothesized that the above-mentioned group differences in sensitivity and phenotype can be extended to explain individual difference in phenotype (Grunt and Young 1952) (e.g., where the most aggressive individuals are those with the most AR, ERα, or AROM in areas of the brain mediating aggressive behavior). However, direct tests of this hypothesis are quite rare. With respect to aggressive behavior, there are only a handful of studies that have related natural behavioral variation to individual differences in the expression of these steroid-binding molecules in unmanipulated animals (Bergeon Burns et al. 2013; Horton et al. 2014; Rosvall et al. 2012; Schlinger and Callard 1989; Silverin et al. 2004; Trainor et al. 2006). To understand how evolution might shape behavior, we need to be able to relate individual differences in sensitivity to phenotype, and we further need to understand whether sensitivity can vary independently of hormone levels in circulation.

Another step in understanding how endocrine mechanisms might evolve is to ask whether the sexes share the same patterns of integration and independence or whether one sex can evolve somewhat independently of the other. High T is thought to be particularly costly for females (Gerlach and Ketterson 2013; Veiga and Polo 2008), owing to the deleterious effects of T on maternal care, mate choice, or a female's ability to yolk an egg (Clotfelter et al. 2004; McGlothlin et al. 2004). Lower T is one solution to avoid some of the costs of T. However, T may be comparatively more beneficial to males than it is to females (e.g., T is required for spermatogenesis), and so T levels in both sexes may reflect a compromise. Indeed, interspecific patterns of T profiles demonstrate that male and female T levels are often positively correlated, consistent with correlated evolution among the sexes (Ketterson et al. 2005; Mank 2007; Møller et al. 2005). A second solution to avoid some of the costs of T is to adjust T sensitivity in particular tissues. For example, to the extent that T suppresses behavior associated with parenting, we might expect selection to favor downregulated (or completely eliminated) sensitivity to T in areas of the brain controlling parental care, particularly in females or species with extensive parental care (Lynn 2008). Applying an individual-based approach can help us to identify whether steroid sensitivity varies among females and among males in a manner that would permit adaptive changes to behavior based upon changes in hormone sensitivity.

Our research in the junco measured neural gene expression for AR, ERα, and AROM in free-living males and females (Rosvall et al. 2012), and our results reveal that individual differences in these measures of neural sensitivity to sex steroids predict natural individual differences in aggression within both sexes. Male and female juncos show largely similar patterns of covariation between the abundance of AR, ERα, and AROM mRNA and naturally existing individual differences in aggressiveness. For example, during staged territorial disputes in both sexes, birds that more frequently dived at a live intruder expressed significantly more AR, ERα, and AROM mRNA in amygdalar tissues. Because changes is gene expression are thought to be a major driver of phenotypic evolution (Carroll 2005), these data suggest that the raw material is in place for evolution to make adaptive shifts in the expression of aggressive behavior, independently of circulating T levels.

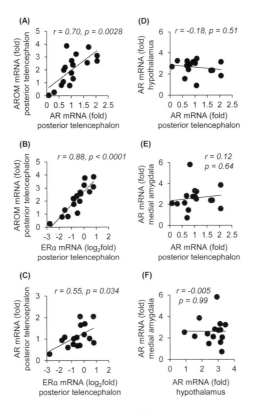

FIGURE 7.2. Measures of sensitivity correlated within a tissue, not among tissues. The abundance of AR, ERα, and AROM mRNA are highly and positively correlated *within a tissue*, shown in (A-C) for the posterior telencephalon. These measures of sex steroid sensitivity are not correlated *among tissues*, shown in (D-F) for AR mRNA abundance in the posterior telencephalon, hypothalamus, and medial amygdala.

With respect to integration or independence among different sex steroid processing molecules, we saw strong, positive correlations between the abundance of AR, ERα, and AROM mRNA *within a tissue* (figure 7.2A-C). In other words, the most aggressive individuals have the potential to process more T via both AR- and ER-mediated pathways, in an integrated fashion. On the other hand, steroid sensitivity in one tissue had no consistent relationship with steroid sensitivity in *other neural tissues*, suggesting that tissue-level patterns of sensitivity can independently regulate specific components of behavior or steroid sensitivity in one area of the brain can vary independently of sensitivity in other areas of the brain (figure 7.2D-F). In principle, these results suggest that different sex steroid-mediated behaviors could evolve independently as steroid sensitivity is upregulated or downregulated in one or another brain area. Thus, theoretically, it ought to be possible for selection to increase singing behavior (a sexually selected signal) via changes in ERα sensitivity in the song control regions of the brain, without necessarily affecting sex steroid sensitivity in areas of the brain that control reproductive or parental behaviors.

Importantly, our research on the junco has shown that circulating T has no consistent relationship with steroid sensitivity across various neural tissues (Bergeon Burns et al. 2013; Rosvall et al. 2012). One notable exception seems to be that sex steroid sensitivity in the hypothalamus tends to be negatively correlated with T titers, possibly because the hypothalamus includes nuclei that are important sites of feedback in the regulation of the HPG axis. The lack of covariation between signal and sensitivity in other tissues suggests that gene expression for these sex steroid sensitive molecules does not simply reflect regulated responses to hormonal inputs from the circulation (e.g., Krey and Mcginnis 1990). Furthermore, circulating T and several measures of sensitivity to sex steroids were both correlated with song rate in male juncos (Rosvall et al. 2012), demonstrating that there is not necessarily a dichotomy between signal and sensitivity. Instead, both of these components of the endocrine system can account for quantitative trait variation. Thus, there appears to be substantial integration in hormone sensitivity within a tissue but substantial independence between tissues. Each tissue appears to be coordinated in its ability to bind and process sex steroids; however, tissues vary independently from one another in this degree of sex steroid sensitivity and whether it predicts phenotypic variation, providing further support for the hypothesis that tissue-specific responses to hormones are likely to be an important component of phenotypic regulation and evolution.

F. Downstream Genomic Responses to Testosterone

It is well known that one of the primary routes by which T affects phenotype is via changes in gene expression that are mediated by nuclear hormone receptors. However, there are only a few studies that have looked at tissue-specific genome-wide effects of hormones, particularly in naturally outbred or other nonmodel systems. The vast majority of research on the downstream effects of hormones has employed experimental manipulations of specific targets in the brain, providing detailed insights into the neural circuits by which hormones influence behavior. This research leaves considerable uncertainty regarding the specific genes whose tissue-specific expression is modified in response to a hormonal input. Identifying T-induced changes in gene expression in tissues throughout the organism is an important step in understanding precisely how hormones bring about coordinated suites of traits, as well as how components of those suites are decoupled in one sex or another, or over evolutionary time.

The genomics era has revealed that sex differences in physiology and behavior are often explained by differences in gene expression between the sexes (Ellegren and Parsch 2007; Naurin et al. 2011), and many of these differences are mediated by hormones (Van Nas et al. 2009). For example, sexually dimorphic gene expression in the mouse brain mediates sex-specific behaviors, including aggression, maternal care, and sexual performance, and hormone levels directly impact both gene expression and behavior (Xu et al. 2012). Dimorphic gene expression is not just limited to the brain but often extends throughout the body (Van Nas et al. 2009; Yang et al. 2006), where it also may contribute to sexual dimorphism. For example, in rodents, gene expression in the liver varies substantially between the sexes (Corton et al. 2012), is responsive to T (Delic et al. 2010), and may contribute to the T-mediated sex difference in immune response (Diodato et al. 2001; Mock and Nacy 1988). Until recently, these genome- and organism-wide approaches were primarily limited to laboratory model systems with mapped genomes (e.g., mice, bees, *Drosophila*), limiting direct applicability to ecology and evolution. Recent advances have made it possible to begin to address genomic questions in ecologically relevant species. The most striking, perhaps, is the emergence of rapid and cost-effective transcriptome sequencing: the sequencing of essentially all of the expressed genes in a given tissue or individual.

Due to extensive field and laboratory studies on both experimental and natural variation in hormones, the junco provides an ideal system for connecting these genomic approaches to decades of previous research on evolutionary endocrinology. We sequenced the transcriptome of the dark-eyed junco and created a custom microarray to assay expression of over 22,000 genes, which amounts to roughly 90 percent of the genes in this species (Peterson et al. 2012). By measuring gene expression in wild, outbred individuals from a natural population, our work sheds light onto the mechanisms by which hormones coordinate behavior and physiology throughout an organism (i.e., hormonal pleiotropy), including those mechanisms that underlie sexual dimorphism and sexual conflict.

Consistent with previous studies in other species, male and female juncos differed in gene expression in ways that were consistent with known phenotypic differences between the sexes, such as muscle development and function. For instance, genes whose products function in the I band and A band portions of contractile muscle fibers were overrepresented among sexually dimorphic genes, including higher expression in males than females of *titin*, which regulates muscle elasticity, and lower expression of *SMAD-related protein 2*, which reduces cellular growth. Whether these sex differences in gene expression are attributable to sex differences in hormones is a key unanswered question, one that we have begun to answer by investigating how each sex responds to T that has been experimentally elevated to the high end of the normal range of variation (Peterson et al. 2013; Peterson et al. 2014).

Not surprisingly, T-implanted and control individuals differed in gene expression related to many biological processes, including hormone balance, cellular communication and signaling, muscle development and metabolism, immune function, and behavior. Summed across the four tissues we have investigated (liver, pectoral muscle, and two brain areas, Peterson et al. 2013; Peterson et al. 2014), testosterone treatment significantly affected expression of roughly 2,500 genes (about 11 percent of expressed genes). AROM, for example, is expressed at a higher level in the hypothalamus of T-implanted compared to control individuals in both sexes. As described above, AROM expression is related to aggression in juncos (Rosvall et al. 2012), and aggression is increased by T implants as well (Zysling et al. 2006). T-implanted females likewise showed reduced expression of amygdalar *monoamine oxidase A* (MAO-A, an enzyme that inhibits serotonergic and dopaminergic signaling), also consistent with the enhancement of aggression in response to T implants. Changes in the expression of several other genes in the hypothalamus were consistent with the known effects of T on metabolism, including increased expression of *cannabinoid receptor 1* (CB_1, which plays a role in hunger/satiation) in T-treated females and increased expression of *melanocortin 4 receptor* (MC4R, which is involved in feeding behavior) in T-treated males, compared to controls. These findings suggest that the genes affected by T implants are likely to be those genes that mediate the known phenotypic effects of T (see chapter 4, this volume), though further experimentation on specific genes and behaviors will test this conclusion more directly.

Most of the genes that responded to T in females did not, however, respond to T in males, and vice versa: only 137 of the roughly 2,500 T-responsive genes were affected in both sexes. Several functional categories of genes responded to T in both sexes, but oftentimes, different genes within each functional category were affected in each sex. For example, T affected gene expression of the growth-related genes *follistatin* and *insulin receptor substrate 4* in the livers of both males and females, but T also affected the expression of three *unique* growth factor receptors in the liver of each sex. Thus, one solution to the potential for sexual conflict regarding T is that T may affect different genes in the two sexes.

Our analyses also revealed that there were some genes whose expression responded to T in opposite directions in the two sexes. Of the 137 genes whose expression was affected by T in both sexes, thirty-one of these genes responded to T in *different* directions in males and females, whereas 106 genes responded to T in the same direction in the two sexes. For example, muscular expression of *smoothelin*, which is a cytoskeletal protein involved in muscle contraction, was upregulated by T treatment in males but depressed by T treatment in females. It is not yet clear why T would upregulate certain genes in females and downregulate (or not affect) those same genes in males, but it is clear that there are intrinsic sex differences in the gene regulatory response to this hormone signal. By identifying the genes that were regulated by T in only one sex and those that were regulated in opposite directions in males and females, our results point towards a short list of the genes that play a key role in the resolution of sexual conflict. Further, the 106 genes that responded similarly in the two sexes are a reminder of the tension between the ability of the sexes to respond independently versus interdependently.

Altogether these results suggest that specific responses to T may be readily decoupled between the sexes and could evolve independently. Elevated T led to somewhat divergent downstream genomic effects in males and females, offering an example of independence between the sexes that might provide the genomic basis for distinct responses to selection in the two sexes (i.e., a solution to sexual conflict over optimal T levels, mediated by sex-specific responses to hormones at a transcriptional level). This independence is particularly striking when layered upon the virtually identical genomes present in males and females of the same species.

Yet another level at which we can explore the continuum between integration and independence is by comparing how different tissues in the brain and body respond to elevated T. Each of the four tissues we have investigated responded differently to the same systemic hormone signal. Some tissues were strongly affected by T (e.g., 10 percent of expressed genes in the liver); others were not (e.g., 1 percent of expressed genes in the hypothalamus were significantly regulated by T). Only about 5 percent of genes that responded to T in any tissue responded in more than one tissue in either sex. However, several functional categories of genes changed expression in response to T treatment in multiple tissues, providing a list of genes that may be responsible for the sweeping and pleiotropic effects of T on the whole organism. For example, in both sexes, T modified the expression of genes related to immune function in every tissue, including *immunoglobulin J* in the muscle of males, *immunoglobulin A heavy chain* in the liver of females, *interleukin 18* in the female hypothalamus and medial amygdala, and *interferon regulatory factor 1* in the male hypothalamus. These data are consistent with the often-observed inverse relationship between T and immune function, and they provide a novel example of the utility of genomic tools to test classic concepts in behavioral endocrinology (e.g., the immunocompetence handicap hypothesis, Boonekamp et al. 2008).

To date, our research on organism- and genome-wide gene expression in the junco has shed new light onto the role of hormones in sexual dimorphism, but more work is needed. One of our immediate goals is to further investigate the transcriptional response to testosterone implants, to determine if the few genes that were affected by testosterone treatment in both sexes might be those that are particularly important in explaining the pleiotropic (and typically masculinizing) effects of testosterone on phenotype. One specific gene illustrates this possibility nicely: gene expression for the rate-limiting enzyme in creatine synthesis (a key step in muscle development) was upregulated in T-treated individuals of both sexes in both liver and muscle tissue, and T is known to have anabolic effects on muscle. Future work also must address which components of the downstream genomic effects of T contribute to individual differences in T-mediated phenotype and fitness. Are the fittest males those that modulate sensitivity in separate tissues to find the ideal balance of the upregulation of sperm manufacturing in the gonad without the downregulation of immune function in the spleen? Does selection favor individuals that use systemic T to effectively coordinate whole-body suites of characteristics, according to the most adaptive combination? These are the sorts of questions about integration and independence that can and will be answered as genomic technology continues to improve our understanding of the mechanisms underlying phenotypic diversity and evolution.

G. Conclusion: What Are the Endocrine Sources of Individual and Sex Differences?

It is crucial for ecological physiologists to find the causes and consequences of individual variation in order to understand how hormone-mediated traits evolve (Bennett 1987). Our research on the junco attempts to meet this challenge by providing a system-wide view of the mechanistic bases for individual differences in testosterone and testosterone-mediated phenotypes. We extend this approach to compare males and females, providing insights into the mechanisms underlying sexual conflict and coevolution. When tied back together with behavioral ecological and evolutionary studies in free-living juncos, these data provide the foundation for two general conclusions regarding the mechanisms by which evolution may alter hormone levels and hormone-mediated traits.

First, regarding individual differences: among the nearly countless interacting components of the endocrine system, our data point to a handful of likely mechanistic sources of individual variation in T and T-mediated traits. Individuals were variable in many components of the HPG axis—from H to P to G, including hormone signal and hormone receptor—but, among the variables we assessed, individual differences in T production were related only to individual differences in properties of the gonad, not the hypothalamus or pituitary. Interindividual variation in levels of hormone signal and sensitivity of hormone targets in the brain were *both* related to variation in behavior, but hormone and target were not usually related to one another. While it remains to be seen if the transcript-level patterns we identified translate to protein, our data clearly show that individual variation in T and T-mediated traits exists at multiple levels within the endocrine system, and many of these components appear to be able to change independently of others. Thus, T may still orchestrate integrated organismal phenotypes (see chapters 4 and 5, this volume), but levels of T in the circulation may not always be the salient feature accounting for individual differences in particular traits. We have not yet addressed whether these individual differences are heritable or whether they vary among species or respond to selection, but these are pressing empirical questions that will take us one step closer to the "holy grail" of uncovering the genetic bases for the evolution of hormone levels and hormone-mediated traits in wild populations of organisms.

Second, regarding sexual conflict: the sexes appear to have found at least partial solutions to sexual conflict over T at many parts of the endocrine system, particularly in testosterone production and the downstream genomic effects of T. The sexes showed remarkable similarity in the sources of interindividual variation in behavior and in T production, but experimentally elevated T had largely different genomic effects in the two sexes. Because sensitivity to hormones did not covary with hormone levels themselves in most cases but did covary with behavioral phenotype, this suggests that evolution could produce changes in phenotype in either sex, independently of circulating hormone levels, by altering hormone sensitivity or the degree to which particular genes are activated by T. Additional genomic tools are becoming more and more tractable for use in nontraditional model organisms like the junco, and these tools will facilitate some critical next steps. With network analyses and RNA-seq, for example, we are not far from being able to identify the specific genes that build or break hormonal pleiotropy and the specific sequence variants that account for functional differences in phenotype.

With these data in mind, can we make predictions about which components of the endocrine system are more likely to covary versus those that are more likely to vary independently from one another? Our data demonstrate that we should not necessarily expect to see covariation across multiple levels of the HPG axis, nor should we expect to see integration between hormone levels, hormone sensitivity, and hormone response. Further, different tissues play different roles within the organism, and tissue-specific hormone sensitivity and downstream genomic effects appear to be mechanisms by which juncos resolve hormone-mediated trade-offs between self-maintenance and reproduction. Still, in multiple datasets, we saw that different measures of hormone sensitivity were tightly linked within a tissue, suggesting that tissues may be heavily integrated in endocrine mechanisms.

An essential next step in developing an understanding of the balance between integration and independence in natural systems will be to compare patterns of covariation across populations or species (see chapters 5, 10, and 11, this volume). These comparisons will help us determine whether there are universal "rules" about the degree of integration of different components of the endocrine system and whether different species or populations have relied on independently varying endocrine components to get to particular phenotypic optima. A comparative approach is also needed to determine whether evolutionary processes maintain intrapopulation patterns of covariation as populations diverge and whether changes in endocrine mechanisms reflect adaptive responses to new environments—these are questions we begin to tackle in greater detail in chapter 11, this volume. Our comparisons among individuals point to sources of variation in the endocrine system that could bring about evolutionary change in testosterone and testosterone-mediated phenotypes, and these comparisons lay the groundwork for interspecific comparisons that will allow us to piece together a view of the mechanisms by which evolution has produced the wide variety of life-history, hormonal, and phenotypic variants seen in nature.

Acknowledgments

The authors are grateful to the collaborators, field/lab assistants, and funding sources that have made this work possible. KAR was supported by the NIH (F32HD068222 and R21HD073583). CMBB was supported by the NSF (predoctoral fellowship and DDIG IOS-0909834). MPP was supported by the NSF (DDIG IOS-1209564). EDK was supported by the NSF (IOS-0820055) and the Indiana METACyt Initiative. All authors were supported by a "Common Themes in Reproductive Diversity" NIH training grant (T32HD049336).

References

Adkins-Regan, E. 2005. *Hormones and Animal Social Behavior.* Princeton: Princeton University Press.

———. 2008. Do hormonal control systems produce evolutionary inertia? *Philosophical Transactions of the Royal Society B-Biological Sciences* 363:1599–1609.

Ball, G. F., and J. Balthazart. 2008. Individual variation and the endocrine regulation of behaviour and phsyiology of birds: A cellular/molecular perspective. *Philosophical Transactions of the Royal Society B-Biological Sciences* 363: 1699–1710.

Bennett, A. F. 1987. Interindividual variability: An underutilized resource. In *New Directions in Ecological Physiology*, edited by M. E. Feder, A. F. Bennett, W. W. Burggren, and R. B. Huey, 147–69. Cambridge: Cambridge University Press.

Bergeon Burns, C. M., K. A. Rosvall, T. P. Hahn, G. E. Demas, and E. D. Ketterson.

2014. Examining sources of variation in HPG axis function among individuals and populations of the dark-eyed junco. *Hormones and Behavior* 65:179–87.

Bergeon Burns, C. M., K. A. Rosvall, and E. D. Ketterson. 2013. Neural steroid sensitivity and aggression: Comparing individuals of two songbird subspecies. *Journal of Evolutionary Biology* 26:820–31.

Boonekamp, J. J., A. H. F. Ros, and S. Verhulst. 2008. Immune activation suppresses plasma testosterone level: A meta-analysis. *Biology Letters* 4:741–44.

Canoine, V., L. Fusani, B. Schlinger, and M. Hau. 2007. Low sex steroids, high steroid receptors: Increasing the sensitivity of the nonreproductive brain. *Developmental Neurobiology* 67:57–67.

Carroll, S. B. 2005. Evolution at two levels: On genes and form. *PLoS Biol* 3:e245.

Clotfelter, E. D., D. M. O'neal, J. M. Gaudioso, J. M. Casto, I. M. Parker-Renga, E. A. Snajdr, D. L. Duffy, V. Nolan, Jr., and E. D. Ketterson. 2004. Consequences of elevating plasma testosterone in females of a socially monogamous songbird: Evidence of constraints on male evolution? *Hormones and Behavior* 46: 171–78.

Cornil, C. A., G. F. Ball, J. Balthazart, and T. D. Charlier. 2011. Organizing effects of sex steroids on brain aromatase activity in quail. *PLoS ONE* 6:e19196.

Corton, J. C., P. R. Bushel, J. Fostel, and R. B. O'lone. 2012. Sources of variance in baseline gene expression in the rodent liver. *Mutation Research-Genetic Toxicology and Environmental Mutagenesis* 746:104–12.

Crews, D., R. Gillette, S. V. Scarpino, M. Manikkam, M. I. Savenkova, and M. K. Skinner. 2012. Epigenetic transgenerational inheritance of altered stress responses. *Proceedings of the National Academy of Sciences of the United States of America* 109:9143–9148.

Darwin, C. 1859. *On the Origin of Species*. Cambridge: Harvard University Press.

Dawson, A., B. K. Follett, A. R. Goldsmith, and T. J. Nicholls. 1985. Hypothalamic gonadotropin-releasing hormone and pituitary and plasma FSH and prolactin during photostimulation and photorefractoriness in intact and thyroidectomized starlings (*Sturnus vulgaris*). *Journal of Endocrinology* 105:71–77.

Delic, D., C. Grosser, M. Dkhil, S. Al-Quraishy, and F. Wunderlich. 2010. Testosterone-induced upregulation of mirnas in the female mouse liver. *Steroids* 75:998–1004.

Diodato, M. D., M. W. Knoferl, M. G. Schwacha, K. I. Bland, and I. H. Chaudry. 2001. Gender differences in the inflammatory response and survival following haemorrhage and subsequent sepsis. *Cytokine* 14:162–69.

Ellegren, H., and J. Parsch. 2007. The evolution of sex-biased genes and sex-biased gene expression. *Nature Reviews Genetics* 8:689–98.

Farner, D. S., and J. C. Wingfield. 1980. Reproductive endocrinology of birds. *Annual Review of Physiology* 42:457–72.

Forlano, P. M., B. A. Schlinger, and A. H. Bass. 2006. Brain aromatase: New lessons from non-mammalian model systems. *Frontiers in Neuroendocrinology* 27: 247–74.

Gerlach, N. M., and E. D. Ketterson. 2013. Experimental elevation of testosterone lowers fitness in female dark-eyed juncos. *Hormones and Behavior* 63:782–90.

Goncalves, D., J. Saraiva, M. Teles, R. Teodosio, A. V. M. Canario, and R. F. Oliveira. 2010. Brain aromatase mRNA expression in two populations of the peacock blenny *Salaria pavo* with divergent mating systems. *Hormones and Behavior* 57:155–61.

Goodson, J. L., L. C. Wilson, and S. E. Schrock. 2012. To flock or fight: Neurochemical signatures of divergent life histories in sparrows. *Proceedings of the National Academy of Sciences of the United States of America* 109 Suppl 1:10685–92.

Grunt, J. A., and W. C. Young. 1952. Differential reactivity of individuals and the response of the male guinea pig to testosterone propionate. *Endocrinology* 51: 237–48.

Hau, M. 2007. Regulation of male traits by testosterone: Implications for the evolution of vertebrate life histories. *Bioessays* 29:133–44.

Hau, M., and J. C. Wingfield. 2011. Hormonally-regulated trade-offs: Evolutionary variability and phenotypic plasticity in testosterone signaling pathways. In *Mechanisms of Life History Evolution: The Genetics and Physiology of Life History Traits and Trade-Offs*, edited by T. Flatt and A. Heyland, 349–61. Oxford: Oxford University Press.

Hews, D. K., and M. C. Moore. 1997. Hormones and sex-specific traits: Critical questions. In *Parasites and Pathogens*, edited by N. Beckage, 277–92. London: Chapman and Hall.

Horton, B. M., W. H. Hudson, E. A. Ortlund, S. Shirk, J. W. Thomas, E. R. Young, W. M. Zinzow-Kramer, and D. L. Maney. 2014. Estrogen receptor α polymorphism in a species with alternative behavioral phenotypes. *Proceedings of the National Academy of Sciences of the United States of America* 111:1443–48.

Huffman, L. S., M. M. Mitchell, L. A. O'Connell, and H. A. Hofmann. 2012. Rising stars: Behavioral, hormonal, and molecular responses to social challenge and opportunity. *Hormones and Behavior* 61:631–41.

Jawor, J. M., J. W. McGlothlin, J. M. Casto, T. J. Greives, E. A. Snajdr, G. E. Bentley, and E. D. Ketterson. 2006. Seasonal and individual variation in response to GnRH challenge in male dark-eyed juncos (*Junco hyemalis*). *General and Comparative Endocrinology* 149:182–89.

Ketterson, E. D., J. W. Atwell, and J. W. McGlothlin. 2009. Phenotypic integration and independence: Hormones, performance, and response to environmental change. *Integrative and Comparative Biology* 49:365–79.

Ketterson, E. D., and V. Nolan, Jr. 1999. Adaptation, exaptation, and constraint: A hormonal perspective. *American Naturalist* 154:S4–S25.

Ketterson, E. D., V. Nolan, Jr., and M. Sandell. 2005. Testosterone in females: Mediator of adaptive traits, constraint on sexual dimorphism, or both? *American Naturalist* 166:S85–S98.

Krey, L. C., and M. Y. McGinnis. 1990. Time-courses of the appearance-disappearance of nuclear androgen plus receptor complexes in the brain and

adenohypophysis following testosterone administration-withdrawal to castrated male rats' relationships with gonadotropin secretion. *Journal of Steroid
Biochemistry* 35:403–8.

Lande, R. 1979. Quantitative genetic analysis of multivariate evolution, applied to
brain-body size allometry. *Evolution* 33:402–16.

———. 1980. Sexual dimorphism, sexual selection, and adaptation in polygenic
characters. *Evolution* 34:292–305.

London, S. E., L. Remage-Healey, and B. A. Schlinger. 2009. Neurosteroid production in the songbird brain: A re-evaluation of core principles. *Frontiers in Neuroendocrinology* 30:302–14.

Lynn, S. E. 2008. Behavioral insensitivity to testosterone: Why and how does testosterone alter paternal and aggressive behavior in some avian species but not
others? *General and Comparative Endocrinology* 157:233–40.

Malisch, J. L., and C. W. Breuner. 2010. Steroid-binding proteins and free steroids
in birds. *Molecular and Cellular Endocrinology* 316:42–52.

Mank, J. E. 2007. The evolution of sexually selected traits and antagonistic androgen expression in actinopterygiian fishes. *American Naturalist* 169:142–49.

Maruska, K. P., and R. D. Fernald. 2011. Plasticity of the reproductive axis caused
by social status change in an African cichlid fish: I. Testicular gene expression
and spermatogenesis. *Endocrinology* 152:291–302.

Maruska, K. P., B. Levavi-Sivan, J. Biran, and R. D. Fernald. 2011. Plasticity of the
reproductive axis caused by social status change in an African cichlid fish: I. Pituitary gonadotropins. *Endocrinology* 152:281–90.

Matsumoto, T., S. Honda, and N. Harada. 2003. Alteration in sex-specific behaviors in male mice lacking the aromatase gene. *Neuroendocrinology* 77:416–24.

McGlothlin, J. W., and E. D. Ketterson. 2008. Hormone-mediated suites as adaptations and evolutionary constraints. *Philosophical Transactions of the Royal
Society B-Biological Sciences* 363:1611–20.

McGlothlin, J. W., D. L. H. Neudorf, J. M. Casto, V. Nolan, Jr., and E. D. Ketterson. 2004. Elevated testosterone reduces choosiness in female dark-eyed juncos
(*Junco hyemalis*): Evidence for a hormonal constraint on sexual selection? *Proceedings of the Royal Society B-Biological Sciences* 271:1377–84.

McGlothlin, J. W., D. Whittaker, S. Schrock, N. Gerlach, J. Jawor, E. Snajdr, and E.
Ketterson. 2010. Natural selection on testosterone production in a wild songbird population. *American Naturalist* 175:687–701.

Mock, B. A., and C. A. Nacy. 1988. Hormonal modulation of sex-differences in resistance to leishmania-major systemic infections. *Infection and Immunity* 56:
3316–19.

Møller, A. P., L. Z. Garamszegi, D. Gil, S. Hurtrez-Bousses, and M. Eens. 2005. Correlated evolution of male and female testosterone profiles in birds and its consequences. *Behavioral Ecology and Sociobiology* 58:534–44.

Naurin, S., B. Hansson, D. Hasselquist, Y. H. Kim, and S. Bensch. 2011. The sex-biased brain: Sexual dimorphism in gene expression in two species of songbirds. *BMC Genomics* 12:37.

Ogawa, S., D. B. Lubahn, K. S. Korach, and D. W. Pfaff. 1997. Behavioral effects of estrogen receptor gene disruption in male mice. *Proceedings of the National Academy of Sciences of the United States of America* 94:1476–81.

Peterson, M. P., K. A. Rosvall, J.-H. Choi, C. Ziegenfus, H. Tang, J. K. Colbourne, and E. D. Ketterson. 2013. Testosterone affects neural gene expression differently in male and female juncos: A role for hormones in mediating sexual dimorphism and conflict. *PLoS ONE* 8:e61784.

Peterson, M. P., K. A. Rosvall, C. A. Taylor, J. A. Lopez, J. H. Choi, C. Ziegenfus, H. Tang, J. K. Colbourne, and E. D. Ketterson. 2014. Potential for sexual conflict assessed via testosterone-mediated transcriptional changes in liver and muscle of a songbird. *Journal of Experimental Biology* 217:507–17.

Peterson, M. P., D. J. Whittaker, S. Ambreth, S. Sureshchandra, A. Buechlein, R. Podicheti, J.-H. Choi, Z. Lai, K. Mockatis, J. Colbourne, H. Tang, and E. D. Ketterson. 2012. De novo transcriptome sequencing in a songbird, the dark-eyed junco (*Junco hyemalis*): Genomic tools for an ecological model system. *BMC Genomics* 13:305.

Pradhan, D. S., A. E. M. Newman, D. W. Wacker, J. C. Wingfield, B. A. Schlinger, and K. K. Soma. 2010. Aggressive interactions rapidly increase androgen synthesis in the brain during the non-breeding season. *Hormones and Behavior* 57:381–89.

Riters, L. V., M. Baillien, M. Eens, R. Pinxten, A. Foidart, G. F. Ball, and J. Balthazart. 2001. Seasonal variation in androgen-metabolizing enzymes in the diencephalon and telencephalon of the male European starling (*Sturnus vulgaris*). *Journal of Neuroendocrinology* 13:985–97.

Rosvall, K. A. 2013. Proximate perspectives on the evolution of female aggression: Good for the gander, good for the goose? *Philosophical Transactions of the Royal Society B-Biological Sciences* 368:20130083.

Rosvall, K. A., C. M. Bergeon Burns, J. Barske, J. L. Goodson, B. A. Schlinger, D. R. Sengelaub, and E. D. Ketterson. 2012. Neural sensitivity to sex steroids predicts individual differences in aggression: Implications for behavioural evolution. *Proceedings of the Royal Society B-Biological Sciences* 279:3547–55.

Rosvall, K. A., C. M. Bergeon Burns, T. P. Hahn, and E. D. Ketterson. 2013. Sources of variation in HPG axis reactivity and individually consistent elevation of sex steroids in a female songbird. *General and Comparative Endocrinology* 194:230–39.

Schlinger, B. A., and G. V. Callard. 1989. Aromatase-activity in quail brain—correlation with aggressiveness. *Endocrinology* 124:437–43.

Schmidt, K. L., D. S. Pradhan, A. H. Shah, T. D. Charlier, E. H. Chin, and K. K. Soma. 2008. Neurosteroids, immunosteroids, and the balkanization of endocrinology. *General and Comparative Endocrinology* 157:266–74.

Silverin, B., M. Baillien, and J. Balthazart. 2004. Territorial aggression, circulating levels of testosterone, and brain aromatase activity in free-living pied flycatchers. *Hormones and Behavior* 45:225–34.

Sperry, T. S., D. W. Wacker, and J. C. Wingfield. 2010. The role of androgen receptors in regulating territorial aggression in male song sparrows. *Hormones and Behavior* 57:86–95.

Spinney, L. H., G. E. Bentley, and M. Hau. 2006. Endocrine correlates of alternative phenotypes in the white-throated sparrow (*Zonotrichia albicollis*). *Hormones and Behavior* 50:762–71.

Staub, N. L., and M. De Beer. 1997. The role of androgens in female vertebrates. *General and Comparative Endocrinology* 108:1–24.

Trainor, B. C., K. M. Greiwe, and R. J. Nelson. 2006. Individual differences in estrogen receptor alpha in select brain nuclei are associated with individual differences in aggression. *Hormones and Behavior* 50:338–45.

Tsutsui, K., G. E. Bentley, L. J. Kriegsfeld, T. Osugi, J. Y. Seong, and H. Vaudry. 2010. Discovery and evolutionary history of gonadotrophin-inhibitory hormone and kisspeptin: New key neuropeptides controlling reproduction. *Journal of Neuroendocrinology* 22:716–27.

Van Doorn, G. S. 2009. Intralocus sexual conflict. *Year in Evolutionary Biology 2009*, special issue of *Annals of the New York Academy of Sciences* 1168:52–71.

Van Nas, A., D. Guhathakurta, S. S. Wang, N. Yehya, S. Horvath, B. Zhang, L. Ingram-Drake, G. Chaudhuri, E. E. Schadt, T. A. Drake, A. P. Arnold, and A. J. Lusis. 2009. Elucidating the role of gonadal hormones in sexually dimorphic gene coexpression networks. *Endocrinology* 150:1235–49.

Vasudevan, N., and D. W. Pfaff. 2008. Non-genomic actions of estrogens and their interaction with genomic actions in the brain. *Frontiers in Neuroendocrinology* 29:238–57.

Veiga, J. P., and V. Polo. 2008. Fitness consequences of increased testosterone levels in female spotless starlings. *American Naturalist* 172:42–53.

Viau, V. 2002. Functional cross-talk between the hypothalamic-pituitary-gonadal and -adrenal axes. *Journal of Neuroendocrinology* 14:506–13.

Voigt, C., and W. Goymann. 2007. Sex-role reversal is reflected in the brain of African black coucals (*Centropus grillii*). *Developmental Neurobiology* 67:1560–73.

Williams, T. D. 2008. Individual variation in endocrine systems: Moving beyond the "tyranny of the golden mean." *Philosophical Transactions of the Royal Society B-Biological Sciences* 363:1687–98.

Wingfield, J. C. 2012. Regulatory mechanisms that underlie phenology, behavior, and coping with environmental perturbations: An alternative look at biodiversity. *Auk* 129:1–7.

Wingfield, J. C., S. E. Lynn, and K. K. Soma. 2001. Avoiding the "costs" of testosterone: Ecological bases of hormone-behavior interactions. *Brain Behavior and Evolution* 57:239–51.

Xu, X., J. K. Coats, C. F. Yang, A. Wang, O. M. Ahmed, M. Alvarado, T. Izumi, and N. M. Shah. 2012. Modular genetic control of sexually dimorphic behaviors. *Cell* 148:596–607.

Yang, X., E. E. Schadt, S. Wang, H. Wang, A. P. Arnold, L. Ingram-Drake, T. A. Drake, and A. J. Lusis. 2006. Tissue-specific expression and regulation of sexually dimorphic genes in mice. *Genome Research* 16:995–1004.

Zysling, D. A., T. J. Greives, C. W. Breuner, J. M. Casto, G. E. Demas, and E. D. Ketterson. 2006. Behavioral and physiological responses to experimentally elevated testosterone in female dark-eyed juncos (*Junco hyemalis carolinensis*). *Hormones and Behavior* 50:200–207.

PART III

Evolutionary Diversification in the Avian Genus *Junco*

Pattern and Process

A motivating objective of this volume is to deepen the connections among scientists who study evolution and organismal biology. As a generalization, evolutionary biologists are more excited by variation in gene sequences and the origin of new species than they are by the particulars of hormone synthesis or cell signaling. Organismal biologists have their own foci as well. But we are all engaged in understanding how phenotypes emerge from genotypes and vice versa, and how this might change over time, so it is essential to "get on with it." We posit that people studying the same organism each from their own perspectives can provide the push that is needed.

The authors of this and previous sections have only recently become acquainted. Around 2005, Milá and McCormack decided to attack the junco's complex evolutionary history with new-at-the-time mitochondrial sequencing techniques. They wrote to people known to be studying juncos to see whether they had any spare DNA to complement their own field sampling. The Ketterson group had lots of DNA from Virginia and sent some to Milá.

About that time Ketterson and Price met at a Research Coordination Network organized by the ESF and NSF entitled Integrating Ecology and Endocrinology in Avian Reproduction. Connected only by their history of studying juncos, they found much to talk about. Price explained to Ketterson the potential role of phenotypic plasticity in population divergence, and Ketterson explained to Price the myriad ways in which hormones might enable plasticity. Grant applications by Price and Ketterson followed, and while the proposals were rejected, the collaboration was begun.

Soon after, Atwell, then a graduate student in the Ketterson-Nolan group, teamed up with Cardoso, who was a postdoc with Price, to study juncos in southern California that had previously been studied by Pamela Yeh when she was a graduate student of Price. Later still when Atwell was conducting research in Spain, he happened to contact Milá on one of his last nights in Madrid to "talk juncos." They immediately hit it off, and the meeting led to an offer from Milá to Atwell to join him on one of his collecting trips to Mexico. In time these connections led to a film and this book. Without their common interest in an organism, the contributors to the synthesis attempted in this volume would not have come together.

Chapters 8 and 9 provide complementary and contrasting views of the juncos' evolutionary history in North and Mesoamerica. As our ability to obtain and analyze massive amounts of information from the genome advances, along with our ability to reconstruct past environments, some of the differences in their interpretations will resolve themselves. However, despite ever-growing knowledge of rates of change in sequence variation and rates of change in environments, a critical limitation may remain. How will we know how the birds of today differ from the birds of the past in their response to climate, their habitat needs, and their patterns of movement? For these questions we will need more knowledge of the mechanisms of behavior and physiology in relation to the environment.

CHAPTER EIGHT

More than Meets the Eye

Lineage Diversity and Evolutionary History of Dark-Eyed and Yellow-Eyed Juncos

Borja Milá, Pau Aleixandre, Sofía Alvarez-Nordström, and John McCormack

The truth is rarely pure, and never simple.
—Oscar Wilde

A. The Striking Pattern of Geographic Variation in the Genus *Junco* as a Model for the Study of Speciation Processes

In his seminal monograph on the geographic variation in juncos, Alden Miller stated that "the genus *Junco* contributes a rather complete exemplification of the stages and processes that lead to the first mile-post in the evolutionary path, the full species" (Miller 1941, 375). Whether or not there are mileposts in the evolutionary process and whether or not we agree on what exactly a species is, the main point made by Miller is well taken, and profound: the range of patterns of geographic variation found in juncos spans the entire continuum from the local population to the species and from complete isolation to freely interbreeding forms, to the point that they illustrate the different stages of lineage divergence and thus can illuminate the process of speciation itself.

Miller identified a total of twenty-one different junco "forms," or geographically discrete phenotypes, that vary in their degree of phenotypic divergence (from slight to strong) and the amount of gene flow among adjacent forms (from none to extensive, and across narrow to broad areas). Within the dark-eyed junco alone, we find various degrees of distinctness and gene flow among forms. Just to name a few, there are completely isolated forms like the one on Guadalupe Island (*insularis*) or those on the sky islands of northern Baja California (*townsendi* and *pontilis*); there are widely divergent allopatric forms such as the slate-colored (*hyemalis*) and gray-headed juncos (*caniceps*); and there are forms that mix with some neighbors but not others, such as Oregon juncos, which meet gray-headed juncos at narrow contact zones in California and Nevada (*thurberi* and *caniceps*) yet interbreed freely with slate-colored juncos across broad areas in western Canada (*montanus*, *hyemalis*, and *cismontanus*). These complex patterns reflect a complex evolutionary history, with lineages of different ages evolving under different conditions and interacting in different ways, and thus hold the promise of unique insight into "that mystery of mysteries," the origin of new species (Darwin 1859).

To be sure, Alden Miller accomplished a formidable task. He examined over ten thousand specimens, described in painstaking detail the phenotypic characteristics of all junco morphs, and, very importantly, mapped the limits of their geographic distributions, carefully documenting the extent of intergradation where different morphs came into contact. Few widespread species in the world, if any, have had their distribution and geographic variation mapped with such precision. His main objective was to thoroughly analyze junco races and species "in order to determine the degree of unity of each, and to trace differentiation from individual variants through successive stages of group differentiation to the species" (173). Yet after a decade's worth of field and museum work, even he might admit that the titanic task remains unfinished. Some of the closing remarks of his monograph cannot conceal a certain frustration: when discussing the relationships among gray-headed, Oregon, and slate-colored juncos, he writes that "the characterization of forms as completely or incompletely differentiated is rather arbitrary" (Miller 1941, 374).

Indeed, the phenotypic diversity in the genus *Junco* has puzzled generations of ornithologists, and the lack of consensus on how that diversity had to be classified has led to a convoluted taxonomic history (Ketterson et al. 2013, chapter 2, this volume). For all the trouble Miller went through, and the ten junco species he settled for in his extensive monograph (one more than the previous assessment by Dwight 1918), the American Ornithologists' Union currently lumps nine of the ten into only two species, the dark-eyed junco (*Junco hyemalis*) and the yellow-eyed junco (*J. phaeonotus*), both including several subspecific taxa (Nolan et al. 2002; Sullivan 1999). They agree with him only on the species status of the divergent volcano junco in Costa Rica (*Junco vulcani*). The focus on eye color for the current vernaculars ignores the striking diversity in plumage and beak color

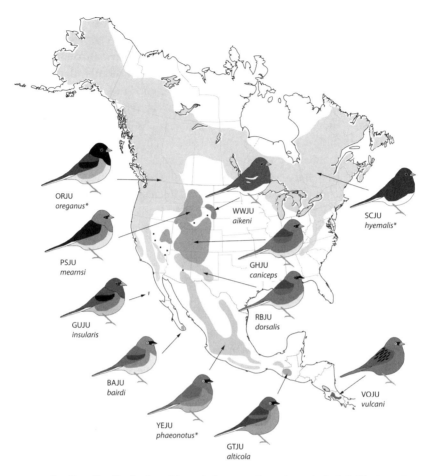

FIGURE 8.1. Breeding distribution of the main junco species and morphs included in the analysis. Four-letter codes for vernacular names are as follows: VOJU, volcano junco; GTJU, Guatemala junco; YEJU, yellow-eyed junco; BAJU, Baird's junco; GUJU, Guadalupe junco; PSJU, pink-sided junco; ORJU, Oregon junco; WWJU, white-winged junco; SCJU, slate-colored junco; GHJU, gray-headed junco; RBJU, red-backed junco. Asterisks indicate that other races exist that are not illustrated, including *palliatus* and *fulvescens* for YEJU; *carolinensis* for SCJU; and *thurberi*, *montanus*, *shufeldti*, *pinosus*, *townsendi*, and *pontilis* for ORJU. Black squares correspond to the approximate location of isolated hybrid populations. Individuals of mixed origin also exist where colored ranges come into contact. See also color plate.

patterns among junco groups (figure 8.1), and disagreements on the group's taxonomy stem from the difficulty of inferring evolutionary history from traits that are under natural or sexual selection. Inferring the evolutionary history of juncos, and thus properly establishing species limits and species names, requires a phylogenetic analysis of neutrally evolving characters.

Prior to DNA sequencing and the ensuing molecular revolution in sys-tematics, taxonomists had to rely solely on phenotypic characters to re-construct the evolutionary history on which to base their taxonomic de-cisions. Because selection is a powerful force, it can modify phenotypic traits dramatically in very short periods of time and in ways that may lead to convergence with distantly related taxa subjected to the same selective pressures. Thus, even though natural and sexual selection are the main drivers of evolution, in order to reconstruct evolutionary history we must rely on phylogenies based on neutrally evolving characters. A phylogeny based on neutral characters provides valuable information on the history of lineages, including the timing and rate of lineage splitting as well as the geographic pattern of lineage formation. In turn, because selection acts on phenotypes, study of neutral genes must go hand in hand with the study of phenotypic traits to understand the factors involved in speciation, and once a robust phylogenetic framework is in place, it can be used to under-stand the evolution of phenotypic characters.

In this chapter we present for the first time a genus-wide molecu-lar phylogeny of juncos featuring the full scope of major phenotypic and geographic diversity. A previous survey of mtDNA variation in the genus (Milá et al. 2007) was limited to continental dark-eyed junco forms and two yellow-eyed forms (mainland Mexico and Guatemala) and used sequence from the mtDNA control region, a fragment with very high mutation rate, which is useful to detect differences among recently diverged lineages yet can sometimes evolve so rapidly that the signal becomes overwritten by new mutations. Here we include data from all main junco forms, including isolated populations in Chiapas (*fulvescens*), Baja California (*bairdi*) and Guadalupe Island (*insularis*) in Mexico, and the highlands of Guatemala (*alticola*) and Costa Rica (*vulcani*). We also use sequence from the cyto-chrome *c* oxidase I (or COI) gene, a coding fragment of the mitochondrial genome that shows a mutation rate well suited to detecting differences among species-level lineages and that has been shown to be congruent with species-level variation in birds (Kerr et al. 2007; Tavares et al. 2011).

TABLE 8.1. **Junco taxa analyzed: Sampling localities, sample sizes, and COI haplotypes per junco population**

Taxon	Code	Locality	n	Haplotypes (freq.)
J. vulcani	VOJU	Chirripó (Costa Rica)	2	V(2)
J. bairdi	BAJU	Sierra de la Laguna (BCS, Mexico)	16	X(12), Y(3), Z(1)
J. insularis	GUJU	Isla Guadalupe (BCN, Mexico)	26	A(1), J(5), K(7), L(12), M(1)
J. alticola	GTJU	Chichim (Huehuetenango, Guatemala)	16	E(2), F(1), G(10), H(3)
J. p. fulvescens	YEJU	S. Cristóbal de las Casas (CHIS, Mexico)	9	B(9)
J. p. phaeonotus	YEJU	Ajuno (MICH, Mexico)	7	A(7)
		Monarca Reserve (MICH, Mexico)	8	A(8)
		Llano del Capulin (OAX, Mexico)	12	A(12)
		La Cima (DF, Mexico)	23	A(22), O(1)
J. p. palliatus	YEJU	Bajío de la Víbora (DGO, Mexico)	22	A(13), P(7), R(1), S(1)
		Cerro de Potosí (NL, Mexico)	4	P(4)
		Sierra del Carmen (COAH, Mexico)	8	A(8)
		Pinaleño Mts. (AZ, USA)	7	A(1), C(6)
J. h. dorsalis	RBJU	White Mts. (AZ, USA)	10	A(10)
J. h. caniceps	GHJU	Uinta Mts. (UT, USA)	10	A(9), I(1)
J. h. oreganus	ORJU	Sawtooth N. F. (ID, USA)	12	A (11), U(1)
		Santa Monica Mts. (CA, USA)	16	A(15), C2(1)
		Great Bear Rainforest (BC, Canada)	14	A(12), A2(1), B2(1)
		Juneau (AK, USA)	15	A(12), N(1), Q(1), T(1)
J. h. mearnsi	PSJU	Shoshone N. F. (WY, USA)	10	A(10)
J. h. hyemalis	SCJU	Great Mountain Forest (CT, USA)	4	A(4)
		Mountain Lake Biol. Station (VA, USA)	12	A(11), W(1)
J. h. aikeni	WWJU	Black Hills N. F. (WY, USA)	10	A(10)

Notes: Four-letter codes for vernacular names are as follows: VOJU, volcano junco; BAJU, Baird's junco; GUJU, Guadalupe junco; GTJU, Guatemala junco; YEJU, yellow-eyed junco; RBJU, red-backed junco; GHJU, gray-headed junco; PSJU, pink-sided junco; ORJU, Oregon junco; SCJU, slate-colored junco; WWJU, white-winged junco.

B. A Preliminary Molecular Phylogeny of the Genus *Junco*

The phylogeny of the genus *Junco* based on the COI gene[1] reveals the existence of five clades: the volcano junco in Costa Rica (*vulcani*), the most divergent taxon; Baird's junco in Baja California (*bairdi*); the Gua-

1. We obtained samples for this study in the field using mist-nets, often attracting territorial males by means of song playbacks. We obtained blood samples by venipuncture of the sub-brachial vein, and we extracted genomic DNA from blood using a DNEasy kit by Qiagen™. The COI gene was amplified using primers BirdF1 and BirdR1 (Hebert et al. 2004), and the resulting 640 base-pair (bp) fragment was purified and sequenced in an ABI 3730XL automated sequencer. Sequences were aligned automatically using SEQUENCHER 4.1 (Gene-

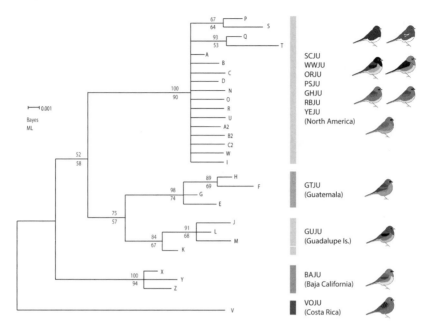

FIGURE 8.2. Preliminary phylogenetic tree of the main forms within the genus *Junco*, based on mtDNA sequence data (COI gene, 690 base pairs). Branch support values correspond to a Bayesian analysis (above branches), and a Maximum Likelihood analysis (below branches). See also color plate.

dalupe junco on Guadalupe Island (*insularis*); the Guatemala junco in the highlands of Guatemala (*alticola*); and a fifth clade including all yellow-eyed and dark-eyed juncos found in mainland Mexico, the United States, and Canada (figure 8.2). Even though the relative divergence of these five

_____

Codes) and polymorphisms were checked visually for accuracy. All sequences translated un-ambiguously into their amino acid sequences, suggesting that they are of mitochondrial ori-gin and not potentially misleading nuclear copies ("numts") or pseudogenes. We identified a total of twenty-nine haplotypes in our sample of 271 juncos from throughout the range (table 8.1). To resolve the phylogenetic relationships among COI haplotypes we used Bayes-ian inference in the program MRBAYES (Huelsenbeck et al. 2001) and a maximum likelihood approach with the program MEGA 5 (Tamura et al. 2011), with a HKY+Γ nucleotide sub-stitution model as estimated with JMODELTEST (Posada 2008). We also constructed a haplo-type network using the median-joining algorithm in the program NETWORK (Fluxus Ltd.). To estimate divergence among groups, we calculated genetic distances among morphs/species corrected for population polymorphism and estimated values of percent divergence among groups using ARLEQUIN 3.5 (Excoffier et al. 2005).

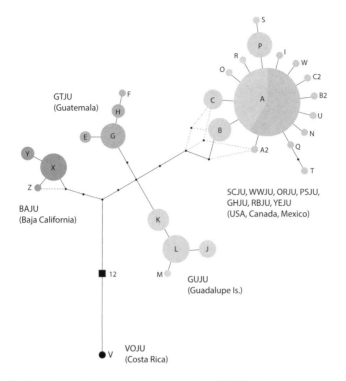

FIGURE 8.3 Haplotype network of cytochrome c oxidase I (COI) data. Circles represent haplotypes, and sizes are proportional to their frequency in the sample. Each branch represents one nucleotide change; black dots represent missing haplotypes. The black square on the branch to volcano juncos represents twelve substitutions. Taxon codes as in figure 8.1.

clades is robust, the relationship among them is more tentative, as most internal nodes received relatively low support, and additional molecular data will be necessary to establish a definitive topology.

The haplotype network reflects the clear divergence among the five junco lineages and also provides insight into their evolutionary history (figure 8.3). The North American clade, which includes mainland dark-eyed juncos (*hyemalis*) and the three mainland yellow-eyed subspecies (*phaeonotus*, *palliatus*, and *fulvescens*), shows a star-like pattern, with the presence of a single high-frequency haplotype ("A") surrounded by closely related, low-frequency haplotypes. This pattern is consistent with results from a previous study using a noncoding region of the mtDNA and genome-wide nuclear markers, which demonstrated that a rapid population expansion was more parsimonious than a selective sweep on mtDNA (Milá et al. 2007).

As shown in table 8.2, percent divergence in COI sequence between the volcano junco and the other junco lineages was about 1.5 to 2 percent, Baird's and Guatemala juncos are about 1 percent divergent from other yellow-eyed juncos, and divergence is negligible within the dark-eyed/ yellow-eyed junco clade. Applying a rough molecular clock calibration of 2 percent divergence per million years (Weir and Schluter 2008), we estimate that except for the volcano junco, which diverged from other juncos about one million years ago, most junco diversification has occurred within the last half million years, in the second half of the Pleistocene.

C. Insights into the Evolutionary History of the Genus *Junco*

A phylogeny based on the COI gene provides a striking, unexpected new view of the evolutionary history of the genus *Junco*. The most obvious result in light of current taxonomy and nomenclature is that dark-eyed and yellow-eyed juncos are not monophyletic groups, and, indeed, iris color shows little congruence with lineage history. Even considering the low support for many nodes deep in the phylogeny, eye color reverted to dark brown (the ancestral, dominant color in Emberizids) at least twice in the history of the group, once in the Guadalupe junco, and recently as the yellow-eyed juncos from Mexico expanded into North America and radiated into the various dark-eyed junco morphs in the United States and Canada.

A second pattern that stands out is the general lack of association between the amount of genetic differentiation and the degree of plumage differentiation. Genetic divergence is clearly associated with geographic isolation, as the most divergent clades are those found in either small isolated mountain ranges (*vulcani, alticola, bairdi*) or an oceanic island (*insularis*). Yet despite their marked genetic divergence, these groups are not particularly differentiated phenotypically, for instance compared to the most divergent plumage patterns found within the dark-eyed junco radiation in mainland North America, which has produced patterns as divergent as those of *caniceps, oreganus,* and *hyemalis*.

Another surprising finding is the divergence and phylogenetic position of the Guadalupe junco (*insularis*). Because of its dark iris and a plumage pattern similar to that of the pink-sided junco (*mearnsi*) in the Rocky Mountains, this junco had always been thought to represent a recent colonization of Guadalupe Island by a dark-eyed mainland relative of

TABLE 8.2 **Genetic distances among junco groups based on COI gene sequence data**

	VOJU	BAJU	GTJU	GUJU	YEJU	GHJU	ORJU	PSJU	WWJU	SCJU
VOJU	*0.00*	15.19	16.44	18.04	19.82	19.95	19.96	20.00	20.00	19.94
BAJU	1.54	*0.05*	7.38	9.23	11.01	11.04	11.15	11.19	11.19	11.13
GTJU	1.66	0.70	*0.08*	6.55	10.14	10.39	10.40	10.44	10.44	10.37
GUJU	1.79	0.86	0.57	*0.14*	7.96	7.80	7.90	7.65	7.65	7.84
YEJU	1.99	1.06	0.95	0.70	*0.11*	0.71	0.87	0.59	0.59	0.74
GHJU	2.04	1.10	1.02	0.72	0.00	*0.03*	0.45	0.15	0.15	0.33
ORJU	2.03	1.10	1.00	0.72	0.00	0.00	*0.06*	0.32	0.32	0.49
PSJU	2.06	1.13	1.04	0.72	0.07	0.00	0.00	*0.00*	0.00	0.19
WWJU	2.06	1.13	1.04	0.72	0.07	0.00	0.00	0.00	*0.00*	0.19
SCJU	2.04	1.11	1.01	0.72	0.00	0.00	0.00	0.00	0.00	*0.04*

Notes: Above diagonal: uncorrected average number of substitutions among populations; along diagonal, in italics: uncorrected average number of substitutions within each population; and below diagonal: percent sequence divergence based on Kimura-2-parameter distances corrected for intrapopulation polymorphism. Taxon codes as in table 8.1.

the *hyemalis* group (Miller 1941; Dwight 1918). However, the phylogeny reveals that *insularis* is an old, highly divergent lineage and suggests that it may be more closely related to yellow-eyed juncos in Guatemala and Baja than to other dark-eyed juncos. In any event, it is certainly not closely related to *mearnsi* of mainland North America. Thus, the similarity in plumage color between *insularis* and *mearnsi* appears to be due to evolutionary convergence and the evolution of its dark iris an independent event from that of other dark-eyed juncos. Consistent with the old age of the *insularis* lineage are its divergent morphology, especially bill size and shape (Aleixandre et al. 2013), and very divergent song (Mirsky 1976).

In addition to the volcano junco in Costa Rica, another highly divergent lineage is Baird's yellow-eyed junco in the mountains of southern Baja California, Mexico. Again, plumage differentiation is a misleading indicator of evolutionary history, as Baird's junco is superficially similar to other yellow-eyed juncos from mainland North America. However, the deep cinnamon of its flanks and back (not found in any other yellow-eyed junco) sets it apart, as well as its highly divergent and complex male song (Pieplow and Francis 2011). The strong isolation of Baird's junco had been predicted by Miller (1941), who keenly observed that "the most complete isolation is that of a species such as *bairdi*, in which there is no chance of mixture through stray migrants" (378).

With respect to the dark-eyed junco radiation in mainland North America, the star-like pattern of the COI haplotype network indicates a recent population expansion that is consistent with previous results from the mitochondrial control region and genome-wide AFLP markers (Milá et al. 2007). This pattern, combined with the deep divergence of lineages in Mexico and Guatemala, strongly suggests a southern origin of the genus. Miller's careful observations had led him to predict the old age of southern lineages using the geographic distribution of what he deemed "primitive characters," those based on prevalent patterns and juvenal plumages in other Emberizids, including dorsal stripes, lack of a hood pattern, slight sexual dimorphism, absence of tail white, dark-colored maxilla, and dark iris. His predictions, tested here with genetic data, turn out to be correct: "The isolated southern areas may properly be viewed as an asylum for antiquated types, but are they centers of origin? [...] The greatest environmental changes in recent geologic history have been in a broad zone bordering the glacial areas. The areas of greatest environmental stability would preserve the primitive types, and these in general must have been the southern and insular areas" (371). Our results to date are consistent with this assessment, as junco lineages in southern and insular areas are not only genetically more divergent but also diverge from a more basal position in the phylogeny, even if several of the weakly supported basal nodes are collapsed. The fact that the younger lineages are concentrated in the temperate and boreal zones might be related to the environmental changes mentioned by Miller and is consistent with studies showing that phenotypic diversification rates are relatively higher in the temperate zone (Martin et al. 2009; Weir and Schluter 2007; Weir and Wheatcroft 2011).

Phylogeny reconstruction depends entirely on the existence of shared molecular substitutions in the ancestors of lineages in existence today (in this case, in ancestral juncos, represented by the internal branches of the phylogeny in figure 8.2). Given the rapid nature of the junco radiation, and the short time frame between speciation events, analysis of a quickly evolving molecular marker like mtDNA is necessary to uncover a signal of this history. The drawback to analyzing a single gene, however, is that no single DNA fragment ever fully represents the history of species because there is considerable random variation in how the different alleles for a given gene sort themselves into descendent species during the process of speciation. Faster speciation results in more randomness in gene sorting, lowering our confidence in results from any single gene. Species trees, which result from analysis of multiple genes in a way that accounts for the randomness of lineage sorting, can substantially increase our statistical power to uncover evolutionary history. A species tree using eight nuclear genes collected via next-generation sequencing in a recent study by McCormack et al. (2012) suggests that nuclear genes do contain sufficient signal to help resolve junco phylogeny, or at least its deeper nodes. Although this study did not include all junco lineages, their phylogeny was similar in most ways to the COI phylogeny presented here, showing *J. vulcani* as the basal taxon to other juncos, followed by *J. bairdi* (figure 8.4). Although support for these relationships was low, similar to the COI phylogeny, these results provide a measure of confidence that a species tree including many nuclear genes in addition to mtDNA will provide added clarity to the evolutionary history of juncos.

D. Diversification Mechanisms in the Juncos

We now know more about how juncos diversified, but what were the evolutionary mechanisms driving diversification? Was speciation in juncos predominantly caused by their physical isolation, or by natural selection

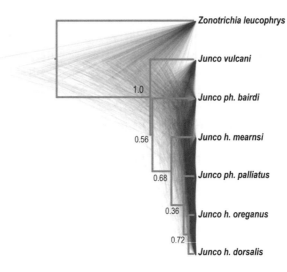

FIGURE 8.4. A species tree of the genus *Junco*, from eight nuclear loci isolated via next-generation sequencing (from McCormack et al. 2012). The thick lines represent the consensus topology with the highest probability, whereas the thin lines represent a "cloudogram" of all possible relationships, revealing high uncertainty in most parts of the tree. Similar to the COI phylogeny, *J. vulcani* branches from the base of the junco phylogeny, followed by *J. bairdi*. Modified from McCormack et al. 2012.

driven by ecological differences, or perhaps by mating differences driven by female choice? A striking pattern that emerges when we combine phylogenetic and phenotypic patterns in juncos is the marked contrast between the more genetically divergent lineages with limited plumage divergence and least genetically divergent lineages with marked plumage divergence. Long-term geographic isolation seems to be the main factor in driving the divergence of Guatemala, Baird's, and Guadalupe juncos. Highland habitats in Mexico and Guatemala are not particularly variable, which could account for the relative lack of morphological divergence among yellow-eyed junco lineages. Further analyses involving ecological niche models of junco habitats should allow us to test this hypothesis (see chapter 9, this volume).

In contrast to the old lineages in the south, the rapid evolution of markedly divergent plumage types among the dark-eyed juncos of mainland North America does suggest a role for natural or sexual selection, or both. The apparent similarity of the high-elevation habitats inhabited by most dark-eyed juncos and the lack of major differences in morphology (for instance in bill morphology) suggest that natural selection might have been less important than sexual selection in the North American lineages,

although the apparent similarities among niches should be tested empiri-
cally. The phenotypic pattern in dark-eyed juncos contrasts with another
recent, rapid North American radiation in the red crossbill complex (Benkman 2003), where different lineages that occupy almost the same geographic ranges as the dark-eyed junco morphs show marked divergence
in morphology (especially beak size and shape) associated with food
types, yet very little in terms of plumage color and pattern. Other North
American bird radiations show remarkable divergence of plumage color
(e.g., *Passerina* buntings and *Spinus* siskins), but these likely reflect mil-
lions of years of evolution, rather than thousands of years, as in the juncos.

Sexual selection has long been considered a potentially important
mechanism in divergence and speciation (Coyne and Orr 2004; Ritchie
2007; West-Eberhard 1983), and has received both theoretical (Lande
1981) and empirical support (Wilson et al. 2000). The marked differences
in plumage pattern and color among dark-eyed junco morphs suggest the
role of sexual selection, and in particular divergent female preferences
for plumage color, yet direct evidence for sexual selection playing a role
in dark-eyed junco diversification is still scant. Some studies at the population level have shown that sexual selection can modify the amount of
white on tail feathers in slate-colored and Oregon juncos (Hill et al. 1999;
McGlothlin et al. 2005; Price et al. 2008; Yeh 2004). At the genus level no
experimental data are available, yet the data collected by Alden Miller
can again be useful. One predictable outcome of sexual selection is the
appearance of sexual dimorphism in secondary sexual characters. Based
on the examination of thousands of junco specimens, Miller (1941) quantified the amount of sexual dimorphism in the different junco taxa as expressed in head color, and he concluded that dimorphism is lacking in
vulcani, alticola, fulvescens, and *bairdi*, nearly lacking in *phaeonotus, palliatus, dorsalis,* and *insularis*, slight in *caniceps*, and present in the different
Oregon and slate-colored taxa, with values (on a scale of 1 to 3) of 1 for
mearnsi, pontilis, townsendi, carolinensis, pinosus and *aikeni*; 2 for *montanus, cismontanus, hyemalis*, and *thurberi*; and 3 for *shufeldti* and *oreganus*
(370). The pattern revealed by these scores is highly congruent with both
a latitudinal and a phylogenetic pattern, with more genetically divergent
lineages (*vulcani, alticola, bairdi*, and *insularis*) showing no or almost no
dimorphism and the highest values being found in the more recently diverged races in northern North America.

Miller (1941) also contributed interesting observations regarding patterns of intrapopulation variation in plumage color traits, which he dem-
onstrates to be highest in the northern races (*hyemalis, oreganus, monta-*

nus, shufeldti, and *thurberi*) (367). There, high amounts of genetic standing variation appear to have been preserved, which result in the recurrent appearance in many races of some traits at low frequencies, such as white wing bars (a trait that became fixed in *aikeni*), or reddish crown feathers. These observations suggest the existence of numerous polygenic traits with high standing variation in the populations, which could confer juncos high "evolvability" of plumage traits (similar to the beak evolvability proposed for finches [Lovette et al. 2002]) and thus an increased potential for divergence by sexual selection. In the case of juncos, a plausible scenario for a role of sexual selection in the origin and divergence of forms is provided by the combined effect of naturally fragmented high-elevation mountain habitats, high standing variation for polygenic plumage color traits, and a profusion of small populations as juncos recolonized the North American continent following the last glacial maximum. Early divergence in these population isolates, helped by small effective population sizes and behavioral imprinting following the emergence of novel traits (Price 1998), could have promoted character fixation and speciation. Advances in our understanding of quantitative traits and the genetic basis of traits under selection, combined with the ever-increasing genome sequencing capacity, will soon allow us to properly test some of these hypotheses in juncos.

Another mode of diversification potentially at play in juncos is hybrid speciation, or the formation of an independent lineage resulting from the hybridization of two distinct parental forms. Phenotypically intermediate yet stable and geographically discrete morphs are suggestive of a hybrid origin. An example is the red-backed junco (*dorsalis*) of central Arizona and New Mexico, which shares plumage color with both the gray-headed junco (*caniceps*) and the northern yellow-eyed junco (*palliatus*), yet has a dark eye like the former and a bicolored bill like the latter. Miller was quite convinced of the hybrid origin of *dorsalis*. An alternate hypothesis is that *dorsalis* is not of recent hybrid origin but rather represents an evolutionary link between dark-eyed and yellow-eyed lineages. Miller also proposed a stable race of hybrid origin in some areas of western Canada (*cismontanus*), which he claimed shows premating isolation from its parental forms, northern Oregon juncos (*montanus*) and slate-colored juncos (*hyemalis*). Recent studies have already documented possible cases of hybrid speciation in birds (Brelsford et al. 2011; Elgvin et al. 2011; Hermansen et al. 2011), and explicit tests will be required to investigate junco taxa of hybrid origin. As with hypotheses involving sexual selection, testing the hybrid origin of these junco forms may soon be within our reach given the increased access to genome-wide molecular markers.

E. Taxonomic Implications and Future Research

The process of lineage splitting and divergence is often a slow, gradual, and messy affair, especially at the outset, when a certain amount of gene flow is to be expected among incipient lineages (see chapter 9, this volume). We tend to agree with the view that sees species as distinct evolutionary lineages and regards differences among so-called "species concepts" as largely differences in criteria for species designation (de Queiroz 2007). The complexity of junco variation is a telling example of how natural variation is a continuum that often cannot be easily, and much less objectively, divided into discrete taxonomic units or Alden Miller's "mileposts in the evolutionary path" (375). In cases of incipient evolutionary lineages such as those observed among some dark-eyed and yellow-eyed juncos, the designation of species is particularly difficult. Indeed, depending on the criteria used, and as Miller himself put it when discussing the relationships among gray-headed, Oregon, and slate-colored juncos, "the characterization of forms as completely or incompletely differentiated is rather arbitrary" (374). We trust that the ongoing quest for additional molecular data will help resolve some relationships and patterns of gene flow among hybridizing junco forms in mainland North America. A reticulate pattern rather than a single bifurcating tree might be an outcome at this early stage of lineage divergence, yet recent work on the cichlid fish radiation of Lake Victoria, known to have also taken place since the last glacial maximum, has revealed reciprocally monophyletic clades when there is a sufficient number of molecular loci, thus showing that lineage sorting in some loci can be reached relatively quickly (Wagner et al. 2013).

Fortunately, criteria for species designation in the case of well-differentiated lineages are far less problematic. Based on the genetic evidence presented here and the phenotypic data provided by previous work (largely Miller's), we propose that the three genetically divergent monophyletic lineages within the genus in addition to the volcano junco (*J. vulcani*) be treated as separate taxonomic species. These include two juncos with yellow irises, Baird's junco (*J. bairdi*) of Baja California and the Guatemala junco (*J. alticola*), and the dark-iris Guadalupe junco (*J. insularis*) from Guadalupe Island in the Pacific. Given the allopatric ranges of these taxa, their compliance with species designation criteria under the biological species concept is, as usual, impossible to establish, yet the proposed taxonomic arrangement is consistent with species designation criteria usually employed by adherents to the evolutionary and phylogenetic species concepts.

With respect to the taxonomy and nomenclature of the taxa included in the clade grouping most mainland forms of dark-eyed and yellow-eyed juncos, the COI data are of little help. New, more variable markers will be necessary to resolve these extremely young junco lineages, and ongoing analysis of genomic data is identifying markers that may be able to determine the amount of gene flow among forms and further clarify taxonomic relationships. In addition, research into chromosomal inversions might identify areas of the genome potentially important in population divergence (see chapter 9, this volume). For now, and taking into account the above recommendations regarding the taxonomy of divergent lineages, we consider the yellow-eyed junco (*J. phaeonotus*) to be composed of three subspecies (*phaeonotus*, *palliatus*, and *fulvescens*), corresponding to northern, central, and southern continental Mexico, respectively, and the dark-eyed junco (*J. hyemalis*) to be composed of the following groups: gray-headed junco (*J. h. caniceps* and *dorsalis*), Oregon junco (*J. h.oreganus, thurberi, pinosus, shufeldti, montanus, townsendi,* and *pontilis*), pink-sided junco (*J. h. mearnsi*), white-winged junco (*J. h. aikeni*), and slate-colored junco (*J. h. hyemalis* and *carolinensis*). Even though our genetic data reveal no mtDNA divergence supporting the separation of *J. phaeonotus* and *J. hyemalis* as different species, we feel that the marked differences in eye color and, especially, song between the two groups (Aleixandre et al. 2012) are sufficient to maintain them as separate species, pending a genome-wide assessment of genetic differentiation.

A better understanding of phenotypic variation within and between junco forms is also necessary to infer evolutionary processes and patterns of lineage divergence in the group. Song variation among populations appears to be limited among *J. hyemalis* forms (see chapter 13, this volume) but is proving to be a useful trait showing marked differentiation among the more divergent lineages, as already demonstrated in the Guadalupe junco (Aleixandre et al. 2013) and Baird's junco (Pieplow and Francis 2011). In-depth, quantitative analysis of plumage color traits across the range is also needed and will be useful in assessing patterns of introgression at contact zones, especially in conjunction with cline analysis of molecular markers. And much remains to be done to unravel the mysteries of mate choice in this complex system, and the relative roles of various visual and chemical cues are just now beginning to be explored (see chapter 12, this volume).

To further advance our understanding of speciation mechanisms in juncos and vertebrates in general, we will have to continue our quest for genomic data that increase our power to resolve phylogenetic relationships and patterns of gene flow, but we also need to work towards unveiling the genetic basis of adaptive variation. Identifying the genes that code for the traits that selection acts upon will allow us to associate polymorphism with levels of divergence and gene flow and infer strength and modes of selection, the number of loci involved, and many other fascinating aspects of this complex evolutionary process. The genus *Junco* provides the possibility of investigating old speciation events as well as striking cases of "speciation-in-action," and we are confident that juncos will remain an illuminating and challenging system in the study of vertebrate diversification.

Acknowledgments

Ellen Ketterson, Jonathan Atwell, Trevor Price, and an anonymous referee provided useful comments that improved the manuscript. Sampling expeditions to localities across the junco range over the years were made possible by the help of many people, including Adán Oliveras, Adrián Gutiérrez, Vicente Rodríguez, Allison Alvarado, Omar Espinosa, Sergio Larios, Allison Lee, Fritz Hertel, Elena Berg, Rich Van Buskirk, Adolfo Navarro Sigüenza, Cesar Rios, Roberto Sosa, Alfonso Aguirre, Julio Hernández Montoya, Ellen Ketterson, Jonathan Atwell, Steve Burns, Christine Bergeon Burns, Jatziri Calderón, Ricardo Rodríguez Estrella, and Thomas B. Smith. Funding sources for this work over the years include a grant from Spain's Ministry of Science and Innovation (CGL2011–25866) to BM, a UC-MEXUS Dissertation Grant to BM, Indiana University, and the Center for Tropical Research at the University of California, Los Angeles.

References

Aleixandre, P., J. Hernández-Montoya, and B. Milá. 2013. Speciation on oceanic islands: Rapid adaptive divergence vs. cryptic speciation in a Guadalupe Island songbird (Aves: *Junco*). *PLoS One* 8(5):e63242.

Benkman, C. W. 2003. Divergent selection drives the adaptive radiation of crossbills. *Evolution* 57:1176–81.

Brelsford, A., B. Milá, and D. E. Irwin. 2011. Hybrid origin of Audubon's warbler. *Molecular Ecology* 20:2380–89.

Coyne, J. A., and H. A. Orr. 2004. *Speciation.* Sunderland, MA: Sinauer Associates, Inc.

Darwin, C. 1859. *The Origin of Species by Means of Natural Selection, or the Preservation of Favoured Races in the Struggle for Life.* London: John Murray.

de Queiroz, K. 2007. Species concepts and species delineation. *Systematic Biology* 56:879–86.

Dwight, J. 1918. The geographical distribution of color and of other variable characters in the genus *Junco*: A new aspect of specific and subspecific values. *Bulletin of the American Museum of Natural History* 38:269–309.

Elgvin, T. O., J. S. Hermansen, A. Fijarczyk, T. Bonnet, T. Borge, S. A. SÆther, K. L. Voje, et al. 2011. Hybrid speciation in sparrows II: A role for sex chromosomes? *Molecular Ecology* 20:3823–37.

Excoffier, L., G. Laval, and S. Schneider. 2005. Arlequin ver. 3.0: An integrated software package for population genetics data analysis. *Evolutionary Bioinformatics Online* 1:47–50.

Hebert, P. D. N., M. Y. Stoeckle, T. S. Zemlak, and C. M. Francis. 2004. Identification of birds through DNA barcodes. *PLoS Biology* 2:1657–63.

Hermansen, J. S., S. A. SÆther, T. O. Elgvin, T. Borge, E. Hjelle, and G.-P. SÆtre. 2011. Hybrid speciation in sparrows I: Phenotypic intermediacy, genetic admixture and barriers to gene flow. *Molecular Ecology* 20:3812–22.

Hill, J. A., D. A. Enstrom, E. D. Ketterson, V. J. Nolan, and C. Ziegenfus. 1999. Mate choice based on static versus dynamic secondary sexual traits in the dark-eyed junco. *Behavioral Ecology* 10:91–96.

Huelsenbeck, J. P., F. Ronquist, R. Nielsen, and J. P. Bollback. 2001. Bayesian inference of phylogeny and its impact on evolutionary biology. *Science* 294:2310–14.

Kerr, K. C. R., M. Y. Stoeckle, C. J. Dove, L. A. Weigt, C. M. Francis, and P. D. N. Hebert. 2007. Comprehensive DNA barcode coverage of North American birds. *Molecular Ecology Notes* 7:535–43.

Lande, R. 1981. Models of speciation by sexual selection on polygenic traits. *Proceedings of the National Academy of Sciences of the United States of America* 78:3721–25.

Lovette, I. J., E. Bermingham, and R. E. Ricklefs. 2002. Clade-specific morphological diversification and adaptive radiation in Hawaiian songbirds. *Proceedings of the Royal Society B-Biological Sciences* 269:37–42.

Martin, P. R., R. Montgomerie, and S. C. Loughheed. 2009. Rapid sympatry explains greater color pattern divergence in high latitude birds. *Evolution* 64:336–47.

McCormack, J. E., J. M. Maley, S. M. Hird, E. P. Derryberry, G. R. Graves, and R. T. Brumfield. 2012. Next-generation sequencing reveals phylogeographic structure and a species tree for recent bird divergences. *Molecular Phylogenetics and Evolution* 62:397–406.

McGlothlin, J. W., P. G. Parker, V. J. Nolan, and E. D. Ketterson. 2005. Correlational selection leads to genetic integration of body size and an attractive plumage trait in dark-eyed juncos. *Evolution* 59:658–71.

Milá, B., J. E. McCormack, G. Castañeda, R. K. Wayne, and T. B. Smith. 2007. Recent postglacial range expansion drives the rapid diversification of a songbird lineage in the genus *Junco*. *Proceedings of the Royal Society B-Biological Sciences* 274:2653–60.

Miller, A. 1941. Speciation in the avian genus *Junco*. *University of California Publications in Zoology* 44:173–434.

Mirsky, E. N. 1976. Song divergence in hummingbird and junco populations on Guadalupe island. *Condor* 78:230–35.

Nolan, V. J., E. D. Ketterson, D. A. Cristol, C. M. Rogers, E. D. Clotfelter, R. C. Titus, S. J. Schoech, et al. 2002. Dark-eyed Junco (*Junco hyemalis*). In *The Birds of North America*, edited by A. Poole and F. Gill, no. 718. Philadelphia: The Birds of North America, Inc.

Pieplow, N. D., and C. D. Francis. 2011. Song differences among subspecies of the yellow-eyed juncos (*Junco phaeonotus*). *Wilson Journal of Ornithology* 123: 464–71.

Posada, D. 2008. jModelTest: Phylogenetic Model Averaging. *Molecular Biology and Evolution* 25:1253–56.

Price, T. 1998. Sexual selection and natural selection in bird speciation. *Philosophical Transactions of the Royal Society B-Biological Sciences* 353:251–60.

Price, T. D., P. J. Yeh, and B. Harr. 2008. Phenotypic plasticity and the evolution of a socially selected trait following colonization of a novel environment. *American Naturalist* 172:S49–S62.

Ritchie, M. G. 2007. Sexual selection and speciation. *Annual Review of Ecology and Systematics* 38:79–102.

Sullivan, K. A. 1999. Yellow-eyed junco (*Junco phaeonotus*). In *The Birds of North America*, edited by A. Poole and F. Gill. Philadelphia: The Birds of North America Inc.

Tamura, K., D. Peterson, N. Peterson, G. Stecher, M. Nei, and S. Kumar. 2011. MEGA5: Molecular Evolutionary Genetics Analysis using maximum likelihood, evolutionary distance, and maximum parsimony methods. *Molecular Biology and Evolution* 28:2731–39.

Tavares, E. S., P. Gonçalves, C. Y. Miyaki, and A. J. Baker. 2011. DNA barcode detects high genetic structure within Neotropical bird species. *PLoS One* 6: e28543.

Wagner, C. E., I. Keller, S. Wittwer, O. M. Selz, S. Mwaiko, L. Greuter, A. Sivasundar, et al. 2013. Genome-wide RAD sequence data provide unprecedented resolution of species boundaries and relationships in the Lake Victoria cichlid adaptive radiation. *Molecular Ecology* 22:787–98.

Weir, J. T., and D. Schluter. 2007. The latitudinal gradient in recent speciation and extinction rates of birds and mammals. *Science* 315:1574–76.

———. 2008. Calibrating the avian molecular clock. *Molecular Ecology* 17: 2321–28.

Weir, J. T., and D. Wheatcroft. 2011. A latitudinal gradient in rates of evolution of avian syllable diversity and song length. *Proceedings of the Royal Society B-Biological Sciences* 278:1713–20.

West-Eberhard, M. J. 1983. Sexual selection, social competition and speciation. *Quarterly Review of Biology* 58:155–83.

Wilson, A. B., K. Noack-Kunnmann, and A. Meyer. 2000. Incipient speciation in sympatric Nicaraguan crater lake cichlid fishes: sexual selection versus ecological diversification. *Proceedings of the Royal Society B-Biological Sciences* 267: 2133–2141.

Yeh, P. J. 2004. Rapid evolution of a sexually selected trait following population establishment in a novel habitat. *Evolution* 58:166–74.

The Potential Role of Parapatric and Alloparapatric Divergence in Junco Speciation

Trevor D. Price and Daniel M. Hooper

Until segregation has proceeded to the point of adding infertility, there is always the possibility of its being broken down when forms are again thrown together in the same region. Even with infertility, unless it be quite absolute, there is some possibility.
—Miller 1941, 380

Introduction

Mitochondrial genetic analyses of the junco indicate five well-marked clades (chapter 8, figs. 8.2 and 8.3, this volume). Four of these clades define relatively small populations at the southern end of the junco range (in Guatemala, Costa Rica, southern Baja California, and on Guadalupe Island, see figure 8.1). The other clade lies to the north and covers a vast area, including most of Mexico (the yellow-eyed junco), and the United States and Canada (the dark-eyed junco). Across this whole distribution, mitochondrial DNA shows little variation (figures 8.2, 8.3) and putatively neutral nuclear genetic markers also appear to be largely homogeneous (figure 8.4). One explanation for the genetic similarity of the northern forms is that these taxa diverged from each other very recently following range expansion, perhaps only since the end of the last glaciation, 10,000 years ago (Milá et al. 2007; chapter 8, this volume). Despite the apparent genetic homogeneity over much of North America, many allopatric and parapatric forms have been described, based on distinctive plumage patterns as well as other characteristics, including eye-color and morphometric measures (see chapters 3 and 8, figures 3.1 and 8.1, color plates 4 and 5, this volume). In fact, phenotypic differences between taxa within this clade can be greater than differences between the four southern taxa (see figure 8.1). Because of this, historically, some of the taxa within this northern clade have themselves been given species status. In this book we have continued this tradition, separating the distinctive Mexican yellow-eyed junco from the dark-eyed junco, despite their genetic similarity (chapter 8, this volume).

A pattern of rapid divergence in color patterns at high latitudes in the New World appears to be quite general in birds (Martin et al. 2010). Rapid divergence at high latitudes also applies to bird song (Weir and Wheatcroft 2011; Weir et al. 2012) and to color divergence in other groups of organisms, such as damselflies (Turgeon et al. 2009). Differences in recent evolutionary rates between low and high latitudes have been attributed to the great disturbance at northern latitudes throughout the Pleistocene, with populations expanding out of southern refugia and diversifying as they do so (Turgeon et al. 2009; Weir and Wheatcroft 2011). However, the northern junco is exceptional, at least among birds, for the apparent youth of the component taxa (Johnson and Cicero 2004; Weir and Schluter 2007; Milá et al. 2007).

In this chapter we consider explanations for why the color patterns are so different among these young northern junco taxa, despite high genetic similarity. One possibility is that the colors represent very rapid sorting of standing variation or even de novo mutations from a single ancestral population that expanded out of a southern refuge following the most recent ice age, a view presented elsewhere (Milá et al. 2007; chapter 8, this volume). A second alternative is that the phenotypic differences have been preserved in different geographic locations over long time periods, but gene flow between locations resulting from both high dispersal and frequent range shifts has led to homogenization across much of the rest of the genome. We present a novel mechanism for this process that involves individuals from migratory (northern) populations settling to interbreed with residents. This unidirectional gene flow has the feature that at neutral markers all populations come to resemble the population sending out migrants, so the total effective population size is estimated to be low. Overall, we suggest an important characteristic of the junco populations is the possibility for "divergence with gene flow," the idea that populations can continue to exchange genes across much of their genome, even as they maintain differences from each other at other places in the genome. We conclude that while it is possible that the color patterns have only recently become established in northern populations, a plausible alternative is that they are more ancient, with similarity at genetic markers such as mitochondrial DNA the result of differences in the way these genetic markers and color patterns introgress across populations.

The development of genetic differences between populations for any trait, including color, requires some evolutionary force to cause the different variants to become established in different populations. The two main ways variation can be introduced are through mutation or hybridization. Once variation has been introduced, genetic drift and sexual and natural selection pressures have all been invoked as agencies of population divergence. Among these processes genetic drift is seen as quite unimportant as a driver of divergence in traits such as color patterns, because it is such a weak force (Price 2008; Price et al. 2010). Indeed, color variants have been shown to be subject to strong selection in contemporary studies (Hoekstra et al. 2001; Siepielski et al. 2011), including some on the junco (McGlothlin et al. 2005; Price et al. 2008). In what follows, we emphasize a role for selection, rather than drift. We consider possibilities for junco population divergence in the face of gene flow, including a novel mechanism based in seasonal migration. Finally, we review what is known about chromosomal inversions in juncos and suggest they may aid in population differentiation within these genomic regions despite less restricted gene flow across others.

A. Divergence with Gene Flow

Under the so-called biological species concept, species are interbreeding populations reproductively isolated from other interbreeding populations (Mayr 1963). Mayr (1963) emphasized the importance of geographical separation of populations, limiting gene flow between them and allowing them to diverge. Geographical separation is clearly important to the vast majority of bird speciation events (Mayr 1963; Price 2008). However, we now recognize that the development of complete reproductive isolation can take a long time, with complete infertility of hybrids taking more than two million years and hybrid inviability much longer (Price and Bouvier 2002). This creates opportunities for some gene exchange even as populations diverge, because geographical ranges have certainly fluctuated over these timescales, potentially bringing populations into contact, and because individual birds may disperse over long distances.

Thus, the development of reproductive isolation between populations may proceed even if the populations do not remain completely allopatric. One possibility is that populations continue to exchange migrants throughout the whole period of divergence, right up to the completion of speciation (to be consistent with population genetics terminology, migrants are defined as individuals which disperse from one population to another, and we use the term "seasonal migration" to refer to annual movements). In this case, immigrants and their offspring become increasingly disadvantaged as divergence proceeds. That is, even though migration between populations continues, gene flow gets curtailed. In the speciation literature, this is termed "parapatric speciation." An alternative to parapatric speciation is that some divergence proceeds in allopatry (i.e., a complete absence of gene flow), but as a result of climate change or other factors affecting distributions, populations come into contact before they are fully reproductively isolated, potentially leading to genetic exchange. The process of differentiation and contact may be repeated multiple times, with each period of contact resulting in some gene flow, until eventually differentiation has proceeded to complete reproductive isolation. This is termed "alloparapatric speciation" (Endler 1977; Coyne and Orr 2004).

B. Gene Flow and the Establishment of Different Color Variants in Different Populations

Whenever migrants from one population mate with residents from another to produce viable and fertile offspring, there is the possibility for gene exchange. However, some genes are more likely to pass from one population to the other (Wu 2001). Any variants brought in by migrants that are favored in the resident population should rapidly become established and differentiation between populations lost. On the other hand, variants that are not favored in the resident population will not become established and differences between populations maintained. In the first case, selection in both populations favors the same variant: selection pressures are parallel. In the second case, selection favors one variant in one population and the other variant in the other: selection pressures are divergent.

Normally elements of both parallel and divergent selection will be involved in establishing genetic (including color) differences between populations, but they are important to distinguish conceptually, because the power of gene flow to prevent or slow differentiation differs greatly under the alternative processes (Price 2008; Nosil and Flaxman 2011). In the divergent selection model, selection can overcome the homogenizing effects of migration and maintain differences between populations (Hedrick 2000; chapter 3 in Price 2008). However, in the parallel selection model, gene flow efficiently prevents population differentiation. We illustrate this in figure 9.1, with the panels distinguishing the case of some gene flow (left) from the case of no gene flow (right). First, a barrier is established, resulting in two geographically separated populations. In one of those populations an attractive trait arises and becomes established relatively soon after the populations become geographically separated, at about Time 1. In the left hand panel, sometime after Time 1, populations come in contact, perhaps for only a short time, leading to gene exchange. The attractive trait spreads rapidly into the second population, erasing the differences between the two populations. In the right panel there is a similar time progression but without any gene flow at all. Now, there is sufficient time for a second, different trait to arise in the other populations. If different attractive traits become established in each population, these traits are considered to mutually interfere with each other's spread (for example, frequency dependent selection may favor the common type in each population, as in Stein and Uy [2006]). In this way differences between the two populations are maintained in perpetuity, even if the populations come into contact and hybridize.

Both in the juncos and more generally, we have little idea if the different color patterns are arbitrary attractive traits that have become established because they arose in different populations that are subject to parallel selection pressures (as suggested by Milá et al. in chapter 8, this volume) or if the respective colors are specifically adapted to the environments in which the populations occur. It has been generally hard to distinguish the alternatives, mainly because few clear correlates of color and environment have been identified (Gomez and Théry 2007); such relationships have not been examined systematically in juncos. If junco color differences have been established in the face of ongoing gene flow, then it seems likely that they became established as a result of divergent selection pressures (i.e. adaptation to different environments in different places). Many of the color differences do seem quite arbitrary, however, and it is worth considering how parallel selection pressures may lead to divergence. First, different color patterns may have arisen and become

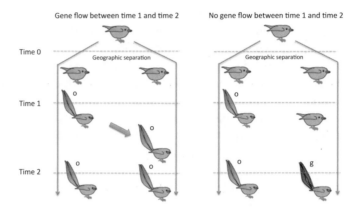

FIGURE 9.1. Example of the importance of allopatry during divergence under parallel selection pressures. At time 0, a single population becomes divided by a physical barrier, leading to two allopatric populations. In the left panel, some gene flow occurs between these two populations at a point between time 1 and time 2. In the right panel the two populations remain allopatric. Consider two alternative attractive signals, a long orange tail (o) and a long green tail (g), and that once established, the common form is favored. This implies that anytime between time 1 and time 2, gene flow between populations (left panel) results in establishment of the orange-tailed form everywhere, and the green-tailed form could not establish. The implication is that long periods without gene flow are particularly conducive to divergence under parallel selection pressures.

fixed in long-separated allopatric populations, under parallel selection pressures as illustrated in the right panel of figure 9.1. The patterns are then ancient and interfere with each other's spread, with subsequent contact between populations resulting in much of the rest of the genome homogenizing.

Other possibilities for differentiation under parallel selection pressures consider more continuous gene flow scenarios. First, the color mutation may be locked up in an ancient chromosomal inversion. Here, even if the color mutation is favored everywhere throughout a species range, it would be unable to spread if alleles at other loci within the inversion are only favored in a portion of the range; we return to inversions in a later section. Second, different mutations in different populations may have arisen sufficiently close together in time that they are simultaneously spreading in each population (Kondrashov 2003), that is, one does not spread through the range before an alternative color form arises somewhere else (cf. figure 9.1). This seems similar to the suggestions of Milá et al. (chapter 8, this volume) for how junco populations have come to differ so quickly, with the added proviso that the new mutations have not only originated close in time but also recently. Potentially, all these scenarios will be testable with genomics methods.

In the next sections we consider divergence with gene flow in the juncos in more detail. First we consider a contemporary study, and second we show a reconstruction of ancestral ranges at the time of the last glaciation, which indicates that range expansions and contractions have likely been a pervasive feature for juncos throughout the Pleistocene.

C. A Present-Day Example of Divergence with Gene Flow in the Junco

Species that migrate seasonally tend to have high natal dispersal distances (Paradis et al. 1998), thereby promoting gene flow across populations. Thus, seasonally migratory species are expected to experience gene flow during divergence more often than species with year-round residency (Winker 2010).

In migratory birds, divergence with gene flow may occur when a few individuals establish a sedentary breeding population in what were previously the wintering grounds. Winker (2010) termed this "heteropatric" speciation because residents and seasonal migrants are in contact for only part of the year, but the process fits the general model of parapatric speciation. The idea is that a new life history resulting from year-round residency subjects the resident population to novel selection pressures. Thus, even if some migrant birds remain and interbreed with residents, as the resident population becomes increasingly differentiated from the migratory one, immigrant individuals and their offspring become increasingly disadvantaged relative to the residents. Note that once a resident population has been founded, to the extent that seasonal migration is heritable, hybrid offspring of immigrants will themselves tend to emigrate (i.e., disperse) and be lost from the population. In this way effective migration is lowered further. The junco system provides the best-studied example of this process in its very early stages.

Juncos breeding in the mountains of southern California (*J. h. thurberi*) are apparently facultative and partial migrants. In the nonbreeding season, small flocks visit the coast and are present in low numbers in San Diego parks and gardens (Unitt 2005; T. Price, pers. obs). Probably in the early 1980s some individuals did not return to their breeding grounds but instead remained to breed on the campus of the University of California at San Diego (UCSD), where a resident population has persisted ever since, holding more or less steady at about seventy breeding pairs. The population has been intermittently monitored since the mid-1990s and was intensively studied by Pamela Yeh for her PhD thesis from 1998 to 2002. Wintering birds continued to visit, and we suspected 3–10 percent of the breeding population each year to have been formed by immigrants (Yeh and Price 2004). Despite this high immigration rate, the population as a whole rapidly changed in reproductive schedules, plumage pattern, morphology, song, social and personality behaviors, nest site selection, migratory behavior (*Zugunruhe*), and hormonal profiles (reviewed in chapter 10, this volume; and Yeh et al. 2007). Based on common garden experiments, at least some of these changes appear to be genetically determined (Yeh 2004; Rasner et al. 2004; Atwell et al. 2012; Atwell et al. 2014; Atwell et al., in review). Both quantitative genetic analyses of trait divergence and molecular genetic (microsatellite) diversity comparisons indicate that drift does not explain the apparent genetic differences. The contrasting environments of the native range (woodland, temperate) and invaded range (urban, Mediterranean) clearly placed multiple novel selective pressures on the population, driving rapid divergence.

Many juveniles fledge, but most disappear before breeding (Yeh and Price 2004), which may partly reflect persistent migratory tendencies. To the extent that some individuals show higher dispersal tendencies than others and dispersal is associated with multiple correlated traits bound together in a "behavioral syndrome" (Dingemanse et al. 2003, 2010; Atwell et al. 2014), other traits will be subject to selection as a result of any emigration. Measurements of surviving juveniles suggest that selection on the plumage pattern, leading to a reduction in the amount of white in the tail, was associated with differential juvenile disappearance (Price et al. 2008). It may simply be that individuals with greater tendency to emigrate from the San Diego population also have more tail white, with the net result that the residents have evolved to have less tail white (Price et al. 2008). Consistent with this, a common garden study suggests that resident UCSD juncos exhibit reduced genetic propensity to migrate compared to ancestral range juncos (Atwell et al., in review).

Under the principle of divergence with gene flow, increasing differentiation of the resident population should lower fitness of immigrants, further reducing gene flow and thereby facilitating more divergence in a feedback mechanism (chapter 4 in Price 2008; Winker 2010). Reduced immigrant fitness could result from reduced mating success of immigrants.

Both plumage (Yeh 2004; Atwell et al. 2014a) and songs (chapter 14, this volume; Newman et al. 2008; Cardoso and Atwell 2011) of the UCSD population differ slightly from populations in the nearby mountains, but we have no evidence that these differences affect mating success. For example, songs from the native range and from the UCSD population elicit comparable aggressive responses by resident males (Newman et al. 2006). Other differences between residents and migrants are surely affecting migrant fitness. One feature of the UCSD population is that reproductive schedules are greatly extended with respect to the native range. Up to four broods are produced each year compared to one to two in the native range (birds begin breeding in February alongside wintering nonbreeding individuals [Yeh and Price 2004]). Each year some breeding San Diego birds are on the native range schedule, breeding late and producing few broods (Price et al. 2008). We do not know if this is a particular feature of recent immigrants, and it may not be genetic, because testosterone profiles of mountain and UCSD populations converged in a common garden experiment (Atwell et al. 2014) and individual repeatability of breeding dates was not detected in a longitudinal study (Yeh and Price 2004). However, late breeding is clearly selected against, because the production of fewer broods leads to fewer surviving offspring the following year (Yeh and Price 2004).

The main importance of these results for the current discussion is that divergence in multiple traits, including plumage color, can take place despite ongoing gene flow. This also implies that established color patterns in different populations should be resistant to homogenization in the face of moderate levels of gene flow, whenever the local color patterns are selectively favored.

D. Range Shifts in the Junco

Pleistocene glacial cycles must have repeatedly caused major changes to the geographic ranges of the different junco lineages. In figure 9.2, we show a reconstruction of the distribution of suitable habitat for breeding dark-eyed juncos at the last glacial maximum ca. 21,000 years ago, compared with the present-day (based on climate data posted at www .worldclim.org). The inference from this reconstruction is that habitat for North American juncos was quite restricted, lying to the west and south, with Alaska an important refuge. Although this reconstruction comes with many uncertainties, permanent ice sheets covered Canada and extended

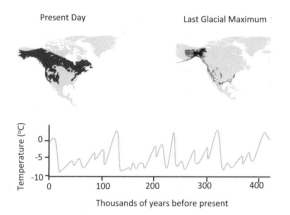

FIGURE 9.2. *Top left*: Breeding distribution of North American dark-eyed juncos, *Junco hy-emalis*, according to Ridgely et al. (2005). *Top right*: "Reconstructed range" at the last glacial maximum (21,000 years ago). *Below*: Estimated temperature climate fluctuations over the last 420,000 years based on oxygen isotope ratios in a polar (Antarctic) ice core, redrawn from Petit et al. (1999) (line has been smoothed and masks smaller variations). To estimate conditions suitable for *Junco hyemalis* at the last glacial maximum reconstruction we used the Worldclim database (worldclim.org), which gives 19 climatic variables at a resolution of 2.5 arc minutes across the world for both the present day and estimated for 21,000 years ago (before present). We included all 19 variables in the analyses. We clipped the dataset to cover the area illustrated in the figure using the R program Raster (Hijmans and van Etten 2011). We then used Maxent (Phillips et al. 2006) to contrast the climate estimates for all cells drawn from the junco's current range *(top left)* with all North American cells, to project probabilities of occurrence 21,000 years before present. The distributions are areas where P(occurrence)>0.66 (black), 0.33>P>0.66 (dark grey), and P<0.33 (pale grey).

south of Chicago (Peltier 2004), so these areas were clearly not occupied by juncos. Junco ranges must have been labile, leading to repeated population separation into refugia followed by contact, associated with the ebb and flow of the ice sheets, which over the last several hundred thousand years operated on an approximately 100,000-year timescale (Petit et al. 1999; see figure 9.2).

E. Mechanisms of Retaining Color Differences Despite Gene Flow

In the classic model of alloparapatric speciation, when differentiated taxa come into contact, neutral and selectively favored alleles spread across the ensuing hybrid zone, while traits favored in one or other taxon are resistant to gene flow (Wu 2001). Given that strong hybrid incompatibilities are unlikely in the North American juncos (Miller 1941; Price and Bouvier 2002), the possibility for periodic episodes of gene exchange in the junco has presumably persisted throughout the Pleistocene. Thus, if different junco colors are favored in different populations, the rest of the genome may homogenize across contact zones, while the color patterns remain restricted to each taxon, even in the face of substantial gene flow. In this case, the hybrid zone moves, except for the color. One excellent example of a moving hybrid zone comes from the hermit warbler (*Setophaga occidentalis*)/Townsend's warbler (*Setophaga townsendi*) hybrid zone in the Pacific northwest, where the Townsend's warbler has apparently moved south from northern Alaska, probably since the last glaciations, replacing the hermit warbler. Mitochondrial DNA of the hermit warbler occurs 2,000 kilometers north of the present zone (in Washington State) in what are phenotypically pure Townsend's populations (Rohwer et al. 2001). In this case, it is mtDNA that has been left behind as the hybrid zone moved south. Similar principles should apply also to color patterns if they are favored in one location and not the other.

Zones of hybridization between different color forms of the dark-eyed junco do not seem to have moved historically since Miller surveyed them in the 1930s (B. Milá, pers. comm.) although this remains to be directly quantified and movement of genetic material other than color would not be detected. Nevertheless juncos have clearly experienced repeated range shifts through the Pleistocene, and it is possible that a hybrid zone moved through a large part of the landscape, homogenizing much of the genome, with color variants selectively retained from ghost populations, whose other genetic differences have now largely been lost. This is one way color patterns could have remained differentiated while the rest of the genome homogenized. It may be contrasted with the argument of Milá et al. in chapter 8 (this volume) that the color patterns are novel variants arising in association with a rapid range expansion out of the south. However, ancestral reconstructions of breeding range argue against *both* these explanations. With respect to the hypothesis of a rapid range expansion, it appears that northerly breeding populations were retained throughout the Pleistocene. With respect to the hypothesis of homogenization by gene exchange, there is no reason to expect effective population sizes to be as small as currently inferred from the genomic data. Indeed, a naturally fragmented range should create conditions for preservation of multiple genetic variants and a large effective population size (see Hedrick 2000). In the next section we consider a possible resolution whereby gene flow can both homogenize much of the genome and also lead to a small population size.

F. A Resolution to the Problem of Low Genetic Diversity and High Color Pattern Differentiation

Following the previous section, we propose that color patterns have been retained by selection over long periods and the genome homogenized by migration; note that only a few immigrants per generation are required to prevent differentiation by genetic drift (e.g., Hedrick 2000). However, we add the feature that gene flow is *unidirectional*, from northern populations to southern ones. By analogy with our review of the San Diego population, we argue that seasonal migrants occasionally remain on their winter quarters and interbreed with residents, but residents rarely, if at all, become migratory, so gene flow from resident to migrant populations does not occur. Consider a northern migratory population of juncos that persisted in Alaska throughout the Pleistocene. A reasonable effective population size for that population may number in the ten thousands. If migrants from that population occasionally interbreed with residents of other populations, the genetic make-up of the other populations rapidly comes to resemble that of the source (see Hedrick 2000) and will have a correspondingly low effective population size as well (Price and Hudson, unpublished observations).

G. Chromosomal Rearrangement-Assisted Divergence with Gene Flow

Chromosomal rearrangements, especially inversions, occur regularly in birds (Shields 1982; Christidis 1990; Price 2008). Here we consider the roles they may play in junco differentiation. We emphasize two features: first, gene flow can actually favor the spread of inversions, and second, inversions can contribute to the build-up of genetic differences between populations even when gene flow is occurring.

Chromosome inversions are structural mutations that occur when a chromosome breaks at two points and the region bounded by those breakpoints is repaired in the opposite orientation. Inversions are found both as polymorphisms within species and as fixed differences between species (King 1993; Price 2008; Faria and Navarro 2010). One importance of an inversion is that when it is paired with the noninverted homologue, recombination between loci on the inverted segment is reduced or completely suppressed (Rieseberg 2001; Noor et al. 2001b). Inverted regions may thus retain and accumulate genetic differentiation, despite gene flow between populations (Navarro and Barton 2003). To see this, suppose that one of two populations exchanging migrants is fixed for the inverted form of a chromosome and one for the original, noninverted form. Now suppose that the inversion carries alleles that are favored in one population but not the other (i.e., is locally adapted). This means that the inversion—and hence all the genetic variation inside the inversion unique to one population—remains restricted to that population. Note that selection does not act upon the inversion itself but rather on the locally adapted genes within it.

Now consider a mutation that is favored throughout the species range (as in the attractive traits of figure 9.1). If the mutation occurs within the region spanned by the inversion it will not pass from one population to the other, because it is held in association with one or more alleles that are favored only in that population, but if it occurs in a region outside the inversion, it can decouple from locally adapted alleles and spread right through the species range. In this way, genetic differences between the genomes of hybridizing populations can build up within chromosomal rearrangements (Navarro and Barton 2003). Critically, when considering population divergence from a genomic perspective, two hybridizing populations may appear indistinguishable, except at loci within rearranged regions of the genome. We give an example in the final paragraph of this section.

One feature of rearrangement-assisted divergence models is that, counterintuitively, differentiation can be promoted by gene flow. This is because inversions themselves are more likely to become established if gene flow is present. Kirkpatrick and Barton (2006) model the situation where an inversion arises in one population and by chance captures a set of at least two locally adapted alleles. The inversion keeps these alleles together in descendants from the mating of a resident with an immigrant. In consequence, local carriers of the inversion have a fitness advantage over those that do not carry the inversion and the inversion increases in frequency in the population where the locally adapted alleles it carries are favored (Kirkpatrick and Barton 2006).

In summary, a role for inversions in differentiation, as outlined here, follows a two-step process, with gene flow integral to divergence. First, an inversion arises and becomes fixed in one population because it carries alleles specifically adapted to the local conditions that population experiences (Kirkpatrick and Barton 2006). Second, once an inversion is established in one population, hybridization is less likely to remove those early stages of divergence in the parallel selection model illustrated in figure

9.1. Thus, if a mutation that is favored throughout the species range arises within the inversion, it does not spread throughout the range but remains confined to the population where the inversion itself is restricted.

The well-studied North American *Drosophila* pair *D. pseudoobscura* and *D. persimilis* offer an excellent example of these scenarios. These two species differ by chromosomal inversions on three of their six chromosomes (Tan 1935). Recombination rates show a nine-fold decrease between the inverted and collinear regions of the rearranged chromosomes and thereby facilitate the maintenance and build-up of divergence between taxa (Machado and Hey 2003; Stevison et al. 2011). In sympatry all forms of reproductive isolation and most other species-specific traits between taxa map to loci within these inversions (Noor et al. 2001a), while in the allopatric *D. pseudoobscura bogotana*, loci responsible for reproductive isolation are spread more evenly across the genome (Brown et al. 2004). A reasonable scenario is that genetic differences that cause reproductive isolation and are present in allopatric populations have been lost when the species hybridize, unless they are held in large linked blocks. Strikingly, when considering only loci found on collinear regions of the genome more than 2.5 Mb away from inversion breakpoints, *D. p. pseudoobscura* and *D. persimilis* would be erroneously designated as sister species and *D. p. bogotana* the outgroup (Machado and Hey 2003; Brown et al. 2004; Machado et al. 2007; McGaugh and Noor 2012). The first avian example of inversions as reproductive barriers has recently come from a comparative genomic assessment of the European hybrid zone between the hooded (*Corvus corone cornix*) and carrion crow (*C. c. corone*). Almost all of the limited genetic differentiation between the two forms (eighty-one of eighty-two fixed differences) are located within a two-megabase inversion difference between crows, and genes within this inversion appear to be involved in regulation of both divergent feather pigmentation and visual perception (Poelstra et al. 2014). These sorts of effects may apply to the juncos too, which, as we describe below, harbor multiple inversions.

H. Chromosomal Rearrangements in the Junco

Juncos offer a natural bird system with which to examine the role of chromosomal rearrangements in facilitating population differentiation despite gene flow. After examining mitotic cell cultures from 246 individuals in six

TABLE 9.1 **Sampling locality and karyotype information about chromosomal inversions in juncos, from Shields (1973)**

Population	‡Collection locality	Individuals	*Chromosome 2	*Chromosome 5
Junco hyemalis hyemalis	Ontario Prov., Canada (N44.83 W78.95)	187	0.68	0.03
Junco h. oreganus	Clark Co., WA, USA (N45.83 W122.23)	19	0.95	0.26
Junco h. aikeni	Custer Co., SD, USA (N43.74 W103.66)	13	1.00	0.15
Junco h. caniceps	Lincoln Co., NM, USA (N33.45 W105.79)	13	0.69	0.31
Junco phaeonotus palliatus	Cochise Co., AZ, USA (N31.40 W110.28)	12	0.67	0.33
Junco insularis	Guadalupe Is., Mexico (N29.05 W118.31)	2	1.00	1.00

‡ Latitude and longitude coordinates are estimated as the midpoint of sampling site locations in each area visited by Shields.
* Proportion of individuals homozygous for ancestral form of the chromosome relative to outgroup sister genus *Zonotrichia* (Thorneycroft 1968; Shields 1973; Lucca and Rocha 1985).

junco populations, Shields (1973) found large polymorphic inversions on two chromosomes (table 9.1). Later, Shields (1976) demonstrated a three-fold decrease in the amount of crossing over between rearranged chromosomes when compared with chromosomes that are colinear, a result of pairing difficulties.

The junco inversions are at different frequencies in different populations (table 9.1, comparing the five North American populations, χ^2 = 11.61, P = 0.02 for the rearranged second largest chromosome and χ^2 = 35.54, P < 0.00001 for the rearranged fifth largest chromosome—hereafter referred to as chromosome two and five, respectively). The inversion on chromosome five may form a cline from high to low frequency, from west to east across North America (correlation of longitude and frequency, r = 0.728, N = 6 populations, including that from Guadalupe island, P = 0.1) and may be correlated with population-specific seasonal migration distance; however, more data are required to assess this (see table 9.1).

Within the single largest sampled population (N=187 slate-colored juncos, *J. h. hyemalis*) both arrangements conform to Hardy-Weinberg equilibrium (chi-square tests, P > 0.2; any deviations from Hardy-Weinberg would indicate nonrandom mating, or perhaps strong selection, e.g., Hedrick 2000). Nor is there evidence for linkage disequilibrium (i.e., correlation) between the two arrangements (chi-square tests, P > 0.2). How-

ever, Rising and Shields (1980) found that within *J. h. hyemalis*, inversions
are correlated with ecologically relevant morphological shape proportions (chromosome two with beak, tarsus, and wing ratios; chromosome
five with beak shape and, in females, with wing and tarsus proportions)
but not with overall body size. They postulated that differences between
inversion-associated morphological variants were important in habitat
partitioning during the winter in association with selection for different
foraging behaviors and resource use. This has yet to be explicitly tested.

   Inversion polymorphisms are common in Passerellidae, the family of
new-world sparrows to which juncos belong, but few have been studied in
depth (Shields 1982). In the closely related white-throated sparrow (*Zonotrichia albicollis*), an inversion polymorphism on chromosome two is associated with both behavioral (i.e., level of aggression, mate choice, and
parental investment) and morphological (plumage color, body size, wing
shape, and beak shape) differences between individuals (Thorneycroft
1966, 1975; Rising and Shields 1980; Tuttle 2003; Maney 2008). The morphology of the uninverted (inferred ancestral) chromosome two of the
junco appears to be identical to that of chromosome three of the white-
throated sparrow (Thorneycroft 1968; Shields 1973; Lucca and Rocha
1985). While the two inversions in the junco (table 9.1) and two in the
white-throated sparrow appear to have arisen independently (Rising
and Shields 1980; Thomas et al. 2008; Huynh et al. 2011), the junco inversions appear identical to the inversion polymorphisms segregating in the
rufous-collared sparrow (Lucca and Rocha 1985). Whether this similarity
is due to an ancient inversion polymorphism in the common ancestor of
both *Junco* and *Zonotrichia* or has more recently introgressed between the
two is unknown.

I. Conclusions and Future Directions

The northern junco complex contains numerous well-marked forms,
based largely on discrete plumage color differences. We still have very
little understanding of how color differences evolve, and an outstanding question is the role of parallel and divergent selection pressures (the
question is outstanding because, as we have argued, the importance of
gene flow is so different in the two modes). In other species, much work
has been devoted to searching for associations of color with the environment, which would be support for the divergent selection model (e.g.,

Marchetti 1993; Gomez and Théry 2007), but strong correlates have been hard to find (Gomez and Théry 2007; Price et al. 2010). Given that the junco complex across North America is so genetically homogeneous, at least at the nuclear and mitochondrial markers so far studied (Milá et al. 2007; chapter 8, this volume), it should be possible to use modern molecular methods to locate those regions in the genome that differentiate one subspecies from the other and potentially identify the loci underlying the phenotypic differences. This would be one contribution toward an understanding of mechanisms of divergence at these loci, and it would also help ascertain whether color differences are much older than the genetic differences at mitochondrial and studied nuclear loci, as we predict.

The southern taxa (from Mexico and Central America) are genetically differentiated from one another (chapter 8, this volume). We attribute the greater genetic divergence in the south to relatively less influence of climate change on range expansions, plus year round residency (i.e., long-term separation in allopatry). By contrast, further to the north greater climate-enforced range shifts in response to habitat gain and loss, as well as increased levels of seasonal migration, are expected to reduce genetic differentiation. With respect to alloparapatric speciation, isolated populations may have been present both during glacial maxima (figure 9.2) as well as during interglacial periods, such as the present. In any event, frequent range changes are certain to have occurred, regularly bringing populations in contact. In addition, seasonal migration in northern populations leads to the expectation of much gene flow during population divergence. Thus, large scale shifts in range and high levels of gene flow between populations likely account for the lower genetic differentiation between northern populations than between southern ones. One important feature is that within the complex of scarcely genetically differentiated forms, gene flow may be much higher from north to south than the reverse, which would mean that all populations come to resemble the source populations, decreasing overall genetic diversity over that expected if migration is bidirectional.

While color differences between populations may have been selectively retained in the face of gene flow, chromosomal rearrangements may be an important part of the puzzle of how northern junco populations are so strongly differentiated in color despite similarity in genetic markers so far studied. Gene flow between populations experiencing different selective pressures promotes the spread of chromosomal inversions if they prevent recombination between locally adapted alleles (Kirkpatrick and Barton 2006). The inversion on the fifth chromosome segregates at highest frequencies in populations towards the northeast and those with greater seasonal movements but is less frequent to entirely absent in southwestern populations (table 9.1). This suggests a possible link between migration, gene flow, and inversions in driving differentiation at some parts of the genome. If the loci responsible for seasonal migration are associated with an inversion, they provide a powerful mechanism for adaptive fixation of the inversion within a population, and rapid transitions between seasonal migration and year-round residency, repeatedly using the same set of alleles. Analogously, in the three-spine stickleback, three chromosomal inversions have repeatedly been involved in the divergence between marine and freshwater populations (Jones et al. 2012).

More generally, if junco genomes are a mosaic of homogeneous and highly diverged regions associated with inversions—as expected from chromosomal models of speciation—then alleles that have accumulated in the inversion would be retained even in the face of limited gene flow between taxa. If this is the case, then we should be able to detect signatures of divergence with gene flow by comparing divergence within the chromosomal inversions to divergence outside. With myriad combinations of parapatric and allopatric populations, stable hybrid zones, and behavioral as well as morphological divergence between junco populations, there are many avenues for studying how chromosomal rearrangements might be involved in junco diversification. For this reason, we anticipate the junco system continuing to hold its place as a model system in the study of speciation.

Acknowledgments

We thank Jonathan Atwell, Borja Milá, Ellen Ketterson, and John McCormack for their multiple careful readings of this chapter.

References

Atwell, J. W., G. C. Cardoso, D. J. Whittaker, S. Campbell-Nelson, K. W. Robertson, and E.D. Ketterson. 2012. Boldness behavior and stress physiology in a novel urban environment suggest rapid correlated evolutionary adaptation. *Behavioral Ecology* 23:960–69.

Atwell, J. W., G. C. Cardoso, D. J. Whittaker, T. D. Price, and E. D. Ketterson. 2014. Hormonal, behavioral, and life-history traits exhibit correlated shifts in relation to population establishment in a novel environment. *American Naturalist* 184:147–60.

Atwell, J.W., R. J. Rice, and E. D. Ketterson. In review. Rapid loss of migratory behavior associated with recent colonization of an urban environment. Manuscript.

Brown, K. M., L. M. Burk, L. M. Henagan, and M. A. F. Noor. 2004. A test of the chromosomal rearrangement model of speciation in *Drosophila pseudoobscura*. *Evolution* 58:1856–60.

Cardoso, G. C., and J. W. Atwell. 2011. Directional cultural change by modification and replacement of memes. *Evolution* 65:295–300.

Christidis, L. 1990. Aves. In *Animal Cytogenetics*, edited by B. John, 116. Berlin and Stuttgart: Gebrüder Bornträger.

Coyne, J. A., and H. A. Orr. 2004. *Speciation*. Sunderland, MA: Sinauer.

Dingemanse, N. J., C. Both, A. J. van Noordwijk, A. L. Rutten, and P. J. Drent. 2003. Natal dispersal and personalities in great tits (*Parus major*). *Proceedings of the Royal Society B-Biological Sciences* 270:741–47.

Dingemanse, N. J., A. J. N. Kazem, D. Reale, and J. Wright. 2010. Behavioural reaction norms: Animal personality meets individual plasticity. *Trends in Ecology & Evolution* 25:81–89.

Endler, J. A. 1977. *Geographic Variation, Speciation, and Clines*. Princeton: Princeton University Press

Faria, R., and A. Navarro. 2010. Chromosomal speciation revisited: Rearranging theory with pieces of evidence. *Trends in Ecology & Evolution* 25:660–69.

Gomez, D., and M. Théry. 2007. Simultaneous crypsis and conspicuousness in color patterns: Comparative analysis of a neotropical rainforest bird community. *American Naturalist* 169:S42–S61.

Hedrick, P. W. 2000. *Genetics of Populations*. 2nd edition. London: Jones and Bartlett.

Hoekstra, H. E., J. M. Hoekstra, D. Berrigan, S. N. Vignieri, A. Hoang, C. E. Hill, P. Beerli, and J. G. Kingsolver. 2001. Strength and tempo of directional selection in the wild. *Proceedings of the National Academy of Sciences of the United States of America* 98:9157–60.

Hijmans, R. J., and J. van Etten. 2011. Raster: Geographic analysis and modeling with raster data. R package, 1.9–92 (1-May-2012). http://CRAN.R-project.org/package praster.

Huynh, L. Y., D. L. Maney, and J. W. Thomas. 2011. Chromosome-wide linkage disequilibrium caused by an inversion polymorphism in the white-throated sparrow (*Zonotrichia albicollis*). *Heredity* 106:537–46.

Johnson, N. K., and C. Cicero. 2004. New mitochondrial DNA data affirm the importance of Pleistocene speciation in North American birds. *Evolution* 58:1122–30.

Jones, F. C., M. G. Grabherr, Y. F. Chan, P. Russell, E. Mauceli, J. Johnson, R. Swofford, et al. 2012. The genomic basis of adaptive evolution in threespine sticklebacks. *Nature* 484:55–61.

King, M. 1993. *Species Evolution.* Cambridge: Cambridge University Press.

Kirkpatrick, M., and N. H. Barton. 2006. Chromosome inversions, local adaptation, and speciation. *Genetics* 173:419–34.

Kondrashov, A. S. 2003. Accumulation of Dobzhansky–Muller incompatibilities within a spatially structured population. *Evolution* 57:151–53.

Lucca, E. J., and G. T. Rocha. 1985. Chromosomal polymorphism in *Zonotrichia capensis* (Passeriformes, Aves). *Brazil Journal of Genetics* 8:71–78.

Machado, C. A., T. S. Haselkorn, and M. A. Noor. 2007. Evaluation of the genomic extent of effects of fixed inversion differences on intraspecific variation and interspecific gene flow in *Drosophila pseudoobscura* and *D. persimilis. Genetics* 175:1289–306.

Machado, C. A., and J. Hey. 2003. The causes of phylogenetic conflict in a classic *Drosophila* species group. *Proceedings of the Royal Society B-Biological Sciences* 270:1193–202

Maney, D. L. 2008. Endocrine and genomic architecture of life history trade-offs in an avian model of social behavior. *General and Comparative Endocrinology* 157:275–82.

Marchetti, K. 1993. Dark habitats and bright birds illustrate the role of the environment in species divergence. *Nature* 362:149–52.

Martin, P. R., R. Montgomerie, and S. C. Lougheed. 2010. Rapid sympatry explains greater color pattern divergence in high latitude birds. *Evolution* 64:336–47.

Mayr, E. 1963. *Animal Species and Evolution.* Cambridge: Belknap Press of Harvard University Press.

McGaugh, S. E., & Noor, M. A. 2012. Genomic impacts of chromosomal inversions in parapatric *Drosophila* species. *Philosophical Transactions of the Royal Society B-Biological Sciences* 367:422–29.

McGlothlin, J. W., P. G. Parker, V. Nolan, and E. D. Ketterson. 2005. Correlational selection leads to genetic integration of body size and an attractive plumage trait in dark-eyed juncos. *Evolution* 59:658–71.

Milá, B., J. E. McCormack, G. Castaneda, R. K. Wayne, and T. B. Smith. 2007. Recent postglacial range expansion drives the rapid diversification of a songbird lineage in the genus *Junco. Proceedings of the Royal Society B-Biological Sciences* 274:2653–60.

Miller, A. H. 1941. Speciation in the avian genus *Junco. University of California Publications in Zoology* 44:173–434.

Navarro, A., and N. H. Barton. 2003. Accumulating postzygotic isolation genes in parapatry: A new twist on chromosomal speciation. *Evolution* 57: 447–59.

Newman, M. M., P. J. Yeh, and T. D. Price. 2006. Reduced territorial responses in dark-eyed juncos following population establishment in a climatically mild environment. *Animal Behaviour* 71:893–99.

————. 2008. Song variation in a recently founded population of the dark-eyed junco (*Junco hyemalis*). *Ethology* 114:164–73.

Noor, M. A. F., L. A. Grams, L. A. Bertucci, Y. Almendarez, J. Reiland, and K. R. Smith. 2001a. The genetics of reproductive isolation and the potential for gene exchange between *Drosophila pseudoobscura* and *D. persimilis* via backcross hybrid males. *Evolution* 55: 512–21.

Noor, M. A. F., K. L. Grams, L. A. Bertucci, and J. Reiland. 2001b. Chromosomal inversions and the reproductive isolation of species. *Proceedings of the National Academy of Sciences of the United States of America* 98:12084–88.

Nosil, P., and S. M. Flaxman. 2011. Conditions for mutation-order speciation. *Proceedings of the Royal Society B-Biological Sciences* 278:399–407.

Paradis, E., S. R. Baillie, W. J. Sutherland, and R. D. Gregory. 1998. Patterns of natal and breeding dispersal in birds. *Journal of Animal Ecology* 67:518–36.

Peltier, W. R. 2004. Global glacial isostasy and the surface of the ice-age earth: The ice-5G (VM2) model and grace. *Annual Review of Earth and Planetary Sciences* 32:111–49.

Petit, J. R., J. Jouzel, D. Raynaud, N. I. Barkov, J.-M. Barnola, I. Basile, M. Bender, et al. 1999. Climate and atmospheric history of the past 420,000 years from the Vostok ice core, Antarctica. *Nature* 399:429–36.

Phillips, S. J., R. P. Anderson, and R. E. Schapire. 2006. Maximum entropy modeling of species geographic distributions. *Ecological Modeling* 190:231–59.

Poelstra, J. W., N. Vijay, C. M. Bossu, H. Lantz, B. Ryll, I. Mueller, V. Baglione et al. 2014. The genomic landscape underlying phenotypic integrity in the face of gene flow in crows. *Science* 344:1410–14.

Price, T. 2008. *Speciation in Birds*. Boulder, CO: Roberts and Co.

Price, T. D., and M. M. Bouvier. 2002. The evolution of F1 postzygotic incompatibilities in birds. *Evolution* 56:2083–89.

Price, T., A. B. Phillimore, M. Awodey, and R. Hudson. 2010. Ecological and geographical influences on the allopatric phase of island speciation. In *From Field Observations to Mechanisms: A Program in Evolutionary Biology*, edited by P. R. Grant and B. R. Grant, 251–81. Princeton: Princeton University Press.

Price, T. D., P. J. Yeh, and B. Harr. 2008. Phenotypic plasticity and the evolution of a socially selected trait following colonization of a novel environment. *American Naturalist* 172:S49–S62.

Rasner, C. A., P. Yeh, L. S. Eggert, K. E. Hunt, D. S. Woodruff, and T. D. Price. 2004. Genetic and morphological evolution following a founder event in the dark-eyed junco, *Junco hyemalis thurberi*. *Molecular Ecology* 13:671–81.

Ridgely, R. S., T. F. Allnutt, T. Brooks, D. K. McNicol, D. W. Mehlman, B. E. Young, and J. R. Zook. 2005. Digital distribution maps of the birds of the Western Hemisphere, version 2.1. Arlington: NatureServe.

Rieseberg, L. H. 2001. Chromosomal rearrangements and speciation. *Trends in Ecology & Evolution* 16:351–58.

Rising, J. D., and G. F. Shields. 1980. Chromosomal and morphological correlates in two New World sparrows (Emberizidae). *Evolution* 34: 654–62.

Rohwer, S., E. Bermingham, and C. Wood. 2001. Plumage and mitochondrial DNA haplotype variation across a moving hybrid zone. *Evolution* 55:405–22.

Shields, G. F. 1973. Chromosomal polymorphism common to several species of *Junco. Canadian Journal of Genetics and Cytology* 15:461–71.

———. 1976. Meiotic evidence for pericentric inversion polymorphism in *Junco* (Aves). *Canadian Journal of Genetics and Cytology* 18:747–51.

———. 1982. Comparative avian cytogenetics: A review. *Condor* 84:45–58.

Siepielski, A. M., J. D. DiBattista, J. A. Evans, and S. M. Carlson. 2011. Differences in the temporal dynamics of phenotypic selection among fitness components in the wild. *Proceedings of the Royal Society B-Biological Sciences* 278:1572–80.

Stein, A. C., and J. A. C. Uy. 2006. Unidirectional introgression of a secondary sexual character: A role for female choice? *Evolution* 60:1476–85

Stevison, L. S., K. B. Hoehn, and M. A. F. Noor. 2011. Effects of inversions on within- and between-species recombination and divergence. *Genome Biology and Evolution* 3:830–41.

Tan, C. C. 1935. Salivary gland chromosomes in the two races of *Drosophila pseudoobscura. Genetics* 20:392–402.

Thomas, J. W., M. Caceres, J. J. Lowan, C. B. Morehouse, M. E. Short, E. L. Baldwin, D. L. Maney, et al. 2008. The chromosomal polymorphism linked to variation in social behavior in the white-throated sparrow (*Zonotrichia albicollis*) is a complex rearrangement and suppressor of recombination. *Genetics* 179:1455–68.

Thorneycroft, H. B. 1966. Chromosomal polymorphism in the white-throated sparrow *Zonotrichia albicollis. Science* 154: 1571–1572.

———. 1968. A cytogenetic study of the white-throated sparrow, *Zonotrichia albicollis* (Gmelin). PhD thesis, University of Toronto.

———. 1975. A cytogenetic study of the white-throated sparrow *Zonotrichia albicollis. Evolution* 29:611–21.

Turgeon, J., R. Stoks, R. A. Thum, J. M. Brown, and M. A. McPeek. 2005. Simultaneous Quaternary radiations of three damselfly clades across the Holarctic. *American Naturalist* 165:E78–E107.

Tuttle, E. M. 2003. Alternative reproductive strategies in the white-throated sparrow: Behavioral and genetic evidence. *Behavioral Ecology* 14:425–32.

Unitt, P. 2005. *San Diego County Bird Atlas.* San Diego: San Diego Natural History Museum Press.

Weir, J. T., and D. Schluter. 2007. The latitudinal gradient in recent speciation and extinction rates of birds and mammals. *Science* 315:1574–76.

Weir, J. T., and D. Wheatcroft. 2011. A latitudinal gradient in rates of evolution of avian syllable diversity and song length. *Proceedings of the Royal Society B-Biological Sciences* 278:1713–20.

Weir, J. T., D. J. Wheatcroft, and T. D. Price. 2012. The role of ecological constraint in driving the evolution of avian song frequency across a latitudinal gradient. *Evolution* 66:2773–83.

Winker, K. 2010. On the origin of species through heteropatric differentiation: A review and a model of speciation in migratory animals. *Ornithological Monographs* 69:1–30.

Wu, C. I. 2001. The genic view of the process of speciation. *Journal of Evolutionary Biology* 14:851–65.

Yeh, P. J. 2004. Rapid evolution of a sexually selected trait following population establishment in a novel habitat. *Evolution* 58:166–74.

Yeh, P. J., M. E. Hauber, and T. D. Price. 2007. Alternative nesting behaviours following colonisation of a novel environment by a passerine bird. *Oikos* 116: 1473–80.

Yeh, P. J., and T. D. Price. 2004. Adaptive phenotypic plasticity and the successful colonization of a novel environment. *American Naturalist* 164:531–42.

PART IV

Mechanisms of Divergence among Populations

Part 1 of this volume conveyed how juncos differ geographically in appearance, reproductive timing, and migratory behavior, and how these differences caught the attention of biologists over a century ago.

Part 2 conveyed how intensive study of a single population revealed variation in behavior, physiology, and morphology that appears to be coordinated via hormonal mechanisms and acted upon by natural selection.

Part 3 addressed how rates of evolution in a polytypic species can be addressed by measuring genetic divergence and the potential role of dynamic changes in geographic ranges in creating and eroding population structure. Part 3 posits that differential permeability of genes for some traits versus others across a hybrid zone can help to explain how closely related forms can look so different yet still interbreed successfully.

The knowledge contained in parts 2 and 3 might remain separate in the absence of active intervention to promote synthesis. Part 4 consists of two chapters that address whether the knowledge gained from the Carolina junco in Virginia can be used to predict phenotypes in other populations and thus provide insight into how hormonal mechanisms evolve and their potential role in enabling occupancy of an extremely broad range of habitats.

The colors, the sounds, the displays that birds produce are always intriguing and are explored in two additional chapters in part 4. Whether these products of sexual selection—differences among individuals in mating success—lead to or follow from reproductive isolation and whether mate choice drives speciation are controversial topics requiring more investigation. These investigations will be able to build on what has already been learned.

As with the other sections, externalities also helped to motivate the research described in part 4. The National Science Foundation's expectation that research be transformational as opposed to incremental encourages scientists to broaden their approaches and collaborate with new colleagues. Personal goals contribute as well. The world is a big place and a desire to see more of it, when combined with enhanced communication and global exchange of scientists, made it feasible to conduct research simultaneously in the Appalachians, the Black Hills of South Dakota, urban and montane habitats in southern California, and even in spots as remote as Guadalupe Island. The pace of discovery promises to be high as these ideas and opportunities continue to expand.

CHAPTER TEN

Shifts in Hormonal, Morphological, and Behavioral Traits in a Novel Environment

Comparing Recently Diverged Junco Populations

Jonathan W. Atwell, Danielle J. Whittaker, Trevor D. Price, and Ellen D. Ketterson

I did not claim that every founder population undergoes a drastic change. All I claimed was that when a drastic change occurs, it occurs in a relatively small and isolated population.
—Ernst Mayr, 1988, *Toward a New Philosophy of Biology: Observations of an Evolutionist*

A. Introduction

One unifying theme of studies in organismal biology, ecology, and evolution is understanding how individual organisms or populations respond to changing environmental conditions. Studies of responses to environmental perturbation, including observations of responses to "natural" environmental phenomena as well as experiments designed to simulate average or extreme values of an environmental state, are evidenced throughout the interdisciplinary research on juncos summarized in this volume. For the behavioral ecologist, changing environments might include alterations in social environment or learning conditions (e.g., chapters 12 and 13, this volume); for the endocrinologist or physiologist, seasonal changes in environmental variables such as day length initiate the hormonal cascades of interest (e.g., chapters 4–7, this volume); and for evolutionary biologists interested in diversification and speciation, glacial cycles and vegetation changes driven by long-term climatic patterns are the focus (e.g., chapters 8 and 9, this volume). Historically, most studies incorporating environmental change have focused on natural processes occurring across various scales of time (e.g., immediate, circadian, seasonal, annual, or geologic) and space (e.g., local, regional, global).

In recent decades, however, there has been an increasing emphasis on both the societal importance and the scientific potential of studying the causes and consequences of human-induced environmental change. The past two centuries—deemed a new geologic era, the "anthropocene," by some scientists (Crutzen 2002)—are typified by rapid rates of habitat alteration and fragmentation, spread of exotic species, pollution, harvesting, and global climate change (Price 2012; Sih et al. 2011). Studies characterizing organismal, ecological, and evolutionary changes in response to these human activities provide not only knowledge critical to address issues of biodiversity, sustainability, and human and animal health but also natural experiments that allow biologists to study phenotypically plastic and evolutionary changes in "real time." Accordingly, the biological literature is exploding with articles, and even entirely new journals, devoted to characterizing the causes, consequences, and future of biological responses to recent and rapid environmental change.

Contemporary colonization events, including biological invasions and range expansions, provide particularly informative scenarios in which populations enter novel environments, often becoming partially or completely isolated from gene flow from the ancestral (source) populations. Different biotic and abiotic environments, small population sizes, and limited gene flow all increase the likelihood for evolutionary changes—via selection or drift or both—to occur. Numerous classic examples of insights gained from studying recent divergence among colonizing populations include historic and contemporary studies of Darwin's finches isolated throughout the Galapagos Islands (Grant and Grant 2014), studies of invasive plant species spreading onto new continents (Cox 2004), lizards introduced and left behind by scientists and forgotten for three decades in Croatia (Herrel et al. 2008), house finches expanding from New England across North America (Adelman et al. 2013; Bonneaud et al. 2011; Williams et al. 2014), house sparrows radiating outwards from contemporary Mumbasa, Kenya (Liebl and Martin 2012; Martin et al. 2014), and European blackbirds expanding their ranges into urban centers in Europe (Evans et al. 2012; Miranda et al. 2013; Partecke et al. 2006; Partecke et al. 2004), among others. Close examination of such systems—including both proximate and ultimate mechanisms underlying phenotypic change—allows for insights into the organismal and evolutionary dynamics of adaptation in populations with relatively well-known origins and natural histories.

In this chapter, we provide an overview of a contemporary *Junco* colonization event in southern California, in which a typically montane-breeding junco subspecies *(Junco hyemalis thurberi)* recently (in approximately1983) established a small, isolated population in an urban and coastal environment on the campus of the University of California, San Diego (UCSD). Comparisons of the colonist UCSD junco population with nearby ancestral-range populations have yielded many insights into behavioral, ecological, and evolutionary processes in the context of a climatically mild and urban environment, when contrasted with the ancestral-range population's temperate and rural surroundings. We compile and discuss in one place the documented differences between colonist versus ancestral-range junco populations, and, using data from field and captive "common garden" studies, consider their plastic versus genetic basis.

An ongoing theme of this volume is that organisms are not random assemblages of traits and that because they are not, efforts to understand evolutionary and developmental responses to environmental change must consider the underpinnings and implications of relationships among multiple traits (i.e., trait correlations; chapters 4–7, this volume). Further, because hormonal mechanisms integrate environmental information and can have pleiotropic effects on multiple behavioral, morphological, and physiological traits, and thus modulate life-history trade-offs, they are uniquely situated to play an important role in constraining or facilitating multitrait evolutionary responses to novel and changing environments (Adkins-Regan 2008; Atwell et al. 2014; Hau 2007; Ketterson et al. 2009). We will discuss, in particular, the causes and consequences of divergence in hormonal mechanisms and hormone-mediated traits.

We conclude by exploring areas for future research, including the scope for expanding the study system to include additional pairs of colonist and ancestral-range populations, as in recent years juncos have been colonizing other urban areas as well as offshore islands in southern California.

B. The Natural History of a Unique Contemporary Junco Colonization Event

B.1. The Historic Range and Disposition of Junco hyemalis thurberi *in Southern California*

As detailed in Miller's monograph on the junco (1941), the breeding range of dark-eyed juncos (*Junco hyemalis*) in North America is typically restricted to high latitudes or altitudes, where mixed forest and woodland habitats and seasonal, temperate climates predominate (see also chapter 3). This generalization holds true for *Junco hyemalis thurberi*, the most widespread and abundant "race" or "subspecies" of the Oregon junco group (see chapters 2, 3, and 8, this volume, for discussions of junco diversity and taxonomy), whose breeding range Miller described as follows: "... [consisting of] two tongues of the Transition and Boreal zones [that] extend southward in California. The larger of these is that of the southern Cascade-Sierra Nevada mountain system, which is extended discontinuously by the mountaintop Boreal "islands" of southern California" (see Miller 1941, 276; and figure 2 in Yeh 2004).

Miller separated *thurberi* from adjacent Oregon junco subgroups including *pinosus* (coastal residents in central California), *shufeldti* (coastal Washington and Oregon), *montanus* (interior British Columbia, Washington, and Oregon), and *pontilis* (Sierra Jaurez, northern Baja California), based on plumage coloration and morphometric traits, in particular: "... a mixture of pigments in the back [feathers] that is found widely and almost exclusively in *thurberi* ... and may be employed arbitrarily as a key character to delimit the races ... The color is Verona, snuff, Sayal, or mikado brown ..." a difference which "... permits 90-per cent segregation of individuals in the interior ranges, and 75-per cent along the coast" (Miller 1941, 276).

San Diego County represents the southern end of the *thurberi* breeding range (see Yeh 2004, figure 2), and at this lower latitude, junco breeding has historically been limited to higher elevation coniferous forests and oak-alder woodlands above 1,500 meters in elevation, occurring either on mountaintops or in shaded canyons at least fifty to seventy kilometers inland from the Pacific coast (figure 10.1; Unitt 2005). The rest of San Diego County consists of desert, chaparral, and coastal sage scrub—habitats unsuitable for junco breeding (figure 10.1; Miller 1941).

Juncos exhibit striking variation in migratory behaviors both among populations and among sex and age cohorts within populations (chapters

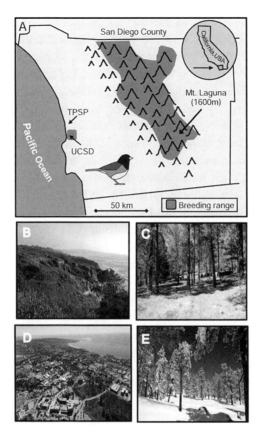

FIGURE 10.1. The breeding range of dark-eyed juncos (*Junco hyemalis thurberi*) in San Diego County, CA, USA, is historically restricted to higher elevation (>1500m) forests and wooded canyons in the central and eastern parts of the county (panel A, adapted from Unitt 2005), for example near Mt. Laguna (plate C). Montane-breeding juncos are facultative altitudinal migrants, with most birds joining flocks and many migrating variable distances to lower elevations during winter, particularly during harsh climatic conditions including winter storms (plate E). In the early 1980s, a small isolated breeding population became established along the coast on the urbanized campus of the University of California, San Diego (UCSD, plate D), an area historically comprised of sage scrub and chaparral not suitable for junco breeding, as currently exhibited at the adjacent Torrey Pines State Park (TPSP, plate B). See also color plate 7.

2 and 3, this volume; Nolan et al. 2002). The *thurberi* group is no exception to this pattern, with a spectrum of migratory variation exhibited throughout its range. Some *thurberi* juncos stay on or near their territories year-round, while others migrate considerable distances (Miller 1941, 290). In San Diego County, *thurberi* juncos' migratory strategy might best be de-

scribed as partial, facultative, differential (with respect to sex and age), and altitudinal (Chapman et al. 2011; Peterson et al. 2013). Even at this lower latitude, cold and snowy conditions are common on mountaintop habitats, favoring seasonal migration to lower latitudes, including coastal areas, but some individuals, particularly males, may remain, and the frequency and duration of migratory behavior may vary with local climatic conditions (J. Atwell, pers. obs.; Yeh 2004; Yeh and Price 2004). Accordingly, flocks of Oregon juncos are regularly seen throughout coastal southern California in winter, including park-like areas in suburban San Diego, with juncos vacating coastal areas in the spring (March and April), returning to their breeding grounds at higher elevations (or higher latitudes) (Unitt 2005).

No *breeding* records for dark-eyed juncos in coastal San Diego County were reported before 1983 (P. Unitt, pers. comm.; Unitt 2005). However, in years prior to and since the juncos' colonization of UCSD, small wintering flocks have been observed lingering late into the spring (e.g., into April), with males beginning to sing as if preparing to breed, for example at Fort Rosecrans National Cemetery on Point Loma (P. Unitt, pers. comm. and T. Price, pers. obs.). However, these lingering juncos have ultimately departed (Yeh 2004; P. Unitt, pers. comm., and T. Price, pers. obs.). Most recently, reports of multiple junco pairs breeding in Balboa Park confirm their affinity for San Diego County's managed park landscapes (R. Hanauer, pers. comm.).

B.2. Establishment of Juncos at UCSD in Approximately 1983

The "discovery" of juncos at UCSD was first reported by an unlikely duo: then-UCSD anthropology professor Stan Walens and his eight-year-old daughter, Rachel. Walens was an avid birdwatcher and amateur ornithologist who frequently spent his lunch hours looking for interesting birds throughout the campus. One day in July 1983, he was joined by Rachel who exclaimed, "Look, Daddy, there's a junco!" to which Walens replied, "Rachel, it's July and juncos are here in the winter." But sure enough, upon closer inspection, Walens observed a breeding pair of juncos.[1]

Although the precise timing of establishment and the composition and

---

1. Quote from S. Walens interview in the film *Ordinary Extraordinary Junco* (2013), chapter 6, "Evolution-in-Action: The Campus Juncos of UCSD." This film chapter provides a summary of the UCSD junco study system and highlights key findings to date, pitched for public and student audiences. Free online: www.juncoproject.org.

size of the colonizing flock are impossible to know, in the years following this initial observation, it became clear to San Diego's ornithological community that a small, stable population of breeding juncos had found a foothold on the UCSD campus (Unitt 2005). Although Walens's first observation of breeding juncos at UCSD in 1983 included only a single pair (S. Walens, pers. comm.), eventual genetic analyses of microsatellite loci revealed a level of allelic diversity that indicated that at minimum a small flock of juncos were part of the founding population, or that it grew via serial recruitment of multiple (genetically) unrelated individuals or pairs (Rasner et al. 2004, see section C1, below).

The first officially published record of the breeding population at UCSD was in 1986 (McCaskie 1986). Unfortunately, however, the campus juncos were not intensively studied until the mid-1990s, when one of the coauthors (Trevor Price) and his students realized their research potential.

B.3. The Origins of the UCSD Colonists?

The taxonomic identity of the colonist juncos at UCSD as *J. h. thurberi* has been evaluated by experts, including San Diego Natural History Museum Curator of Birds and Mammals Dr. Phil Unitt, based on the characteristic plumage coloration and morphology that distinguishes *thurberi* from other forms, including the sedentary *J. h. pinosus* of the central California coast (see above). This means that the colonist juncos at UCSD almost certainly originated from montane-breeding *thurberi* populations, arriving at UCSD as wintering migrants from mountain populations to the east, the northeast, or possibly farther north (i.e., as opposed to coastal *J. h. pinosus* dispersing southwards; see fig. 2 in Yeh 2004). Based on the prior observations of (1) regular wintering flocks at UCSD, (2) lingering and singing flocks in the springtime in this region (e.g., at Pt. Loma, see above), and (3) the characteristic flocking of juncos during migration and wintering (Nolan et al. 2002), the most likely establishment scenario involves a wintering flock remaining to breed.

As Price describes it, first in jest and then in earnest:

It's pretty obvious why juncos decided to stay here. They came down here from the mountains, deeply covered in snow, they took one look at me drinking a cappuccino, and said, "This is a good deal, I'm going to stay here . . ."

. . . But what probably actually happened was that the males were singing getting ready to leave, and sufficient males were singing to stimulate the females enough to start building nests, somehow generating a critical mass action that took off.[2]

Although a lingering (wintering) flock remaining to breed may be the most likely scenario, other alternatives cannot be ruled out, for example that early spring snowstorms in the nearby mountains drove a flock *back down* to a milder, lower elevation site like the UCSD campus where they subsequently began nesting. In either scenario, arguments could be made whether hatch year (i.e., dispersing juveniles) versus after-hatch year (i.e., older adults that had previously bred elsewhere) would be more likely members of a founding transient flock, as songbirds are known to exhibit philopatry to natal sites, breeding sites, and wintering sites—all of which could play a role in facilitating or constraining establishment of a novel breeding population (Berthold 1995; Berthold et al. 2001).

Similarly, although it cannot be explicitly ruled out that *thurberi* juncos from farther afield to the north (e.g., from the Sierra Nevada range or the southern Cascades) were among the colonist flock, the most southerly *thurberi* breeders would be expected to exhibit (1) reduced and/or more flexible migratory propensities, (2) extended and/or more flexible breeding phenology, and (3) relatively local natal origins—all of which would make the "hurdle" of natal or breeding philopatry (i.e., affinity to return to breeding sites elsewhere) more easily "overcome" by the initial colonists if they were hatched or had nested previously in the most southerly parts of the range. The departure of wintering flocks from coastal San Diego County (late March and early April) also corresponds generally with the onset of breeding (late April, early May) in the nearby southern California mountains (Yeh 2004; J. Atwell, pers. obs.). Additionally, morphological characteristics of the regular wintering visitors to campus and Pt. Loma suggest a more southerly *thurberi* origin of the UCSD colonists (Yeh 2004): plumage coloration of winter visitors to UCSD and other coastal areas in San Diego County most closely resembles those *thurberi* that breed on the southern California mountaintops, as detailed below.

It was Miller (1941) who first noticed and quantified variation in head (hood) black coloration and the amount of white in the tail feathers among *thurberi* populations across their range, and he documented a latitudinal gradient wherein southerly breeders exhibit the most white

2. Quote from T. Price interview in the *Ordinary Extraordinary Junco* film (2013), chapter 6, "Evolution-in-Action: The Campus Juncos of UCSD."

in the tail feathers and the most black in the head feathers (see figure 17, p. 279; figure 19, p. 283). As part of her studies of the evolution of the UCSD colonists (see below), Yeh (2004) sampled *thurberi* tail white across the subspecies range and confirmed Miller's latitudinal gradient, also demonstrating that average male tail white in the UCSD population (approximately 36 percent white) is a marked outlier from all California *thurberi*—significantly lower than even the northernmost *thurberi* populations (approximately 40 percent white), suggesting the shift in tail white has occurred post-establishment (see table 1 and figure 3 in Yeh 2004). Also, birds sampled from wintering flocks at UCSD and Pt. Loma (i.e., arguably flocks representing those similar to the most likely colonists) exhibit large amounts of tail white (e.g., males at approximately 44 percent), similar to that observed in the southern-breeding males (see table 1 and figure 3 in Yeh 2004; Yeh and Price 2004; J. Atwell, unpublished data)—consistent with a southern origin for the colonists, followed by a subsequent loss in tail white via selection.

Nevertheless, variance in plumage and morphological traits within junco populations is substantial, and overlap in morphological trait distributions among junco populations is not uncommon (Miller 1941). Therefore, despite inferences based on phenology, geography, and plumage, it is important to acknowledge that the specific geographic origins or phenotypic composition of the UCSD colonists remain undetermined. Attempts to use genetic markers (microsatellites and AFLP restriction enzymes) have failed because populations across the *thurberi* range are genetically so similar (Rasner et al. 2004). Emerging population genomic tools provide greater power, and it may be possible to identify unique alleles that diagnose different populations in the future

B.4. Colonists Isolated on a Biogeographic Island

Beginning in 1998, Price, PhD student Pamela Yeh, and two master's students (Caylor Rasner and Melissa Newman) began monitoring the UCSD juncos intensively. For six years (1998–2003), Yeh, assisted by a small army of undergraduate assistants, surveyed the entire breeding population, mapped territories, and counted and banded most breeding pairs and their offspring (Yeh and Price 2004). This systematic, sustained effort allowed for a detailed understanding of population size, stability, and general demographic parameters as well as the biogeographic extent of the colonist population. Careful consideration of the study system also revealed a range of striking differences between the environments inhab-

ited by the UCSD colonists when compared to those *thurberi* breeding in
the nearby ancestral-range montane forests (summarized in section B.5
and table 10.1 below). These differences would form the basis for subsequent research on morphological, behavioral, and physiological responses
to the novel environment.

Yeh and Price (2004) estimated the population size to be relatively
small, consisting of approximately seventy to eighty-five breeding pairs
(approximately 140–170 individuals; see table 1 in Yeh and Price 2004),
with the population confined almost exclusively to portions of the UCSD
and Veteran's Administration Medical Center–San Diego campuses, cov-
ering an area of approximately 2.5km^2 and situated on the coastal bluff
adjacent to the Pacific Ocean (figure 10.1). This area includes a few campus apartment neighborhoods east of Interstate 5, plus a few pairs/terri-
tories extending into the University Town Center (UTC) shopping area
to the south, as well as the residential neighborhoods just west of UCSD
across La Jolla Shores Drive, with infrequent pairs nesting in residential
Del Mar to the north. However, the consistent, core population range has
been limited primarily to UCSD campus property and adjacent neighborhoods. The lack of geographic expansion of the UCSD population during
the approximately 3 decades since its establishment—perhaps a function
of limited suitable habitat—has been hard to explain, as Yeh and Price
(2004) observed no evidence for density-dependent recruitment.

During the six seasons (1998–2003) of intensive monitoring, the UCSD
population size and geographic extent remained remarkably stable (Yeh
and Price 2004). An additional two years of intensive monitoring (2006–
2008) by Atwell and Cardoso, as well as periodic fieldwork by several different researchers from 2009 to present, has confirmed the general population size and geographic scope at UCSD to be consistent since at least
1998 (J. Atwell, unpublished data). Years of particularly extensive monitoring, in which most pairs and their offspring at UCSD were banded
(e.g., 2000–2002), allowed for calculations of possible immigration rates
(i.e., recruitment of migrants or dispersers) into the local breeding population, which was found to be 10 percent or less (see appendix in Yeh and
Price 2004). Although our initial perception was that UCSD juncos might
breed at a higher density than in the native range, an analysis of nearest
neighbor nest distances revealed no significant difference in population
density when compared to occupied areas in nearby ancestral-range forest habitats at Mt. Laguna (Atwell et al. 2014).

Given its geographically isolated nature, the UCSD population effectively exists as an island during the breeding season, isolated from the other native southern California breeding ranges by a sea of unsuitable breeding habitat (fig. 10.1). However, considering the influx of migrant *Junco hyemalis* into coastal areas during the winter—predominately *thurberi* (though other less frequent races of Oregon junco and even the occasional slate-colored junco [*J. hyemalis hyemalis*] have been observed on campus in winter, P. Yeh and J. Atwell, pers. obs.)—the overall distribution of junco populations in southern California in relation to UCSD can also be described as "heteropatric," a situation in which populations exist in both allopatry and sympatry during different parts of the annual cycle (chapters 2 and 3, this volume; Winker 2010). This framework may be especially appropriate given probable ongoing immigration (i.e., "recruitment") into the UCSD population (Yeh and Price 2004; see also chapter 9, section C, this volume, for an additional discussion of "divergence with gene flow" in this study system).

Occasional nesting junco pairs have also been formally or informally reported throughout other parts of suburban San Diego, for example at Balboa Park, the San Diego State University campus, the suburban UTC neighborhoods adjacent to UCSD, and residential and commerical areas of Del Mar. However, these populations seem to be smaller and more ephemeral, with "microcolonizations" and "microextinctions" of these localized establishments appearing and disappearing with some regularity (P. Unitt, pers. comm.; S. Walens, pers. comm.; also citizen reports to J. Atwell and E. Ketterson). Systematic and longitudinal surveys conducted at a fine geographic scale would be needed to confirm the size, stability, and dynamics of these additional suburban San Diego junco establishments as well as their potential demographic relationship to the UCSD population.

At a wider geographic scale, juncos seem to be expanding their range throughout southern California, including into additional urban and coastal areas, for example onto the campus of the University of California, Los Angeles (UCLA) and the Channel Islands (see section E below), thus increasing the probability that metapopulation dynamics could begin to be a factor in structuring the demography of coastal urban breeding populations of *J. h. thurberi* like the one at UCSD.

B.5. A Novel, Climatically Mild, and Urban Environment

Arising from species-typical montane forest-breeders, the *J. h. thurberi* juncos at UCSD breed in a habitat that is markedly different from the

TABLE 10.1 **Environmental and ecological differences between UCSD (colonist) and Mt. Laguna (ancestral range) habitats**

Mt. Laguna (ancestral range)	Environment	UCSD (colonist)
32°52'N, 116°25'W	Location (Coordinates)	32°52'N, 117°10'W
Semicontiguous, sky islands	Biogeographic context	Small, isolated population (~2.5km^2)
1700m	Elevation	30m
Seasonal, cold winter, hot summers	Temperature	Mild, rarely freezes (winter), mild in summer
Similar annual, but incl. snow (winter)	Precipitation	Similar annual, but addl. fog and irrigation
Mixed woodland, oak-savannah	Ecotones	(Sub)urban, ornamental gardens, lawns
Conifers, oaks, hardwoods, grasses	Vegetation	Eucalyptus (groves), gardens, lawns
Grasses, banks, road/trail edges	Substrates (nest/forage)	Ivy, mulch, leaf litter (eucalyptus), planters
Sparse, mostly dirt	Roads / trails / paths	Dense, paved, incl. parking and walkways
Infrequent, limited recreational use	Human activity	Extensive, frequent, university campus
Limited, infrequent vehicles, aircraft	Anthropogenic noise	Ongoing, vehicle traffic, machinery, aircraft
Very limited, low traffic highway	Anthropogenic light	Extensive, street & building lights, vehicles
Squirrels, chipmunks, crows, snakes	Observed predators	Crows, ants[1], rats, humans (incidental)
Infrequent fire / fuel management	"Disturbance regimes"	Frequent, landscaping, mowing, construction

[1]The discovery of invasive Argentine ants predating junco nests at UCSD is described in Suarez et al. 2005.

whole of the rest of its range with respect to several key environmental variables (table 10.1), which reflect (1) climate and (2) urbanization. With respect to climate, coastal San Diego is known for its mild weather, with the adjacent Pacific Ocean and its coastal fog moderating daily highs and lows. In contrast, at the nearest junco breeding habitat approximately seventy kilometers to the east near Mt. Laguna, winter temperatures regularly dip below freezing, with periodic storms bringing (occasionally heavy) snow and ice through the winter and into early spring. In summer, while temperatures at UCSD rarely rise above 25 degrees Celsius and coastal fog and irrigation provide sources of water, the high elevation (approximately 1,500–1,800 meter) montane forests at Mt. Laguna become very hot and dry, with temperatures regularly exceeding 30 degrees Celsius. Wildfires (and controlled burns) are common in southern California woodlands due to the hot and dry conditions combined with abundant fuel in the form of trees and grasses. Juncos are known to be among the first birds to return to breed in postburn forests (Nolan et al. 2002), although our study site at Mt. Laguna has not burned since at least 1998. The stark differences in temperature and moisture availability between UCSD and Mt. Laguna provide the ecological context for several of the observed biological differences between colonist and ancestral range populations. As such, the system provides a model for temperate versus tropical comparisons, as well as for examining generally how changing climate leads to organismal and evolutionary responses.

In the early 1900s, thousands of Eucalyptus trees (*Eucalyptus cladocalyx*) were planted across the area that is now UCSD (Henter 2005). These groves of tall trees, alongside numerous multistory campus buildings and interspersed with ornamental plantings, gardens, and lawns, collectively provide the characteristic shade, perches, and nesting and foraging substrates that juncos use in their typical breeding habitats. Further, the heavily irrigated, manicured, and managed vegetation at UCSD provides a year-round food supply in the form of both seeds as well as the insects attracted by the lush plantings and athletic fields—in contrast to the highly seasonal pulse of plant growth, flowering, and insect abundances observed in the southern California mountains.

In addition to marked differences in climate and vegetation between UCSD and the ancestral-range habitat, UCSD is a highly "urbanized" environment in which parking lots, streets, concrete walkways, and buildings occupy a large proportion of the campus, and human and vehicle traffic are ever present. Juncos are not typically considered among the "synanthropic" breeding bird species that thrive in urban habitats (Johnston 2001), and until recently, reports of juncos breeding in such urbanized habitats were rare. Adapting to the urban habitat at UCSD has almost certainly provided both challenges and opportunities for the colonists: juncos have been observed foraging on crumbs from cafeterias and nesting in potted plants, but they are also exposed to anthropogenic noise and light that may disrupt communication or alter physiology (Gaston et al. 2014; Slabbekoorn and Peet 2003) and novel predators or pesticides that may pose altered mortality risks to adults or offspring. The San Diego junco system thus also provides an opportunity to study the myriad ways in which urbanization affects the biology of organisms and populations.

C. A Decade of Insights from the UCSD Juncos

Beginning with Price, Yeh, and colleagues, and continuing to the present day, studies of the UCSD population in relation to populations from the nearby native range have provided insights into a number of subfields in biology, including physiological and behavioral ecology, speciation, and evolutionary genetics. To date, the UCSD study system has been the central feature of more than twenty published articles. Research by members of the Ketterson Lab at Indiana University is ongoing.

A primary goal of these studies has been to document phenotypic differences between native and UCSD juncos and to explore the organismal, evolutionary, and ecological basis of such responses. Documented differences are summarized in table 10.2, categorized by the types of traits (10.2A to 10.2G), and whether the observed differences appear to have a plastic or genetic basis, as assessed by common garden studies of captive birds. Table 10.2 also includes a summary of observed molecular genetic differences between populations (10.2H).

C.1. Phenotypic Divergence: Plasticity, Drift, Selection, or "Immigrant Selection"?

Several factors, including developmental plasticity, random genetic drift including a founder effect, and natural and social selection, could all lead to phenotypic divergence in a novel environment. Such processes are not mutually exclusive. For example, plasticity may facilitate initial population persistence and adaptation to a novel selective landscape for a given set of traits, with genetic evolution following via genetic assimilation (Crispo 2007; Price et al. 2003). Similarly, while the frequencies of genotypes or phenotypes in the founding colonists might be random with respect to most traits, it is possible that certain behavioral or physiological phenotypes were "probabilistically predisposed" to be exhibited by founders or persisting individuals in the novel environment, a phenomenon that can be described as "immigrant selection" (Brown and Lomolino 1998). Experimental, observational, and analytical approaches, summarized below, have been employed to evaluate the relative importance of these evolutionary mechanisms in giving rise to the phenotypic and genetic divergences observed in the UCSD population.

A major question is whether phenotypic differences have a plastic or

TABLE 10.2 **Summary of phenotypic differences by trait category (A–G) and molecular genetic differences (H), as documented to date between the colonist (UCSD; UC) versus ancestral-range (Mt. Laguna; ML) junco populations**

	Trait(s)	Difference	Sexes	Plastic/Genetic	Reference(s)
A. Life History	Breeding phenology	UC: Feb–Aug (6 mos.) ML: May–Jul (3 mos.)	Pairs	Plastic	Yeh and Price 2004; Atwell et al. 2014
	Migratory strategy	UC: Sedentary ML: Altitudinal, Partial	M & F	Genetic / Early development	Atwell et al., in review
B. Morphology	Wing length and tail length	UC: Shorter	M & F	Genetic / Early development	Rasner et al. 2004
	Head black plumage (socially selected)	UC: reduced homogeneous black/grey	M & F	Genetic / Early development	Atwell et al. 2014
	Tail white plumage (socially selected)	UC: reduced prop. of white in tail feathers	M & F	Genetic / Early development	Yeh 2004; Atwell et al. 2014
C. Social Behaviors	Territorial aggression (song playbacks)	UC: reduced singing & flyovers response	only M tested	Unknown	Newman et al. 2006
	Parental care (focal watches & videos)	UC: males feed more frequently	M only; F, ns	Unknown	Atwell et al. 2014
	Extrapair mating (incidence / frequency)	UC: fewer extrapair offspring / nests	Pairs	Unknown	Atwell et al. 2014
D. Vocal Behavior	Minimum song frequency (pitch)	UC: higher min. frequency	Only M sing	Plastic *and* genetic	Slabbekoorn. et al. 2004; Cardoso and Atwell 2011; Atwell et al., in prep;
	Frequency of song mistakes / deviations	Mistakes more frequent at UC	Only M sing	Unknown	Ferreira et al., in review
E. 'Boldness & Novelty' Behaviors	Exploratory boldness (aviary common garden)	UC: faster and more extensive exploration	M & F	Genetic / Early development	Atwell et al. 2012
	Flight-initiation distance (FID)	UC: shorter FID	M & F	Unknown	Atwell et al. 2012
	Food resource use	UC: more diverse, adults & nestlings	M & F	Unknown	Atwell, unpublished data
	Nesting microhabitat	UC: novel off-ground nesting common	Pairs	Unknown	Yeh et al. 2007

continued

TABLE 10.2 **continued**

	Trait(s)	Difference	Sexes	Plastic/Genetic	Reference(s)
F. Hormones & Physiology	Testosterone ("initial" & GnRH-induced)	UC: "Longer" profile but lower peak	Only M tested	Plastic	Atwell et al. 2014
	Corticosterone	UC: attenuated CORT response	M & F	Genetic / Early development	Atwell et al. 2012
	Hematocrit	UC: lower	M & F	Genetic / Early development	Atwell, unpublished data
	Seasonal fat deposition (subcutaneous)	UC: reduced winter / migratory fattening	M & F	Genetic / Early development	Atwell et al., in review
G. Disease & Immunity	Haemosporidian infection (e.g., Plasmodium spp.)	UC: reduced incidence & diversity	M & F	Unknown	Sheets et al., in review
	Cociddia infection	> Incidence at UCSD vs. Mt. Laguna	Only M tested	Unknown	Hanauer et al., in prep.
	Pox virus lesions	UC: ~3% incidence ML: not detected	M & F	Unknown	Hanauer et al., in prep.
	Feather mites	UC: Fewer birds infested vs. ML	Only M tested	Unknown	Haunauer et al., in prep.
	Feather mite incidence (*Acarinus* spp.)	UC: greater % of birds infested (tail feathers)	M & F	Unknown	Hanauer et al., in prep.
	White blood cell counts (WBCs)	UC: Higher counts	Only M tested	Unknown	Hanauer et al., in prep.
H. Genes	"Neutral" microsatellite loci	UC: less variable, divergent mtns.	Pop.level	Genetic	Rasner et al. 2004 Whittaker et al. 2012
	MHC class II alleles	UC: divergent from ML;	Pop. level	Genetic	Whittaker et al. 2012

a genetic basis. One time-consuming way to address this question is to ask whether juncos from different locations that are taken from the wild in early life and housed under identical aviary conditions in a "common garden" retain the differences seen in the field. Yeh (2004) and Rasner et al. (2004) conducted common garden studies focusing on morphological differences based on chicks taken from the nest at an early age and hand reared, and Atwell, Ketterson, and colleagues (2012 and 2014) used a similar approach to compare behavior and physiology based on capture of young fledglings. In brief, the evidence from the common garden studies suggests that some traits have a genetic or very early developmental basis, as evidenced by persistent population differences across multiple seasons in captivity, while other traits have converged in the common garden, providing evidence for plasticity (Atwell et al. 2012; Atwell et al. 2014; Rasner et al. 2004; Yeh 2004).

Another approach is to consider the degree to which the UCSD population might exhibit molecular genetic divergence; for example, drift and founder effects imply very small effective population sizes. Analyses of five polymorphic microsatellite loci among *J .h. thurberi* populations have revealed two key insights with respect to the evolutionary history of the UCSD population. First, the UCSD junco population is moderately differentiated from all other *thurberi* sampled throughout the breeding range (n = 12 populations, ranging from southern Oregon southwards to Mt. Laguna), while virtually no genetic structure was detected when the other sampled *thurberi* populations were compared to one another (Rasner et al. 2004). Second, genetic diversity is lower at UCSD than in ancestral populations (Rasner et al. 2004), but the difference is small, suggesting that at least a small flock (e.g., n = 8–20 birds) gave rise to the UCSD population (Rasner et al. 2004; Yeh 2004). A founding population as large as this makes it unlikely that all of the observed phenotypic differences between UCSD and other populations are attributable to a founder effect or subsequent genetic drift (Yeh 2004; Rasner et al. 2004). This conclusion is bolstered by quantitative genetic theory (Lande 1976), which employs known or estimated measures of trait heritability and time (i.e., generations) since population establishment to calculate the *minimum* effective population size that is needed for a trait to have had a reasonable probability of diverging by random genetic drift (Yeh 2004; and see below).

In sum, the multiple approaches (microsatellite genetic data on population structure, common garden experiments, comparative genetic diversity, and quantitative genetic theory) suggest that while founder effects could contribute to some of the genetic and phenotypic divergence observed at UCSD, they are unlikely to explain most of the colonists' phenotypic responses summarized below.

Both phenotypic plasticity and differential selection pressures likely contribute to the divergence of many traits. One interesting possible form of selection is "immigrant selection" (see above; Brown and Lomolino 1998), which can be considered adaptive evolution insofar as it both entails a change in gene frequencies and invokes adaption via differential suitability for a novel selective landscape. Although it is challenging to distinguish between selection postcolonization versus immigrant selection scenarios, the latter might be especially likely to explain divergence for particular traits in the UCSD junco population, such as greater behavioral boldness or neophilia, or less migratory (i.e., more sedentary) dispositions, traits that might have predisposed nonrandom subsets of individuals to establish or persist in the novel urban and climatically mild environment at UCSD. In a similar scenario, Price et al. (2008) argue that differences in dispersal tendency between different phenotypes (e.g., tail white) result in evolution of these phenotypes simply because the more dispersive types leave UCSD and never return.

C.2. Extension of the Breeding Season and Loss of Migration

One of the most striking behavioral shifts observed in the UCSD junco population is the near doubling of both the breeding season duration and the average number of nesting attempts when compared to Mt. Laguna and other areas of the native range. Yeh and Price (2004) used a demographic analysis to show that the shift towards extended breeding is an essential adjustment for the colonist population's persistence in the face of lower realized reproductive output per nest (recruitment of surviving juveniles is much lower at UCSD than in the native range).

Presumably facilitated by the mild temperatures and year-round food and water availability, some UCSD juncos begin regular territorial singing in December and commence breeding in February, with some pairs continuing well into August and raising 3 to 4 broods. In contrast, Mt. Laguna breeders rarely begin nesting before the end of April (prior to which wintery storms are likely), with almost all pairs raising only a single brood and beginning to molt (and hence terminate breeding activities) by mid-July, when hot and dry conditions predominate. This dramatic shift in both the onset and termination of the breeding season at UCSD is likely both the consequence of and a contributor to a range of physiological and behavioral adjustments, with cascading effects on other traits.

Breeding season length is generally a highly plastic trait (Hahn et al. 1997), with photoperiodic cues initiating the transition from nonreproductive to reproductive condition, and supplementary cues such as temperature and rainfall fine-tuning reproductive development to synchronize with local conditions (Wingfield 2008). A common garden study of reproductive development including testosterone profiles supported the assumption that the shift in breeding phenology at UCSD is plastic, as there was no difference in seasonal testosterone profiles among one-year-old birds originating at UCSD versus Mt. Laguna, even though differences are present in the field (Atwell et al. 2014). Instead, in the mild conditions of captivity, the Mt. Laguna birds upregulated their hypothalamic-pituitary-gonadal axes and produced testosterone much earlier (and downregulated later in the season) than in the field studies, suggesting that it is a release at UCSD from inhibitory supplementary cues experienced at montane sites that leads to the extended breeding schedule (Atwell et al. 2014). It is also important to note that although on average UCSD birds exhibit an expanded breeding season, reproductive synchrony is much lower at UCSD, with some pairs still exhibiting shorter, more mountain-like breeding seasons by starting later than others (Yeh and Price 2004). This perhaps suggests condition dependence in the onset of breeding, or that genetic variation for threshold sensitivity to supplementary inhibitory cues is maintained within the UCSD population (Atwell et al. 2014).

Perhaps as a corollary to the extended breeding season at UCSD, or as an underlying foundational component, the UCSD juncos have also become entirely sedentary, remaining near their territories year-round. Although they can at times be observed mingling with migrant and juvenile flocks, UCSD juncos are typically observed foraging on their breeding territories with their mates in the middle of winter. Meanwhile, both wintering migrants at UCSD as well as any lingering Mt. Laguna breeders up on the mountain generally exhibit winter flocking behavior and abandon territorial defense (J. Atwell, pers. obs.). As with the altered duration of the breeding season, this is a remarkable and dramatic shift in seasonal biology associated with a host of physiological and behavioral adjustments.

In contrast to the plasticity of breeding phenology, there is a rich literature indicating a strong genetic component to migratory propensities, including evidence of substantial heritability, strong response to selection in artificially selected lines, and evolutionary lability within and among genera (Berthold et al. 2001; Zink 2011). In keeping with this generalization, a common garden study of migratory biology suggested that the shift towards sedentariness at UCSD does include a genetic basis because captive juncos from UCSD exhibited reduced intensity of spring migratory restlessness behavior (i.e., *Zugunruhe*, Berthold et al. 2001) in comparison to those from Mt. Laguna, despite being raised under identical conditions since early life and not being sampled for migratory behavior until their *second* spring in captivity (Atwell et al., in review). Amounts of seasonal subcutaneous fat deposition were also lower in the captive UCSD juncos when compared to the Mt. Laguna juncos under the common garden conditions, a difference that persisted across multiple seasons in captivity (Atwell et al., in review).

C.3. Morphological Differences: Wing and Tail Length and Socially Selected Plumage

To date, at least four morphological differences have been documented between UCSD and Mt. Laguna juncos: UCSD juncos have shorter wings and tails by about two to three milimeters (about 5 percent shorter) on average, and reduced proportions of tail white (about 25 percent less white) and head black feather plumage (about 30 percent less black), both of which are assumed to be socially selected plumage ornaments based on experimental studies of other junco populations in eastern North America. For example, experimentally enhancing the amount of black in the hoods of juncos confers higher dominance status (Holberton et al. 1989), and males with augmented tail white are preferred by females in mate choice experiments (Hill et al. 1999) and exhibit greater dominance status in flocks (Holberton et al. 1989). Longer wings are also preferred by females and confer greater social dominance (Hill et al. 1999). Hence, the observed divergence in these socially selected plumage traits and body size at UCSD suggests that altered patterns of social selection in the novel environment play a role in reducing the elaboration of socially selected tail white and head black. Another nonexclusive hypothesis is that shorter wings would be favored by selection within a sedentary population inhabiting an urban environment because shorter wings likely allow for more efficient and maneuverable short flights (Alatalo et al. 1984). Finally, longer-winged birds, with more white in the tail, may be part of a "behavioral syndrome," also exhibiting greater dispersal tendencies and therefore leaving UCSD more often (Price et al. 2008).

Two lines of evidence suggest that selection has shaped the genetic evolution of morphological traits in this system. First, although environmental factors (e.g., lasting maternal effects) cannot be ruled out, the differences do appear to be genetically based. Although diet can play a role in the development of plumage color or morphometry, with better nutrition associated with greater tail white and body size (e.g., McGlothlin et al. 2007), in common garden studies (Rasner et al. 2004; Yeh 2004, 2012; Atwell et al. 2014), morphological differences in tail white, head black, and wing and tail lengths have persisted across multiple years and through at least one molt—providing strong evidence of a genetic basis underlying divergence in these physical traits. This is consistent with the substantial heritability of both tail white and wing length observed in a pedigree study of juncos (McGlothlin et al. 2005).

With respect to drift, Yeh (2004) and Rasner et al. (2004) provide a quantitative genetic consideration of possible *random* founder effects underlying variation in tail white and wing length, respectively, following a method established by Lande (1976) described above. These analyses indicate that given the magnitude of apparent trait divergence and the number of generations since population establishment, populations would have to have been maintained at extremely small sizes (e.g., n < 10 birds) for drift to be responsible, which is in contrast with both the molecular genetic estimates (Rasner et al. 2004) and census population sizes of about 140 (Yeh and Price 2004).

Additionally, Price et al. (2008) found physiological and demographic patterns suggesting selection may be acting to reduce tail white at UCSD: late-hatching juveniles were in better body condition than those hatched early, but among surviving juveniles recruited into the population, late-hatched survivors had less tail white. This pattern could be detected perhaps because higher tail white is socially selected against in competitive flocks with earlier-hatched males, or because these individuals emigrated from UCSD (Price et al. 2008). Attempts to measure viability or sexual selection on morphometrics and plumage within UCSD (Price et al. 2008; Atwell et al., unpublished data) and Mt. Laguna (Atwell et al., unpublished data) have been generally inconclusive.

C.4. Differences in Social and Reproductive Behaviors

Given the marked shifts in breeding phenology, migratory strategy, and plumage ornaments, we made several predictions regarding potential differences in social and reproductive behaviors between colonist and

ancestral-range populations. General life-history theory predicts a trade-
off between mating effort and parental care (Magrath and Komdeur
2003), as well as a relationship between breeding synchrony and extrapair
paternity (Spottiswoode and Moller 2004; Stutchbury and Morton 1995).
In particular, territorial aggressive behavior, parental care, and the frequency of extrapair mating were examined within the UCSD and Mt. Laguna populations, with the general prediction that colonist juncos might
exhibit a shift towards parental care and reduced extrapair mating. Further, the sedentary life history and persistent pair bonds would predict reduced intensity of social selection favoring reduced aggressive responses.

C.4.1. TERRITORIAL AGGRESSION.   In accordance with these general hypotheses, Newman et al. (2006) documented that UCSD males responded
less vigorously to territorial song playbacks on their territories during
spring and summer, as measured by combining long-range songs, contact
calls, hops/flights, swoops, and distance-to-speaker into composite aggressiveness measures (i.e., principal components). The differences in overall
aggressiveness scores were driven primarily by more frequent songs and
more swoops performed by Mt. Laguna males in response to the song
playback (Newman et al. 2006). In a subsequent assessment of territorial
aggression using a simulated-territorial intrusion (STI) protocol that included a caged lure bird, mist-net setup, and playback speaker, Atwell
et al. (2014) observed different patterns of territorial response, with
UCSD males approaching the experimental apparatus closely but singing infrequently in comparison to Mt. Laguna males, who sang more frequently but seemed wary to approach the experimental setup. Hence in
this latter study, it was difficult to tease apart possible differences in territorial aggressiveness per se versus those associated with neophobia or
boldness behaviors (see below).

C.4.2. PARENTAL CARE.   When considered across the annual cycle, given
the longer breeding season and multiple broods, junco parents at UCSD
must invest much more annual effort in parental care than Mt. Laguna
juncos, and there is also a trend towards slightly larger clutch sizes at
UCSD when compared to mountain populations (Yeh and Price 2004).
Atwell et al. (2014) compared levels of parental investment at the level
of individual nesting attempts, using both focal watches and video record-
ings to quantify the frequency of parental feeding visits. After correcting
for possible confounds such as brood size, date, age of nestlings, and age
of parents, clear differences in male parental care were found with UCSD

males feeding almost twice as frequently as Mt. Laguna males; the frequency of maternal visits did not differ (Atwell et al. 2014). These data also suggest that UCSD males may be investing more in parental care, as would be predicted by a shift away from extrapair mating effort and territorial defense.

c.4.3. EXTRAPAIR PATERNITY. Many birds, including juncos, are socially monogamous but exhibit varying degrees of "extrapair mating," in which offspring are produced outside the social pair bond. Several ecological factors predict higher or lower rates of extrapair behavior among individuals, populations, or species, including breeding synchrony, breeding density, levels of parental care, clutch sizes, or adult mortality rates (Griffith et al. 2002; Westneat and Stewart 2003). The frequency of extrapair mating is directly proportional to the potential strength or importance of sexual selection, as variation in male mating success is a key driver of the opportunity for sexual selection in most vertebrates (Wade and Arnold 1980).

In the UCSD colonist junco population, we measured a lower frequency of extrapair paternity when compared with the native range during a combined two years of study, driven chiefly by different rates in one year with a larger sample size (Atwell et al. 2014). Taken together with the reduced territorial responses and increased parental care observed in UCSD males, these results fit with the overall pattern of a reduced emphasis on mating effort and territoriality at UCSD, and an increased emphasis on parental care. Possible drivers of this behavioral shift include the extended breeding season, the shift to sedentariness (and hence relative territorial stability), or increased breeding asynchrony at UCSD.

The degree to which the shifts in social behaviors at UCSD represent plastic adjustments versus evolved differences with a genetic basis remains unknown, in part because these behavioral traits are difficult or impossible to assay under a common garden study. While juncos thrive in captivity and behave naturally in many respects, they do not nest reliably in captivity, and thus traits such as territorial ownership or mating behavior are impossible to assess.

C.5. Vocal Behavior

Upward shifts in the frequency (pitch) of birdsong is now widely documented in urban habitats, with the typical finding of increased frequencies as a putative adaptation for signalers to overcome low frequency background noise (Slabbekoorn and Ripmeester 2008) or simply a result of having to sing more loudly (Brumm and Slabbekoorn 2005; Nemeth and Brumm 2010). A comparison of UCSD and Mt. Laguna junco songs was among the first studies to document both the phenomenon of an upward shift in minimum song (trill) frequency and an associated reduced transmission efficiency of lower frequency sound in an urban habitat (Slabbekoorn et al. 2007). UCSD males sing their long-range songs with a higher minimum frequency when compared to Mt. Laguna males, although song lengths, trill rates, maximum frequencies, and peak frequencies do not differ (Slabbekoorn et al. 2007). While experimental studies in other systems have demonstrated that birds can make "real-time" shifts in vocal behavior in response to a noise stimulus, the potential roles of development, cultural evolution, or spectral trait correlations have received little attention. Three studies in the UCSD versus Mt. Laguna junco system have addressed this topic.

First, Cardoso and Atwell (2011a) examined the population-level mechanisms by which junco song types have shifted at UCSD, in particular which aspects of cultural evolution facilitate changes. They found that modification of existing types (i.e., "mutation" in the framework of cultural evolution) as well as creation of new song types, or selection for high frequency types, appear to play an important role (Cardoso and Atwell 2011a). Second, common garden studies of juncos captured from early life (about twenty-five to forty days posthatch) in UCSD versus Mt. Laguna populations and raised under identical captive conditions revealed (1) plastic shifts in response to the acoustic environment of captivity (all song frequencies were shifted down in captivity), (2) shared song types between meme pools of field populations and captive populations (apparently learned during early life), and (3) persistent population differences in minimum trill frequencies—together suggesting that both plasticity and early learning may play an important role in maintaining spectral shifts (Atwell et al., in prep.).

An analysis of covariation between song amplitude (i.e., volume) and frequency within and among UCSD and Mt. Laguna junco trills *did not* find evidence to support an emerging hypothesis that frequency shifts in birdsong in urban habitats are a mechanistic byproduct of louder singing (Cardoso and Atwell 2011b; Nemeth and Brumm 2010).

C.6. Boldness and Response to Novelty Behaviors

One typical observation of urban-dwelling wildlife, including birds, is the degree to which they become tame in the presence of humans, vehicles, novel predators, or pets, as well as generally increased boldness (cf. shyness) and a lack of neophobia (aversion to novel objects) (Martin and Fitzgerald 2005; Møller 2008; Shochat et al. 2006). Several types of behavioral tests have been used to assay the general category of boldness or novelty behaviors, variously described as tameness, flight-initiation distances, exploratory boldness, risk-taking behavior, neophobia/neophilia, and others, each with slightly different contexts and connotations, many of which are highly repeatable within individuals. Repeatable correlations among these types of behavioral traits within individuals and across behavioral contexts are now termed "animal personality" or "behavioral syndromes" (Sih et al. 2004; Wilson et al. 1994).

Among the most widely used boldness assays, first established in studies of great tits (*Parus major*), is "early exploratory boldness," a measure of how quickly and extensively an individual explores a novel arena (Dingemanse et al. 2004; Drent et al. 2003; van Oers et al. 2004). Tameness and boldness are also commonly evaluated using flight-initiation distance, a measure of the distance at which birds take off when approached by an animal (typically a human) or object (Carrete and Tella 2010; Møller 2008). Given the UCSD juncos' relatively urbanized habitat with constant anthropogenic activity and stimuli, as well as a more dynamic and diverse potential resource base, a clear prediction was that UCSD juncos would exhibit tamer, bolder, more exploratory, less neophobic, and greater risk-taking types of behaviors.

In field studies, both incubating females and foraging individuals of both sexes were found to have a much closer flight-initiation distance at UCSD when compared to birds at Mt. Laguna (Atwell et al. 2012). Additionally, juncos at UCSD have been shown to utilize novel above-ground nesting substrates much more frequently than in the native range (Yeh et al. 2007), and UCSD junco parents deliver a wider variety of natural and anthropogenic food sources to nestlings (J. Atwell, unpublished data). In the common garden study, tests of early exploratory boldness revealed sex differences (males were "bolder") and persistent population differences in both sexes, with birds originating from UCSD exploring a novel space more quickly and more extensively. Exploratory boldness in the common garden was also negatively correlated with stress-induced corticosterone levels (Atwell et al. 2012). Together, these studies indicate an overall shift towards bolder, tamer, and less neophobic behavioral strategies in the UCSD population, with evidence for a genetic (or early developmental) basis underlying these behavioral shifts, apparently as the result of either natural selection postcolonization and/or "immigrant selection" during the populations' founding (Atwell et al. 2012). The latter scenario (immigrant selection) may be particularly likely for traits related to lack of neophobia or increased boldness, as a flock of founders might be more likely to represent a nonrandom sample with respect to such traits.

C.7. Disease, Immune Function, and MHC Loci

Contemporary global change—including both climatic warming and urbanization—are contributing to altered pathogen distributions and dynamics across animal populations (Bradley and Altizer 2007; Lafferty 2009). Additionally, urban habitats contain pollutants, pesticides, or other stressors that might depress or alter immune function (Martin et al. 2010). Hence, in characterizing divergence in other behavioral and physiological traits at UCSD, efforts were made also to assess the prevalence and diversity of parasites and to characterize basic aspects of immune function.

Three parasites are more frequent at UCSD when compared to Mt. Laguna, including feather mites (*Acarina* sp.), avian pox virus (*Avipoxvirus*), and intestinal coccideans (*Isospora* sp.) (Hanauer et al., in prep.). These results mirror those from a similar study of house finches (*Haemorhous mexicanus*) in which both pox virus and coccidians were more prevalent in more populated and disturbed (urban) habitats, and similar findings are reported for avian gastrointestinal parasites in general (Giraudeau et al. 2014). All three of these types of parasites can be transmitted horizontally without vectors, which may suggest that urban habitats with more dense avian assemblages or more concentrated food, perching, or nesting resources may facilitate increased transmission of horizontally transmitted pathogens (Giraudeau et al. 2014). This explanation may fit the UCSD junco habitat; although junco nesting densities were not exceptionally high on the campus (Atwell et al. 2014), other robust populations of urban breeding birds share foraging and perching substrates with juncos (e.g., house finches, song sparrows, crows). Further, juncos from multiple adjacent breeding territories (and winter flocks) are routinely observed foraging in common areas including courtyards, cafés, and the edges of athletic fields—particularly in winter, when junco flocks are present at UCSD and few juncos are on the native breeding grounds (J. Atwell, pers. obs.).

In contrast to feather mites, pox, and coccidians, two classes of vector-transmitted blood parasites, the Haemosporidians (e.g., avian malaria, i.e., *Plasmodium* sp.) and Microfilaria (nematodes), are at lower prevalence among juncos of the UCSD population when compared to Mt. Laguna

(Hanauer et al., in prep.; Sheets et al., in review). These findings are once again mirrored in other studies of vector-borne pathogens in urban avian populations, perhaps due to fewer Dipteran vectors such as mosquitoes and flies in the relatively xeric suburban and urban habitats, where pest control measures may also be employed (Giraudeau et al. 2014).

Finally, male UCSD juncos appear to have higher concentrations of circulating leukocytes (i.e., total white blood cell counts) (Hanauer et al., in prep.), suggesting a more generally activated immune system. However, such activation could be due to acute infection, chronic infection, greater investment in preventative defenses, or represent a by-product of another physiological process (Martin et al. 2011), so the possible functional significance remains unclear.

Another way to compare immune function between the two populations is at the genetic level, specifically focusing on the major histocompatibility complex (MHC). This group of genes plays an important role in mobilizing the adaptive immune response in vertebrates and may be critical for the survival of small or declining populations (Aguilar et al. 2004; Klein 1986). These highly variable genes can affect an individual's ability to detect pathogens, and thus heterozygosity is generally considered to be adaptive. A comparison of MHC Class IIB genes (important in detecting extracellular pathogens) in juncos at UCSD and Mt. Laguna revealed that both populations maintained similar levels of diversity (Whittaker et al. 2012), in contrast to significantly lower diversity at UCSD for neutral microsatellite markers (Rasner et al. 2004; Whittaker et al. 2012). Furthermore, no divergence at the sequence level was detected in MHC loci between these two populations (Whittaker et al. 2012). These findings suggest that, although the founding of the UCSD population led to some loss of genetic variation at neutral loci, environmental factors at UCSD may have selected for high diversity at MHC loci. The observation of higher rates of pox and coccidian infections at UCSD, as well as higher immune activation in UCSD birds, support this hypothesis.

D. Hormonal Mechanisms as "Mediators" of Multitrait Population Divergence?

A core theme of this volume is the central role that hormonal mechanisms may play in mediating variation among seasons, sexes, individuals, populations, or species for suites of interrelated traits, including mediating life-history trade-offs and transitions between stages of the annual cycle (chapters 4–7). As central interfaces between genomes and phenomes, hormonal mechanisms modulate individuals' plastic behavioral and physiological responses to environmental variation, but they can also evolve in response to selection (Ketterson et al. 2009).

Consistent hormone-phenotype linkages observed for suites of traits in experimental and observational studies of both the endocrine hypothalamic-pituitary-gonadal (HPG) axis and hypothalamic-pituitary-adrenal (HPA) axes have led to the hypothesis that testosterone- or corticosterone-mediated traits may evolve as a unit, perhaps constraining adaptive responses or evolutionary diversification (e.g., "phenotypic integration" [Ketterson et al. 2009]; "evolutionary constraint" [Hau 2007]; "evolutionary inertia" [Adkins-Regan 2008]). In fact a small handful of comparative studies among species or populations that vary in latitude or life history suggest that modulation of circulating levels of testosterone and corticosterone may facilitate life-history and behavioral diversification. In contrast, because plastic or genetic modulation of phenotypes via hormonal systems can be expressed at multiple mechanistic levels, it has also been suggested that individual trait values may adjust independently from the hormone signal in response to new or changing environments (e.g., "phenotypic independence" [Ketterson et al. 2009]; "evolutionary potential" [Hau 2007]). That is, phenotypes are affected not only by circulating hormone "signal" concentrations but also by proteins associated with hormone transport, enzymes associated with hormone conversions, receptor densities or sensitivities at target tissues, and cofactors associated with hormone-mediated gene transcription. To date, few studies have examined patterns of response for hormone levels and correlated phenotypic traits both *within* and *among* populations in the context of "microdivergence" or adaptation to environmental change.

Given the observed differences in breeding season length and social and boldness behavior between the UCSD and native range populations, we hypothesized that underlying shifts in testosterone or corticosterone expression might play a central role in mediating patterns of adaptive divergence for multiple traits. Over the course of two field seasons, and in a subsequent common garden study, we measured levels of baseline and gonadotropin-releasing-hormone (GnRH)-induced maximum levels of testosterone, along with baseline and stress-induced maximum levels of corticosterone across the season in birds from both UCSD and Mt. Laguna populations (Atwell et al. 2012; Atwell et al. 2014). Many of the same individuals sampled were also assayed for associated behavioral and morphological traits, including territorial aggression, parental care, and plumage ornamentation (in relation to testosterone, Atwell et al. 2014), and exploratory boldness behavior (in relation to corticosterone, Atwell et al. 2012).

For both testosterone- and corticosterone-mediated traits, patterns of behavioral trait variation were generally predicted by circulating hormone levels both among and within UCSD and Mt. Laguna populations, suggesting integrated patterns of trait divergence (Atwell et al. 2012; Atwell et al. 2014). In the field studies, the UCSD population exhibited an earlier seasonal upregulation and later downregulation in both initial and GnRH-induced testosterone but with lower peak levels, as would be predicted by the extended breeding season and shift towards an emphasis on increased parental care and reduced territorial aggression and extrapair mating (Atwell et al. 2014). The differences in seasonal profiles or peak levels of testosterone did not persist in captivity (Atwell et al. 2014), suggesting plastic responses underlie divergence.

Population differences in stress-induced corticosterone levels among free-living juncos were lower at UCSD in accordance with the prediction of attenuated stress responsiveness in relation to urbanization. In contrast to testosterone, these differences persisted across at least two years in the common garden studies, suggesting a genetic basis for divergence in the endocrine HPA axis (Atwell et al. 2012). Similar findings with respect to plastic HPG differences and persistent HPA microdivergence have been documented in other systems, for example among urban versus wildland European blackbird populations (Partecke et al. 2006; Partecke et al. 2004; Partecke et al. 2005).

The testosterone- and corticosterone-mediated traits in the UCSD versus Mt. Laguna populations provide an integrated dataset to consider hormone-phenotype divergence for a range of interrelated traits in response to environmental change. However, the studies of these two hormonal axes were pursued separately with respect to sampling of individual males and hence provide limited opportunity to consider the potential for "functional cross talk" that may occur between gonadal (HPG) and adrenal (HPA) axes, owing to the interactive effects of sex steroids and glucocorticoids in regulating physiological or behavioral changes (Viau 2002). Ongoing studies address potential HPG-HPA interactions and their role in regulating phenotypic divergence in this system (M. Abolins-Abols, pers. comm.).

E. Future Directions: Additional Populations, Origins, and Emerging Tools

Since 1998, the UCSD junco study population, in particular its comparisons to the nearby ancestral-range population at Mt. Laguna, has provided many insights into the organismal and evolutionary processes by which individuals and populations adjust in the face of environmental change. However, the un-replicated nature inherent in comparisons of only two populations (i.e., only one pair of colonist versus ancestral-range populations) has limited our ability to generalize (Garland and Adolph 1994).

However, since work was begun at UCSD, other reports of recent urban or coastal junco establishments in southern and central California have emerged, providing a scope for replicating studies in this system among additional pairs of colonist versus ancestral-range junco populations. For example, a small but stable population of juncos has begun breeding on the previously unoccupied UCLA campus (about 150 kilometers to the north and west of UCSD), with nearby montane-forest breeders in the San Gabriel Mountains identified as a likely ancestral-range source. Studies of behavior and neuroendocrinology of the UCLA and San Gabriel juncos, including a simultaneous parallel study of the UCSD versus Mt. Laguna system, are currently underway (M. Abolins Abols and R. Hanuaer, unpublished data). A similar southern California suburban colonist junco population has been reported in Santa Barbara, though its size, stability, and geography are not well characterized. Further, reports of small populations of Oregon juncos breeding on the undeveloped Channel Islands, including confirmed breeding on Santa Cruz Island across multiple seasons (S. Sillet, pers. comm.), provide a scope for additional and unique colonist versus ancestral-range population comparisons in future studies. A major shortcoming of the UCSD study system is that it was not followed from inception, and it is to be hoped that newly founded populations can be studied from their earliest stages. It may even be possible to induce breeding in suitable locations by use of playbacks and provision of suitable food. An initial attempt to do this in Balbao Park by Price and Yeh in 1999 failed.

Another general shortcoming of these studies is that we do not know the explicit ancestral source of the UCSD birds. A new era of population genomic tools—including both broader and deeper sequencing as well as increasingly high-throughput and lower-cost methods—provides hope that such tools could be applied to match the colonists to a most-likely ancestral range source based on private alleles. As such, a SNP-generation analysis is currently underway to compare genomic similarity between the UCSD population and more than twelve candidate ancestral range populations from throughout the *thurberi*, *pinosus*, and *montanus* ranges. In addition to addressing evolutionary history and population connectivity, next-generation sequencing approaches, including sequence capture of genomic regions as well expressed (gene) transcripts in focal tissues, can be applied to understand the genomic and physiological basis of phenotypic divergence among the UCSD and ancestral-range populations.

Advances in migratory tracking technologies, such as the creation of archival GPS and light-level dataloggers that weigh less than 0.5 grams (Bridge et al. 2013), as well as technical and analytical advances in stable isotope and genetic approaches to connectivity (Rundel et al. 2013), also provide a framework for examining the biogeographic connectivity of juncos among California's *thurberi* populations in more detail. For example, such emerging technologies could be employed to test the hypothesis that the Mt. Laguna and nearby southern-breeding populations do, in fact, winter on the coast in San Diego County. Similarly, such technologies could be deployed to assist in evaluating the breeding origins of juncos at UCSD, especially given modest overwintering return rates observed for juncos flocking on wintering grounds (about ten to twenty percent, Nolan et al. 2002).

F. Conclusions

In this chapter, we reviewed for the first time nearly two decades of research centered on the unique urban colonist junco population at UCSD and its comparison to nearby native range juncos. Studies of the UCSD junco system join a growing number of examples evaluating organismal and evolutionary mechanisms underlying behavioral and physiological divergence in response to contemporary urbanization and climate change. Common garden studies revealed evidence for both plastic and genetic underpinnings of trait divergence. Studies of hormones and associated behavioral phenotypes revealed generally integrated patterns of response for both testosterone- and corticosterone-mediated suites of traits. Ongoing and future research on juncos in southern California is focused on expanding the study system to examine additional urban-coastal-island colonist populations in comparison to ancestral-montane-wildland populations, effectively replicating the initial population comparison. Emerging (genomic and transcriptomic) sequencing technologies as well as advances in migratory tracking technologies provide opportunities to further elucidate the evolutionary history and the mechanisms underlying phenotypic divergence in this study system.

Acknowledgments

We would like to thank the numerous and diverse individuals and institutions that have played critical roles in supporting the research of juncos at UCSD and Mt. Laguna: Stan Walens and his daughter Rachel, who made the initial discovery; Phil Unitt, who has provided foundational expertise and background data on birds and their habitats in San Diego County; then-UCSD graduate students, in particular Pamela Yeh, as well as Melissa Newman and Cayler Rasner, and their tireless teams of undergraduate field assistants; then-postdoc Gonçalo Cardoso, who worked countless hours alongside our team in the field, lab, and office, as did undergraduates Ediri Metitiri, Angela Kemsley, Allison Miller, Russel Nichols, Meelyn Pandit, Faye Parmer, Sarah Puckett, Rebecca Rice, Kyle Robertson, Kimberly Roth, and Beth Schultz. The work summarized here was primarily supported by the National Science Foundation (IOS:0820055 and DEB:0808284, grants to Ketterson and Atwell, respectively) and a National Institutes of Health "Common Themes in Reproductive Diversity" Training Grant to Indiana University (T32HD049336), which supported Atwell's work in this system. We also thank Hopi Hoekstra, Karen Marchetti, and the Descanso Ranger District of the Cleveland National Forest for logistical support supporting the work summarized here.

References

Adelman, J. S., L. Kirkpatrick, J. L. Grodio, and D. M. Hawley. 2013. House finch populations differ in early inflammatory signaling and pathogen tolerance at the peak of *Mycoplasma gallisepticum* infection. *American Naturalist* 181:674–89.
Adkins-Regan, E. 2008. Do hormonal control systems produce evolutionary inertia? *Philosophical Transactions of the Royal Society B-Biological Sciences* 363: 1599–1609.

Aguilar, A., G. Roemer, S. Debenham, M. Binns, D. Garcelon, and R. K. Wayne. 2004. High MHC diversity maintained by balancing selection in an otherwise genetically monomorphic mammal. *Proceedings of the National Academy of Sciences of the United States of America* 101:3490–94.

Alatalo, R. V., L. Gustafsson, and A. Lundbkrg. 1984. Why do young passerine birds have shorter wings than older birds? *Ibis* 126:410–15.

Atwell, J. W., G. C. Cardoso, D. J. Whittaker, S. Campbell-Nelson, K. W. Robertson, and E. D. Ketterson. 2012. Boldness behavior and stress physiology in a novel urban environment suggest rapid correlated evolutionary adaptation. *Behavioral Ecology* 23:960–69.

Atwell, J. W., G. C. Cardoso, D. J. Whittaker, T. D. Price, and E. D. Ketterson. 2014. Hormonal, behavioral, and life-history traits exhibit correlated shifts in relation to population establishment in a novel environment. *American Naturalist* 184:E147–E160.

Atwell, J. W., D. G. Reichard, G. C. Cardoso, M. M. Pandit, T. D. Price, and E. D. Ketterson. In prep. The ontogeny of birdsong divergence in response to urban noise: Insights from a common garden study. Manuscript.

Atwell, J. W., R. J. Rice, and E. D. Ketterson. In review. Rapid loss of migratory behavior associated with recent colonization of an urban environment. Manuscript.

Berthold, P. 1995. *Control of Bird Migration*. London: Chapman and Hall.

Berthold, P., H. G. Bauer, and V. Westhead. 2001. *Bird Migration: A General Survey*. New York: Oxford University Press.

Bonneaud, C., S. L. Balenger, A. F. Russell, J. Zhang, G. E. Hill, and S. V. Edwards. 2011. Rapid evolution of disease resistance is accompanied by functional changes in gene expression in a wild bird. *Proceedings of the National Academy of Sciences of the United States of America* 108:7866–71.

Bradley, C. A., and S. Altizer. 2007. Urbanization and the ecology of wildlife diseases. *Trends in Ecology & Evolution* 22:95–102.

Bridge, E. S., J. F. Kelly, A. Contina, R. M. Gabrielson, R. B. MacCurdy, and D. W. Winkler. 2013. Advances in tracking small migratory birds: A technical review of light-level geolocation. *Journal of Field Ornithology* 84:121–37.

Brown, J. H., and M. V. Lomolino. 1998. *Biogeography*. Sunderland, MA: Sinauer.

Brumm, H., and H. Slabbekoorn. 2005. Acoustic communication in noise. *Advances in the Study of Behavior* 35:151–209.

Cardoso, G. C., and J. W. Atwell. 2011a. Directional cultural change by modification and replacement of memes. *Evolution* 65:295–300.

———. 2011b. On the relation between loudness and the increased song frequency of urban birds. *Animal Behaviour* 82:831–36.

Carrete, M., and J. L. Tella. 2010. Individual consistency in flight initiation distances in burrowing owls: A new hypothesis on disturbance-induced habitat selection. *Biology Letters* 6:167–70.

Chapman, B. B., C. Brönmark, J. A. Nilsson, and L. A. Hansson. 2011. The ecology and evolution of partial migration. *Oikos* 120:1764–75.

Cox, G. W. 2004. *Alien Species and Evolution: The Evolutionary Ecology of Exotic Plants, Animals, Microbes, and Interacting Native Species.* Washington, DC: Island Press.

Crispo, E. 2007. The Baldwin effect and genetic assimiliation: Revisiting two mechanisms of evolutionary change mediated by phenotypic plasticity. *Evolution* 61:2469–79.

Crutzen, P. J. 2002. Geology of mankind. *Nature* 415:23.

Dingemanse, N. J., C. Both, P. J. Drent, and J. M. Tinbergen. 2004. Fitness consequences of avian personalities in a fluctuating environment. *Proceedings of the Royal Society B-Biological Sciences* 271:847–52.

Drent, P. J., K. van Oers, and A. J. van Noordwijk. 2003. Realized heritability of personalities in the great tit (*Parus major*). *Proceedings of the Royal Society B-Biological Sciences* 270:45–51.

Evans, K. L., J. Newton, K. J. Gaston, S. P. Sharp, A. McGowan, and B. J. Hatchwell. 2012. Colonisation of urban environments is associated with reduced migratory behaviour, facilitating divergence from ancestral populations. *Oikos* 121: 634–40.

Garland, T., and S. C. Adolph. 1994. Why not to do 2-species comparative-studies: Limitations on inferring adaptation. *Physiological Zoology* 67:797–828.

Gaston, K., J. Duffy, S. Gaston, J. Bennie, and T. Davies. 2014. Human alteration of natural light cycles: Causes and ecological consequences. *Oecologia* 176:917–31.

Giraudeau, M., M. Mousel, S. Earl, and K. McGraw. 2014. Parasites in the city: Degree of urbanization predicts poxvirus and coccidian infections in house finches (*Haemorhous mexicanus*). *PLoS ONE* 9:e86747.

Grant, P. R., and B. R. Grant. 2014. *40 Years of Evolution: Darwin's Finches on Daphne Major Island.* Princeton: Princeton University Press.

Griffith, S. C., I. P. F. Owens, and K. A. Thuman. 2002. Extrapair paternity in birds: A review of interspecific variation and adaptive function. *Molecular Ecology* 11:2195–212.

Hahn, T., T. Boswell, J. Wingfield, and G. Ball. 1997. Temporal flexibility in avian reproduction. *Current Ornithology* 14:39–80.

Hanauer, R. E., J. W. Atwell, M. Wurzelman, W. Anderson, and E. D. Ketterson. In prep. Parasitism differs between an urban and a nonurban population of a songbird, the dark-eyed junco. Manuscript.

Hau, M. 2007. Regulation of male traits by testosterone: Implications for the evolution of vertebrate life histories. *Bioessays* 29:133–144.

Henter, H. 2005. Tree wars: The secret life of eucalyptus. *UCSD Alumni Magazine* 2.

Herrel, A., K. Huyghe, B. Vanhooydonck, T. Backeljau, K. Breugelmans, I. Grbac, R. Van Damme, et al. 2008. Rapid large-scale evolutionary divergence in morphology and performance associated with exploitation of a different dietary resource. *Proceedings of the National Academy of Sciences of the United States of America* 105:4792–95.

Hill, J. A., D. A. Enstrom, E. D. Ketterson, V. Nolan, and C. Ziegenfus. 1999. Mate choice based on static versus dynamic secondary sexual traits in the dark-eyed junco. *Behavioral Ecology* 10:91–96.

Holberton, R. L., K. P. Able, and J. C. Wingfield. 1989. Status signalling in dark-eyed juncos, *Junco hyemalis*: Plumage manipulations and hormonal correlates of dominance. *Animal Behaviour* 37:681–89.

Johnston, R. 2001. Synanthropic birds of North America. In *Avian Ecology and Conservation in an Urbanizing World*, edited by J. Marzluff, R. Bowman, and R. Donnelly, 49–67. New York: Springer.

Ketterson, E. D., J. W. Atwell, and J. W. McGlothlin. 2009. Phenotypic integration and independence: Hormones, performance, and response to environmental change. *Integrative and Comparative Biology* 49:365–79.

Klein, J. 1986. *Natural History of the Major Histocompatibility Complex*. New York: Wiley.

Lafferty, K. D. 2009. The ecology of climate change and infectious diseases. *Ecology* 90:888–900.

Lande, R. 1976. Natural-selection and random genetic drift in phenotypic evolution. *Evolution* 30:314–34.

Liebl, A. L., and L. B. Martin. 2012. Exploratory behaviour and stressor hyper-responsiveness facilitate range expansion of an introduced songbird. *Proceedings of the Royal Society B-Biological Sciences* 279:4375–81.

Magrath, M. J. L., and J. Komdeur. 2003. Is male care compromised by additional mating opportunity? *Trends in Ecology & Evolution* 18:424–30.

Martin, L. B., and L. Fitzgerald. 2005. A taste for novelty in invading house sparrows, *Passer domesticus. Behavioral Ecology* 16:702–7.

Martin, L. B., D. M. Hawley, and D. R. Ardia. 2011. An introduction to ecological immunology. *Functional Ecology* 25:1–4.

Martin, L. B., W. A. Hopkins, L. D. Mydlarz, and J. R. Rohr. 2010. The effects of anthropogenic global changes on immune functions and disease resistance. *Annals of the New York Academy of Sciences* 1195:129–48.

Martin, L. B., A. L. Liebl, and H. J. Kilvitis. 2015. Covariation in stress and immune gene expression in a range expanding bird. *General and Comparative Endocrinology* 211:14–19.

McCaskie, G. 1986. Wood warblers to finches. *American Birds* 40:1257.

McGlothlin, J. W., D. L. Duffy, J. L. Henry-Freeman, and E. D. Ketterson. 2007. Diet quality affects an attractive white plumage pattern in dark-eyed juncos (*Junco hyemalis*). *Behavioral Ecology and Sociobiology* 61:1391–99.

McGlothlin, J. W., P. G. Parker, V. Nolan, and E. D. Ketterson. 2005. Correlational selection leads to genetic integration of body size and an attractive plumage trait in dark-eyed juncos. *Evolution* 59:658–71.

Miller, A. H. 1941. Speciation in the avian genus *Junco. University of California Publications in Zoology* 44:173–434.

Miranda, A. C., H. Schielzeth, T. Sonntag, and J. Partecke. 2013. Urbanization and its effects on personality traits: A result of microevolution or phenotypic plasticity? *Global Change Biology* 19:2634–44.

Møller, A. P. 2008. Flight distance of urban birds, predation, and selection for urban life. *Behavioral Ecology and Sociobiology* 63:63–75.

Nemeth, E., and H. Brumm. 2010. Birds and anthropogenic noise: Are urban songs adaptive? *American Naturalist* 176:465–75.

Newman, M. M., P. J. Yeh, and T. D. Price. 2006. Reduced territorial responses in dark-eyed juncos following population establishment in a climatically mild environment. *Animal Behaviour* 71:893–99.

Nolan, V., Jr., E. D. Ketterson, D. A. Cristol, C. M. Rogers, E. D. Clotfelter, R. C. Titus, S. J. Schoech, and E. Snajdr. 2002. Dark-eyed junco (*Junco hyemalis*). In *The Birds of North America*, edited by A. Poole and F. Gill, no. 716. Philadelphia: The Birds of North America, Inc.

Partecke, J., I. Schwabl, and E. Gwinner. 2006. Stress and the city: Urbanization and its effects on the stress physiology in European blackbirds. *Ecology* 87: 1945–52.

Partecke, J., T. Van't Hof, and E. Gwinner. 2004. Differences in the timing of reproduction between urban and forest European blackbirds (*Turdus merula*): Result of phenotypic flexibility or genetic differences? *Proceedings of the Royal Society B-Biological Sciences* 271:1995–2001.

———. 2005. Underlying physiological control of reproduction in urban and forest-dwelling European blackbirds *Turdus merula*. *Journal of Avian Biology* 36:295–305.

Peterson, M., M. Abolins-Abols, J. Atwell, R. Rice, B. Milá, and E. Ketterson. 2013. Variation in candidate genes CLOCK and ADCYAP1 does not consistently predict differences in migratory behavior in the songbird genus Junco [v1; ref status: indexed, http://f1000r.es/11p%5D]. *F1000Research* 2:115.

Price, T. D. 2012. Eaglenest Wildlife Sanctuary: Pressures on biodiversity. *American Naturalist* 180:535–45.

Price, T. D., A. Qvarnström, and D. E. Irwin. 2003. The role of phenotypic plasticity in driving genetic evolution. *Proceedings of the Royal Society B-Biological Sciences* 270:1433–40.

Price, T. D., P. J. Yeh, and B. Harr. 2008. Phenotypic plasticity and the evolution of a socially selected trait following colonization of a novel environment. *American Naturalist* 172:S49–S62.

Rasner, C. A., P. Yeh, L. S. Eggert, K. E. Hunt, D. S. Woodruff, and T. D. Price. 2004. Genetic and morphological evolution following a founder event in the dark-eyed junco, *Junco hyemalis thurberi*. *Molecular Ecology* 13:671–81.

Rundel, C. W., M. B. Wunder, A. H. Alvarado, K. C. Ruegg, R. Harrigan, A. Schuh, J. F. Kelly, et al. 2013. Novel statistical methods for integrating genetic and stable isotope data to infer individual-level migratory connectivity. *Molecular Ecology* 22:4163–76.

Sheets, J., C. Bergeon Burns, J. Atwell, E. Ketterson, and G. Spellman. In review.

Epidemiology of Haemosporidia in the dark-eyed junco (*Junco hyemalis hyemalis*): Investigating ecological patterns of infection across founder and parent populations. Manuscript.

Shochat, E., P. S. Warren, S. H. Faeth, N. E. McIntyre, and D. Hope. 2006. From patterns to emerging processes in mechanistic urban ecology. *Trends in Ecology & Evolution* 21:186–91.

Sih, A., A. Bell, and J. C. Johnson. 2004. Behavioral syndromes: An ecological and evolutionary overview. *Trends in Ecology & Evolution* 19:372–78.

Sih, A., M. C. O. Ferrari, and D. J. Harris. 2011. Evolution and behavioural responses to human-induced rapid environmental change. *Evolutionary Applications* 4:367–87.

Slabbekoorn, H., and M. Peet. 2003. Ecology: Birds sing at a higher pitch in urban noise. *Nature* 424:267–67.

Slabbekoorn, H., and E. A. P. Ripmeester. 2008. Birdsong and anthropogenic noise: Implications and applications for conservation. *Molecular Ecology* 17:72–83.

Slabbekoorn, H., P. Yeh, and K. Hunt. 2007. Sound transmission and song divergence: A comparison of urban and forest acoustics. *Condor* 109:67–78.

Spottiswoode, C., and A. P. Møller. 2004. Extrapair paternity, migration, and breeding synchrony in birds. *Behavioral Ecology* 15:41–57.

Stutchbury, B. J., and E. S. Morton. 1995. The effect of breeding synchrony on extra-pair mating systems in songbirds. *Behaviour* 132:675–90.

Unitt, P. 2005. *San Diego County Bird Atlas*. San Diego: San Diego Natural History Museum.

van Oers, K., P. J. Drent, P. de Goede, and A. J. van Noordwijk. 2004. Realized heritability and repeatability of risk-taking behaviour in relation to avian personalities. *Proceedings of the Royal Society B-Biological Sciences* 271:65–73.

Viau, V. 2002. Functional cross-talk between the hypothalamic-pituitary-gonadal and -adrenal axes. *Journal of Neuroendocrinology* 14:506–13.

Wade, M. J., and S. J. Arnold. 1980. The intensity of sexual selection in relation to male sexual behaviour, female choice, and sperm precedence. *Animal Behaviour* 28:446–61.

Westneat, D. F., and I. R. K. Stewart. 2003. Extra-pair paternity in birds: Causes, correlates, and conflict. *Annual Review of Ecology, Evolution, and Systematics* 34:365–96.

Whittaker, D. J., A. L. Dapper, M. P. Peterson, J. W. Atwell, and E. D. Ketterson. 2012. Maintenance of MHC Class IIB diversity in a recently established songbird population. *Journal of Avian Biology* 43:109–18.

Williams, P. D., A. P. Dobson, K. V. Dhondt, D. M. Hawley, and A. A. Dhondt. 2014. Evidence of trade-offs shaping virulence evolution in an emerging wildlife pathogen. *Journal of Evolutionary Biology* 27:1271–1278.

Wilson, D. S., A. B. Clark, K. Coleman, and T. Dearstyne. 1994. Shyness and boldness in humans and other animals. *Trends in Ecology & Evolution* 9:442–46.

Wingfield, J. C. 2008. Organization of vertebrate annual cycles: Implications for control mechanisms. *Philosophical Transactions of the Royal Society B-Biological Sciences* 363:425–41.

Winker, K. 2010. On the origin of species through heteropatric differentiation: A review and a model of speciation in migratory animals. *Ornithological Monographs* 69:1–30.

Yeh, P. J. 2004. Rapid evolution of a sexually selected trait following population establishment in a novel habitat. *Evolution* 58:166–74.

Yeh, P. J., M. E. Hauber, and T. D. Price. 2007. Alternative nesting behaviours following colonisation of a novel environment by a passerine bird. *Oikos* 116: 1473–80.

Yeh, P. J., and T. D. Price. 2004. Adaptive phenotypic plasticity and the successful colonization of a novel environment. *American Naturalist* 164:531–42.

Zink, R. M. 2011. The evolution of avian migration. *Biological Journal of the Linnean Society* 104:237–50.

CHAPTER ELEVEN

A Physiological View of Population Divergence

Comparing Hormone Production and Response Mechanisms

Christine M. Bergeon Burns and Kimberly A. Rosvall

... Rarely did two individuals of the same species agree throughout in their measurements ... These individual differences differ in amount to a surprising degree in various species & in various groups of species, one part or organ being affected in one species or group, & the same part being very constant in another set of species. Some forms are extremely constant in their whole organisation others as variable, causing to the naturalist an odious amount of perplexity. Generally the characters which individually vary, are of slight physiological importance, but this is not always the case ....

But here arises a perplexing question; are these individual differences of the same order & have they the same origin as those other differences, either greater, more permanent, or less closely linked together, which separate recognised varieties.
—Stauffer, R. C., ed., 1975, *Charles Darwin's Natural Selection: Being the Second Part of His Big Species Book Written from 1856 to 1858* (Cambridge: Cambridge University Press)

A. Introduction

Complex phenotypes are the result of several interconnected traits working well together, and correlations among multiple traits within an organism can be referred to as phenotypic integration (Pigliucci 2003). Hormones can provide a mechanism by which such phenotypic integration is achieved. Hormones secreted by an endocrine gland can circulate throughout the body, potentially interacting with multiple tissues to give rise to various morphological, physiological, and behavioral outcomes. Thus by simultaneously mediating multiple traits, hormones produce organism-level responses to the environment, serving as a mechanistic link that contributes to unity of brain and periphery (Pfaff 2010). The gonadal steroid testosterone (T) is one such hormone that has multiple effects on many phenotypic traits. In seasonally breeding birds, T is typically elevated during the breeding season and mediates many male reproductive trade-offs, simultaneously promoting traits such as sperm production, aggression, sexual behaviors, and ornaments at the expense of immunity and parental care (Folstad and Karter 1992; Ketterson and Nolan, Jr. 1992; Wingfield et al. 2001). Such phenotypic integration is thought to be adaptive, as the traits mediated by hormones are most useful when coexpressed.

However, when a species inhabits a new or changing environment, selection may act to alter the expression of complex phenotypes. The uncoupling of hormone-mediated traits might be expected to occur, if selection were sufficiently strong and a particular combination of traits was no longer beneficial (Finch and Rose 1995; Hau 2007; Ketterson and Nolan, Jr. 1999; Ketterson et al. 2009; McGlothlin and Ketterson 2008; Nijhout 2003). Comparative studies of closely related bird species suggest that species differences in life history may be accompanied by variation in the degree to which certain phenotypic traits are linked to circulating levels of T (Lynn 2008). The idea that traits under selection may break free from their previous hormonal control and thus evolve independently of other hormone-mediated traits is referred to as "phenotypic independence" (Ketterson et al. 2009). Such independence might arise via alterations in the degree to which particular traits are influenced by a given level of hormone (e.g., developing hormone insensitivity) or via shifts toward tissue-specific hormone synthesis (Schmidt et al. 2008).

Rather than being mutually exclusive hypotheses, phenotypic integration and phenotypic independence represent ends of a continuum to explain hormone-mediated phenotypic evolution. However, the circumstances under which we should expect to see tight integration versus independence have received very little study and remain unclear. Importantly, the extent of integration may vary depending on the degree to which multiple traits are linked at the level of mechanism (Adkins-Regan 2008; Hau 2007; Ketterson et al. 2009; Williams 2008). Chapter 7 of this volume reviewed the work that has been conducted on dark-eyed juncos to understand how different components of hormone production and response mechanisms do or do not work together to produce the phenotypic variation among individuals upon which selection may act. In this chapter, we build on this discussion by comparing T-mediated behavior and morphology, as well as aspects of the underlying hormone production and response pathways among closely related but phenotypically distinct populations of junco. We discuss how selection may act on existing variation to utilize or, alternatively, to circumvent the constraints of hormone-mediated phenotypic integration as populations begin to diverge.

B. Mechanisms of Testosterone Action: Perception, Transduction, and Response

There are many potential sources of variation along T production and response pathways (figure 11.1). The hypothalamic-pituitary-gonadal (HPG) axis controls T secretion. Environmental cues, such as long-day photoperiodic information indicating the coming spring, are relayed to neurosecretory cells of the hypothalamus in the brain, eliciting an increase in the release of the stimulatory peptide gonadotropin-releasing hormone (GnRH). Hormones including GnRH act on the nearby anterior pituitary, causing secretion of gonadotropins such as luteinizing hormone (LH), which binds to the G-protein coupled luteinizing hormone receptor (LHR) in the testes or ovaries, triggering synthesis and release of T and other steroids by the gonads. While the HPG axis is activated in a hierarchical, top-down manner, it is known to be sensitive to feedback from multiple hormones at each endocrine gland and the brain, contributing to the dynamic and precise regulation of hormone secretion. The axis also interacts with other endocrine systems (e.g., hypothalamic-pituitary-adrenal axis), providing further levels of integration (Adkins-Regan 2008; Hadley and Levine 2007; Wingfield 2012).

T can be regulated via autocrine feedback by acting on the very cells that produce it, can influence spermatogenesis in nearby cells via paracrine action, and can be carried by transport proteins in the bloodstream, leading to many endocrine effects throughout the brain and periphery. Like LH, T must be transduced by receptors to have these downstream effects. T typically acts via nuclear receptors, either by binding directly to androgen receptors (ARs) or by being converted by the enzyme aromatase to 17β-estradiol (E2), which binds to estrogen receptors (ERs). In addition to receptors, binding likelihood is influenced by availability of particular cofactors and enzymes. If bound, the receptors function as transcription factors to regulate gene expression, leading to various physiological, morphological, or behavioral outcomes (Hadley and Levine 2007; Wingfield 2012; see also Cornil 2009). For example, reproductive and ag-

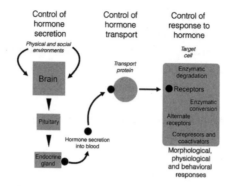

FIGURE 11.1. Mechanisms of testosterone action. One, some, or all of the components involved in this hormone production and response pathway may evolve in response to selection on morphological, physiological or behavioral trait(s). Reprinted from Wingfield (2012) with permission from the American Ornithologists' Union. See also plate 6.

gressive behaviors in birds have been associated with T and its metabolites in brain areas including the medial preoptic area, lateral septum, anterior hypothalamus, lateral ventromedial hypothalamus, and the extended medial amygdala, which is composed of the nucleus taeniae (the avian medial amygdala) and the medial bed nucleus of the stria terminalis (Ball and Balthazart 2004; Cheng et al. 1999; Goodson and Evans 2004; Panzica et al. 1996; Schlinger and Callard 1990; Thompson et al. 1998).

C. Endocrine Mechanisms under Selection

This complex endocrine pathway between environmental stimulus and phenotypic responses provides many targets on which selection may act. For example, a population with a greater aggressive response to a standardized territorial challenge might have a stronger neural signal to secrete T relayed along the HPG axis, a gonad with greater T production response to stimulation, or more receptors for T in the target cell, among other possibilities (figure 11.1). Whether selection acts at one or another component in the pathway (e.g., on the upstream steroid production axis vs. on downstream measures of sensitivity in a particular target cell) may have very different consequences for the number of morphological, behavioral or physiological traits that are ultimately affected (sensu Ramsay et al. 2009). An evolutionary change in hormone production may affect a number of different phenotypes concurrently when more or less hormone becomes available in circulation to a variety of target cells throughout the body, consistent with the phenotypic integration hypothesis (Ketterson et al. 2009). In contrast, a change in a target tissue-specific variable (e.g., receptor abundance) might affect a more limited scope of phenotypes, circumventing constraints posed by phenotypic integration of T-mediated traits and potentially allowing traits to evolve more freely. Thus multiple levels of the pathway are potential sources of divergence in phenotype, and they must be examined collectively.

Yet another consideration is the degree to which components of the pathway are coordinated in their expression over evolutionary time. At one extreme is an endocrine system marked by linkage and coexpression of multiple endocrine components that may respond collectively under selective pressure; at the other extreme is a pathway where each endocrine component is an independent unit of selection (Adkins-Regan 2008; Hau 2007). Where natural systems fall between these alternatives could have important implications for predicting responses to selection. Seasonally, the coexpression of multiple components is well recognized. For example, most birds that breed in the temperate zone demonstrate increased density and abundance of neural sex steroid receptors, as well as aromatase activity/mRNA expression, concurrently with elevated circulating T levels during the breeding season (Canoine and Gwinner 2002; Foidart et al. 1998; Riters et al. 2000; Soma et al. 2003). The tendency for these components to covary under selective pressure is less clear. A change in hormone receptor expression may be one important mechanism facilitating change in expression of some traits independent of variation in hormone levels in circulation (Canoine et al. 2007; Voigt and Goymann 2007).

For example, male tropical spotted antbirds (*Hylophylax n. naevioides*) are territorially aggressive year-round despite seasonal breeding (Wikelski et al. 2000), and T appears to modulate aggression (Hau et al. 2000; Wikelski et al. 1999). During the nonbreeding season, T concentrations are lower (Wikelski et al. 2000), yet ERα and AR mRNA show increased expression in brain areas involved in the regulation of reproductive and aggressive behavior. This seasonal upregulation of ERα and AR may function to increase brain sensitivity to lower levels of circulating steroids, and it is thought to be an adaptive route by which these males can maintain elevated aggression during the nonbreeding season (Canoine et al. 2007). Another example of an evolutionary shift in hormone response mechanisms comes from a sex-role reversed species, the African black coucal (*Centropus grillii*). Female black coucals have significantly more

AR mRNA expression than males in brain areas regulating aggression, suggesting that their male-typical territorial behavior may be mediated by androgens to some degree and modulated via changes in receptor expression. Further, circulating androgens do not covary with receptor mRNA expression, indicating independent regulation of hormone and receptor in this species (Voigt and Goymann 2007).

Examples like these illustrate that changes in T-mediated trait expression can occur via one or another component of the testosterone perception-transduction-response pathway (Wingfield 2012). Evolutionary shifts in hormone response may be particularly adaptive because they circumvent constraints posed by the multiple effects of T across seasons or sexes (e.g., deleterious effects of T on some traits, such as immune function or stress response, but not other traits; see chapter 4, this volume). Among other phylogenetically or ecologically divergent groups, phenotypic variation has also been associated with changes in hormone sensitivity, specifically differential expression of AR, ER, or AROM in behavioral centers of the brain (Gonçalves et al. 2010; Young et al. 2006). However, most comparisons to date involve groups that are distantly related or have quite divergent life histories (e.g., tropical vs. temperate, typical vs. reversed sex roles) and often involve just one or a few endocrine components (but see Kitano et al. 2011). Thus considerable uncertainty remains about the tempo and ease with which these mechanisms evolve. Further, few comparative studies of this kind have also examined individual variation alongside population or species variation, yet a focus on individual variation is essential because evolution requires individual variation to produce change (Bennett 1987; Williams 2008).

D. Variation between Dark-Eyed Junco Subspecies in T-Mediated Phenotype

The junco system provides an excellent tool for diving deeper into the mechanisms by which hormone-mediated traits evolve because relationships between testosterone (T) and phenotype have been so well characterized among individuals within a population (reviewed in chapter 6, this volume), yet the junco species complex is made up of closely related groups (chapter 8, this volume) that have diverged in many of the traits associated with T. Our approach has been to contrast the well-studied Carolina junco (*Junco hyemalis carolinensis*) breeding in the Appalachian mountains of Virginia to the closely related white-winged junco subspecies (*J. h. aikeni*) that is endemic to the Black Hills of South Dakota. We began by asking whether phenotypes varied between populations and then explored the extent to which phenotypic variation was predicted by mechanisms of T production or response. Critically, we also moved beyond group differences, investigating the mechanisms underlying phenotypic variation among individuals within each subspecies (chapter 7, this volume). By combining a comparative approach with an individual-based approach, we were able to use the junco to ask whether the outcomes of past evolutionary processes mirrored the patterns of individual covariation upon which selection may currently act. For example, did the endocrine measure(s) (e.g., hormone levels, hormone receptor) that covaried with phenotype among individuals of a population also differ between populations according to phenotype? Further, did we identify a mechanism whereby multiple traits may diverge simultaneously (e.g., from a change in circulating hormone), or independently (e.g., from a change in receptor abundance in a particular tissue)? Were multiple endocrine measures correlated in their expression among individuals or populations?

We focused first on subspecies differences in traits previously identified as relating to T in Carolina juncos. McGlothlin and colleagues had shown that both territorial aggressiveness and the amount of white ornamentation in the tail feathers in male Carolina juncos were positively correlated to T levels (Ketterson et al. 2009; McGlothlin et al. 2008; McGlothlin et al. 2007). Using field studies with comparable methodology, we demonstrated that territorial aggression (Bergeon Burns et al. 2013) and ornamentation (Bergeon Burns et al., unpublished; figure 11.2A) were each expressed at significantly higher levels in the larger white-winged junco as compared to the Carolina junco. However, contrary to the predictions of phenotypic integration, free-living white-winged juncos did not have significantly higher T levels than Carolina juncos (Bergeon Burns et al., unpublished) when sampled after a standardized injection of gonadotropin-releasing hormone (i.e., GnRH challenge) that stimulates the HPG axis to produce maximal levels of T (i.e., $T_{potential}$ [Goymann et al., 2007]). Similarly, the two subspecies did not significantly differ in T levels measured after a short simulated territorial intrusion in the wild (Bergeon Burns et al. 2013; figure 11.2B), though the white-winged juncos were significantly more variable in T levels than the Carolina juncos, suggesting a potential subspecies difference in some component of the control of T secretion.

With respect to variation among individuals, we found that a relationship between ornamentation and T appears conserved between these two subspecies. That is, in white-winged juncos, males that were more ornamented with white plumage showed a greater ability to elevate testosterone (Bergeon Burns et al., unpublished), just as had been demonstrated in Carolina juncos. In contrast, a relationship between T and aggression was not always conserved across subspecies. Individual variation in aggression was not related to $T_{potential}$ in the white-winged junco (Bergeon Burns et al., unpublished) as it had been in Carolina juncos. However, song rate during a brief simulated territory intrusion did positively correlate with T sampled immediately afterward in both subspecies (Bergeon Burns et al. 2013). Our results indicated some conserved and some variable hormone-phenotype relationships between subspecies. These results were not consistent with tight phenotypic integration via circulating levels of T (Ketterson et al. 2009), as we did not observe a T-mediated change in a suite of traits as one unit.

The contrast between subspecies in the patterns of covariation between aggression and T could perhaps be attributed to plastic responses to environmental differences in South Dakota versus Virginia. We experimentally tested this hypothesis by examining T response to GnRH in Carolina and white-winged juncos in a common aviary, where abiotic and biotic environment were held constant (Bergeon Burns et al. 2014). Similar to our studies on free-living birds, we found that both unmanipulated and post-GnRH challenge T levels were indistinguishable between these two subspecies (figure 11.2C), despite persistent subspecies differences in morphology. This finding suggests that the lack of subspecies variation in T production was not simply a consequence of variable environments. These findings provide some of the first evidence that even very closely related groups exhibit phenotypic independence marked by altered relationships between hormone and traits and suggest that hormone-mediated traits may become decoupled from systemic hormone signal over relatively rapid evolutionary time.

E. Looking Upstream of T: Integration or Independence along HPG Axis

Past research has shown that gonadal T production is a repeatable individual "trait" in juncos: after administration of an injection of GnRH, some individuals reliably produce a lot of T, whereas others reliably pro-

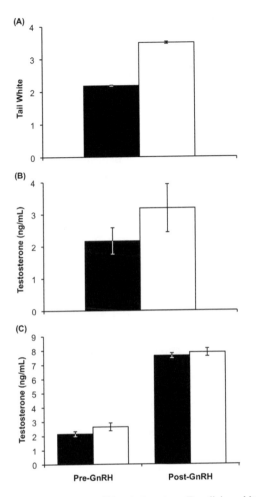

FIGURE 11.2. Subspecies comparisons of T and phenotype. Free-living white-winged juncos (open bars) have significantly more tail white than Carolina juncos (closed bars; A: Bergeon Burns et al., unpublished). Despite this and other phenotypic differences, the subspecies do not differ in T following a short simulated territory intrusion in the wild (B: Bergeon Burns et al. 2013), nor before or following a GnRH challenge in a controlled captive environment (Bergeon Burns et al. 2014).

duce less. We have begun to identify the sources of this individual variation in HPG axis responsiveness by looking *upstream* of the gonadal T response. Importantly, we have also explored differences in functioning of the HPG axis in divergent junco populations to gain insight into the degree to which components of the endocrine system evolve separately versus as a unit. Determining sources of individual and population variation, including the extent to which multiple components of the endocrine pathway are linked in their expression as populations diverge, has important and understudied evolutionary implications, as we discussed above.

We measured the amount of LH produced in response to an injection of GnRH and the amount of T produced in response to an injection of LH. We also measured gonadal sensitivity to LH (i.e., mRNA abundance for gonadal LHR) and hypothalamic sensitivity to negative feedback (i.e., mRNA abundance for AR and AROM in the rostral hypothalamus). Importantly, we measured all these components of the HPG axis in the same individuals, in both white-winged and Carolina juncos. Thus, our experiment is among the first to examine variation among individuals and between closely related taxa along several interrelated components of the HPG axis. We demonstrated that subspecies were similar in LH response to GnRH and in T response to GnRH. However, Carolina juncos expressed greater transcript abundance for LHR in the gonad than white-winged juncos. This result was surprising at first, because greater abundance of LHR could be seen as a mechanism for enhanced output of T by the gonad in response to LH stimulation from the pituitary. Yet we did not detect subspecies differences in either T or LH in circulation (Bergeon Burns et al. 2014). It is possible that subspecies differences in LHR transcript abundance were not translated into differences in the actual number of receptors present to bind LH (e.g., due to posttranscriptional modification). However, additional observations suggest that this seemingly puzzling relationship may reflect subspecies differences in negative feedback.

For instance, we found greater levels of AR transcript in the rostral hypothalamus in Carolina males, potentially indicating increased sensitivity of the hypothalamus to gonadal testosterone (Bergeon Burns et al. 2014). These results make intuitive sense when we reconsider that the greater sensitivity to LH at the level of the gonad appears to be coupled with apparently greater negative feedback at the level of the hypothalamus (figure 11.3A). In other words, the HPG axis of the Carolina juncos may be more dynamic than the HPG of the white-winged junco, potentially producing similar *peak* hormone levels as white-winged juncos when measured at a single time point, as we observed, but returning more quickly to baseline owing to greater feedback inhibition (figure 11.3B). If true, then white-winged juncos may experience greater exposure to T over time, possibly accounting for the enhanced expression of a number of T mediated traits in this subspecies compared to the Carolina

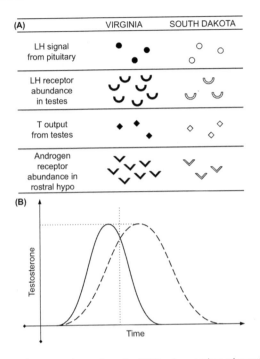

FIGURE 11.3. Subspecies comparisons along the HPG axis, overview schematic. (A) Carolina juncos from Virginia have significantly more mRNA for luteinizing hormone (LH) receptor in the testes and androgen receptor in the hypothalamus when compared to white-winged juncos from South Dakota, yet the subspecies do not differ in circulating LH or T levels (Bergeon Burns et al. 2014). (B) Greater LH sensitivity in the testes could lead to a faster rise in T, while greater feedback sensitivity to T in the hypothalamus could result in a faster fall to baseline in Carolina juncos (solid line) as compared to white-winged juncos (dashed line). Such a scenario could result in similar peak T levels, yet potentially large differences in duration of T elevation across subspecies.

subspecies. More work is needed to directly compare the subspecies for the time course of HPG axis reactivity and the mechanisms that might underlie any differences (e.g., variation in gonadotropin-inhibitory hormone or cross talk with HPA axis; Bentley et al. 2009; Viau 2002). Our results nevertheless point to a pattern of altered hormone sensitivity at multiple levels of the HPG axis across subspecies. An important future direction will be to determine whether additional components of HPG axis sensitivity change concurrently (e.g., pituitary sensitivity to GnRH, gonadal sensitivity to negative feedback, etc.), as these data will yield further insight into the mechanisms of divergence in the functioning of many interrelated components of a dynamic endocrine axis.

Interestingly, in both subspecies, we found that the amount of T a male produced in response to a GnRH injection (which directly stimulates the pituitary) was remarkably similar to the amount of T produced in response to an LH injection (which directly stimulates the gonad), but neither LH levels nor LHR abundance predicted these repeatable individual differences in T responses. Thus our findings indicate that T responses are repeatable but are not driven by variation at the level of the pituitary. Rather, our results implicate testis mass as a significant predictor of individual T responses (Bergeon Burns et al. 2014; see chapter 7 for elaboration). We conclude from these studies that, *within* each subspecies, meaningful, functional variation among individuals in T production resides primarily in just one level of the HPG axis: the gonad. On the contrary, variation *between* subspecies is found at multiple levels of the HPG axis, yet at each level, receptor abundance may more readily diverge than circulating hormone levels.

F. Looking Downstream of T: Integration or Independence of T Response Mechanisms

There is a large literature that suggests that hormone-mediated traits, including aggressive behavior, may be a function of variation in the brain's ability to process sex steroids (Adkins-Regan 2005; Grunt and Young 1952; Shaw and Kennedy 2002; Young et al. 2006). As described above, we previously identified subspecies differences in the degree to which individual variation in some aggressive measures related to the ability to produce T. We also found that white-winged juncos were more aggressive than Carolina juncos on average, but there was no significant difference between subspecies in average T levels. One possible explanation for this phenotypic independence (i.e., change in hormone-mediated phenotype without a change in circulating hormone) is that the subspecies differ in neural sensitivity to T. Thus, we returned to the field, and we looked *downstream* of the gonadal T signal, at neural targets involved in aggressive behavior across individuals of these two divergent subspecies (Bergeon Burns et al. 2013).

We hypothesized that our subspecies difference in aggression would be seen in the relationships between neural sensitivity to T and aggressive behavior. We therefore asked whether the subspecies showed similar or divergent neuroendocrine mechanisms of aggression among individu-

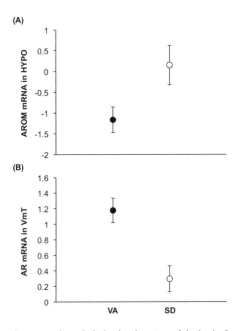

FIGURE 11.4. Subspecies comparisons in behavioral centers of the brain. Less aggressive Carolina juncos from Virginia have (A) significantly less AROM mRNA in the hypothalamus and (B) significantly more AR mRNA in the ventromedial telencephalon than white-winged juncos from South Dakota. Source: Bergeon Burns et al. 2013

als and whether subspecies differences in measures of neural sensitivity to sex steroids predicted subspecies differences in aggression. Unexpectedly, we found that Carolina and white-winged juncos differed in abundance of mRNA for sex steroid-processing molecules in the opposite direction than would be predicted by their behavioral differences: the *more* aggressive white-winged subspecies had *lower* sensitivity to steroids in tissues containing the avian medial amygdala, a brain area associated with aggression, but *greater* sensitivity to steroids in the hypothalamus, a potential site of negative feedback (figure 11.4).

Despite these surprising subspecies differences, we found evidence for shared neuroendocrine mechanisms of aggression among individuals within each of the two populations. For example, among males of each subspecies, we saw significant negative correlations between hypothalamic steroid sensitivity and both T and aggressive behavior, and we also found positive correlations between T and song rate. Our analyses therefore show that both subspecies have meaningful individual variation in T signal and in T sensitivity, suggesting that phenotypic evolution could theoretically occur via changes in either signal or sensitivity (Bergeon Burns et al. 2013), though our subspecies comparison once again identified divergence in sensitivity but not in signal. Combined, our work points to the importance of understanding the many links in the complex mechanistic pathway from environmental stimulus to phenotypic response in order to better understand how phenotypes evolve as populations diverge.

G. Implications for the Evolution of Endocrine Mechanisms

An important finding emerging from these studies is that mean differences between populations in endocrine parameters and phenotypes did not predict patterns of within-population covariation between endocrine parameter and phenotype. For example, we found that several different measures of target tissue sensitivity to steroids varied between these phenotypically divergent populations, but the same endocrine measures were not predictive of differences in phenotype among individuals. These findings are consistent with population genetic theory, as natural selection on a trait limits future additive genetic variation for that trait within a population (Fisher 1930). However, the opposite pattern was also true: current patterns of within-population covariation (on which selection may act) did not necessarily map onto differences among populations (which may reflect outcomes of past evolutionary pressures). For example, T predicted aggression among individuals, but T did not vary between subspecies that differed in aggression. The incongruence between patterns of covariation at the subspecies versus individual level provides a cautionary tale about overinterpreting or overextending results at either level of analysis. However, just as the pathway to understanding evolution should unite both organismal and mechanistic perspectives, we argue that combining the study of individual variation upon which selection may act with an identification of outcomes of past selection via the comparative method allows for richer insights into the mechanisms of divergence.

Specifically, these combined perspectives hold implications for the process by which these junco populations might have diverged. Might the disparity between inter- and intrapopulation patterns of covariation have resulted from drift or founder effects (rather than natural selection) occurring during the rapid postglacial divergence of the genus *Junco* (Milá et al. 2007)? If divergence between white-winged and Carolina subspecies resulted at least in part from random evolutionary processes, perhaps it is not surprising that patterns of individual variation largely did not extend to population differences. In addition, our results—that two subspecies differing in T-mediated traits do *not* differ in T—contrast with findings of divergence in both phenotype *and hormone levels* in the recently diverged California populations (chapter 10, this volume). Thus an additional explanation for our results is that initial divergence might be facilitated by shifts in hormone signal, whereas the multiple changes in hormone sensitivity that we documented might require longer divergence times. Few studies have explored comparisons of this kind among multiple related populations that vary in degree of divergence. In closing, we urge further investigation of intra- and interpopulation variation in these multiple levels of testosterone production and response. We suggest that the junco species complex provides an extraordinary opportunity to understand variation in the mechanisms underlying hormone-mediated phenotypic evolution.

Acknowledgments

Funding for this work was provided by the NIH (F32HD068222 to KAR and T32HD049336 training grant to both authors) and the NSF (predoctoral fellowship and DDIG IOS-0909834 to CMBB). We are grateful for the assistance of many individuals in both the field and laboratory that helped make the research reviewed here possible.

References

Adkins-Regan, E. 2005. *Hormones and Animal Social Behavior*. Princeton: Princeton University Press.

———. 2008. Do hormonal control systems produce evolutionary inertia? *Philosophical Transactions of the Royal Society B-Biological Sciences* 363:1599–1609.

Ball, G., and J. Balthazart. 2004. Hormonal regulation of brain circuits mediating male sexual behavior in birds. *Physiology & Behavior* 83:329–46.

Bennett, A. F. 1987. Interindividual variability: An underutilized resource. In *New Directions in Ecological Physiology*, edited by M. E. Feder, A. F. Bennett, W. W. Burggren, and R. B. Huey. New York: Cambridge University Press.

Bentley, G. E., T. Ubuka, N. L. Mcguire, R. Calisi, N. Perfito, L. J. Kriegsfeld, J. C. Wingfield, and K. Tsutsui. 2009. Gonadotrophin-inhibitory hormone: A multifunctional neuropeptide. *Journal of Neuroendocrinology* 21:276–81.

Bergeon Burns, C. M., K. A. Rosvall, T. P. Hahn, G. E. Demas, and E. D. Ketterson.

2014. Examining sources of variation in HPG axis function among individuals and populations of the dark-eyed junco. *Hormones and Behavior* 65:179–87.

Bergeon Burns, C. M., K. A. Rosvall, and E. D. Ketterson. 2013. Neural steroid sensitivity and aggression: Comparing individuals of two songbird subspecies. *Journal of Evolutionary Biology* 26:820–31.

Canoine, V., L. Fusani, B. Schlinger, and M. Hau. 2007. Low sex steroids, high steroid receptors: Increasing the sensitivity of the nonreproductive brain. *Developmental Neurobiology* 67:57–67.

Canoine, V., and E. Gwinner. 2002. Seasonal differences in the hormonal control of territorial aggression in free-living European stonechats. *Hormones and Behavior* 41:1–8.

Cheng, M. F., M. Chaiken, M. Zuo, and H. Miller. 1999. Nucleus taenia of the amygdala of birds: Anatomical and functional studies in ring doves (*Streptopelia risoria*) and European starlings (*Sturnus vulgaris*). *Brain Behavior and Evolution* 53:243–70.

Cornil, C. A. 2009. Rapid regulation of brain oestrogen synthesis: The behavioural roles of oestrogens and their fates. *Journal of Neuroendocrinology* 21:217–26.

Finch, C. E., and M. R. Rose. 1995. Hormones and the physiological architecture of life-history evolution. *Quarterly Review of Biology* 70:1–52.

Fisher, R. 1930. *The Genetical Theory of Natural Selection.* Oxford: Clarendon Press.

Foidart, A., B. Silverin, M. Baillien, N. Harada, and J. Balthazart. 1998. Neuroanatomical distribution and variations across the reproductive cycle of aromatase activity and aromatase-immunoreactive cells in the pied flycatcher (*Ficedula hypoleuca*). *Hormones and Behavior* 33:180–96.

Folstad, I., and A. J. Karter. 1992. Parasites, bright males, and the immunocompetence handicap. *American Naturalist* 139:603–22.

Gonçalves, D., J. Saraiva, M. Teles, R. Teodósio, A. V. M. Canário, and R. F. Oliveira. 2010. Brain aromatase mRNA expression in two populations of the peacock blenny *Salaria pavo* with divergent mating systems. *Hormones and Behavior* 57:155–61.

Goodson, J. L., and A. K. Evans. 2004. Neural responses to territorial challenge and nonsocial stress in male song sparrows: Segregation, integration, and modulation by a vasopressin v-1 antagonist. *Hormones and Behavior* 46:371–81.

Goymann, W., M. M. Landys, and J. C. Wingfield. 2007. Distinguishing seasonal androgen responses from male-male androgen responsiveness—revisiting the challenge hypothesis. *Hormones and Behavior* 51:463–76.

Grunt, J. A., and W. C. Young. 1952. Differential reactivity of individuals and the response of the male guinea pig to testosterone propionate. *Endocrinology* 51:237–48.

Hadley, M. E., and J. E. Levine. 2007. *Endocrinology.* Upper Saddle River, NJ: Pearson Prentice Hall.

Hau, M. 2007. Regulation of male traits by testosterone: Implications for the evolution of vertebrate life histories. *Bioessays* 29:133–44.

Hau, M., M. Wikelski, K. K. Soma, and J. C. Wingfield. 2000. Testosterone and year-round territorial aggression in a tropical bird. *General and Comparative Endocrinology* 117:20–33.

Ketterson, E. D., J. W. Atwell, and J. W. McGlothlin. 2009. Phenotypic integration and independence: Hormones, performance, and response to environmental change. *Integrative and Comparative Biology* 49:365–79.

Ketterson, E., and V. Nolan, Jr. 1992. Hormones and life histories: An integrative approach. *American Naturalist* 140:S33–S62.

———. 1999. Adaptation, exaptation, and constraint: A hormonal perspective. *American Naturalist* 154:4–25.

Kitano, J., Y. Kawagishi, S. Mori, C. L. Peichel, T. Makino, M. Kawata, and M. Kusakabe. 2011. Divergence in sex steroid hormone signaling between sympatric species of Japanese threespine stickleback. *PLoS ONE* 6:e29253.

Lynn, S. E. 2008. Behavioral insensitivity to testosterone: Why and how does testosterone alter paternal and aggressive behavior in some avian species but not others? *General and Comparative Endocrinology* 157:233–40.

McGlothlin, J. W., J. M. Jawor, T. J. Greives, J. M. Casto, J. L. Phillips, and E. D. Ketterson. 2008. Hormones and honest signals: Males with larger ornaments elevate testosterone more when challenged. *Journal of Evolutionary Biology* 21: 39–48.

McGlothlin, J. W., J. M. Jawor, and E. D. Ketterson. 2007. Natural variation in a testosterone-mediated trade-off between mating effort and parental effort. *American Naturalist* 170:864–75.

McGlothlin, J. W., and E. D. Ketterson. 2008. Hormone-mediated suites as adaptations and evolutionary constraints. *Philosophical Transactions of the Royal Society B-Biological Sciences* 363:1611–20.

Milá, B., J. E. Mccormack, G. Castaneda, R. K. Wayne, and T. B. Smith. 2007. Recent postglacial range expansion drives the rapid diversification of a songbird lineage in the genus *Junco*. *Proceedings of the Royal Society B-Biological Sciences* 274:2653–60.

Nijhout, H. F. 2003. Development and evolution of adaptive polyphenisms. *Evolution & Development* 5:9–18.

Panzica, G. C., C. Viglietti-Panzica, and J. Balthazart. 1996. The sexually dimorphic medial preoptic nucleus of quail: A key brain area mediating steroid action on male sexual behavior. *Frontiers in Neuroendocrinology* 17:51–125.

Pfaff, D. W. 2010. Hormone/brain relations serving the unity of the body. *Physiology & Behavior* 99:149–50.

Pigliucci, M. 2003. Phenotypic integration: Studying the ecology and evolution of complex phenotypes. *Ecology Letters* 6:265–72.

Ramsay, H., L. H. Rieseberg, and K. Ritland. 2009. The correlation of evolutionary rate with pathway position in plant terpenoid biosynthesis. *Molecular Biology and Evolution* 26:1045–53.

Riters, L., M. Eens, R. Pinxten, D. Duffy, J. Balthazart, and G. Ball. 2000. Seasonal

changes in courtship song and the medial preoptic area in male European starlings (*Sturnus vulgaris*). *Hormones and Behavior* 38:250–61.

Schlinger, B. A., and G. V. Callard. 1990. Aromatization mediates aggressive-
behavior in quail. *General and Comparative Endocrinology* 79:39–53.

Schmidt, K. L., D. S. Pradhan, A. H. Shah, T. D. Charlier, E. H. Chin, and K. K.
Soma. 2008. Neurosteroids, immunosteroids, and the Balkanization of endocrinology. *General and Comparative Endocrinology* 157:266–74.

Shaw, B. K., and G. G. Kennedy. 2002. Evidence for species differences in the pattern of androgen receptor distribution in relation to species differences in an
androgen-dependent behavior. *Journal of Neurobiology* 52:203–20.

Soma, K. K., B. A. Schlinger, J. C. Wingfield, and C. J. Saldanha. 2003. Brain aromatase, 5 alpha-reductase, and 5 beta-reductase change seasonally in wild male
song sparrows: Relationship to aggressive and sexual behavior. *Journal of Neurobiology* 56:209–21.

Thompson, R. R., J. L. Goodson, M. G. Ruscio, and E. Adkins-Regan. 1998. Role of
the archistriatal nucleus taeniae in the sexual behavior of male Japanese quail
(*Coturnix japonica*): A comparison of function with the medial nucleus of the
amygdala in mammals. *Brain Behavior and Evolution* 51:215–29.

Viau, V. 2002. Functional cross talk between the hypothalamic-pituitary-gonadal
and -adrenal axes. *Journal of Neuroendocrinology* 14:506–13.

Voigt, C., and W. Goymann. 2007. Sex-role reversal is reflected in the brain of
African black coucals (*Centropus grillii*). *Developmental Neurobiology* 67:
1560–73.

Wikelski, M., M. Hau, and J. Wingfield. 1999. Social instability increases plasma testosterone in a year-round territorial neotropical bird. *Proceedings of the Royal
Society B-Biological Sciences* 266:551.

———. 2000. Seasonality of reproduction in a neotropical rain forest bird. *Ecology* 81:2458–72.

Williams, T. D. 2008. Individual variation in endocrine systems: Moving beyond the
"tyranny of the golden mean." *Philosophical Transactions of the Royal Society
B-Biological Sciences* 363:1687–98.

Wingfield, J. C. 2012. Regulatory mechanisms that underlie phenology, behavior,
and coping with environmental perturbations: An alternative look at biodiversity. *Auk* 129:1–7.

Wingfield, J. C., S. E. Lynn, and K. K. Soma. 2001. Avoiding the "costs" of testosterone: Ecological bases of hormone-behavior interactions. *Brain Behavior and
Evolution* 57:239–51.

Young, L. J., P. K. Nag, and D. Crews. 2006. Species differences in estrogen receptor
and progesterone receptor-mRNA expression in the brain of sexual and uni-
sexual whiptail lizards. *Journal of Neuroendocrinology* 7:567–76.

CHAPTER TWELVE

Mate Choice in Dark-Eyed Juncos Using Visual, Acoustic, and Chemical Cues

Danielle J. Whittaker and Nicole M. Gerlach

I judge people on how they smell, not how they look.
—Jennifer Lopez

A. Introduction

Sexually reproducing organisms need to identify suitable mates in order to avoid wasting energy in a mating attempt that may not result in viable offspring. In addition, sexual selection theory predicts that individuals should prefer the highest quality mates possible and that "indicator" traits should convey information about an individual's health or genetic quality (Andersson 1994). Thus, each instance of mate choice involves three processes: *species recognition*, in which the reproductive individual distinguishes between heterospecifics and conspecifics; *mate recognition*, in which a potential mate is determined to be the compatible sex and in breeding condition; and *mate assessment*, in which the quality of the potential mate is judged (Johansson and Jones 2007).

The various traits that are assessed for these functions are known as "cues" and can include morphological, behavioral, acoustic, and olfactory phenotypes (Johnstone 1997). Cues that have been shaped by natural selection for the purpose of communication are known as "signals" (Johnstone 1997). Species recognition cues must be divergent across species and conserved within species and are thought to be under strong stabilizing selection. Mate recognition cues must differ between the sexes and across the breeding cycle. Mate assessment cues must advertise the sender's individual identity and quality and are often under strong directional selection so that, for example, higher quality males will have more exaggerated ornaments (Johansson and Jones 2007; Ptacek 2000). Despite these differing requirements, these three functions are not mutually exclusive and can be served by a single cue or set of cues; it has been suggested that species recognition may emerge as an epiphenomenon of female choice for cues indicating high quality (Andersson 1994; Johansson and Jones 2007; Ptacek 2000).

The majority of songbirds are socially monogamous—that is, they form male/female pairs that work together over the course of the breeding season to defend a territory and/or care for offspring (Lack 1968; Mock 1985). While social monogamy is common in songbirds, genetic monogamy is not; in most species studied, at least some broods have offspring sired by a male other than the one socially paired to their mother (Griffith et al. 2002; Westneat et al. 1990). The presence of these extra-pair offspring suggests that females in these social systems must make mate choice decisions in multiple contexts. A female must choose a social mate that will help defend and care for her offspring, but she may make a different choice for the male that will sire her young. While the species recognition and mate recognition cues are likely to be the same for both types of mates, females may use very different cues to select social and genetic mates.

In this chapter, we will explore the traits used for mate recognition and assessment in the dark-eyed junco. We will first review what is known about mate preferences for visual and acoustic traits in juncos and discuss the relative importance of each, as well as the use of multiple cues in mate choice. We will then consider a third dimension of communication that is often completely overlooked when considering mate choice in birds: olfaction. Finally, we will consider the various contexts of mate choice in a socially monogamous system, the limits of female mate choice, how mate choice can be assessed by experimental and observational means, and how it may affect sexual selection and population divergence.

B. Traits and Preferences in Dark-Eyed Juncos

Juncos are socially monogamous songbirds, typically forming pairs that last for one breeding season or longer and providing biparental care once

nestlings have hatched (Nolan et al. 2002). Juncos also engage in apprecia-
ble levels of extrapair behavior: in all populations studied to date, about
20–30 percent of offspring were sired by extrapair males (Atwell et al.
2014; Ferree 2007b; Gerlach et al. 2012a; Bergeon Burns and Gerlach,
unpublished data). Thus, junco courtship and mate choice occur in the
context of both within-pair and extrapair mating. These and other characteristics of junco biology and behavior are shared by many species of
songbirds, and understanding aspects of mate choice in juncos may more
generally inform our knowledge of the mate choice process of songbirds
in general.

Male courtship typically involves several visual and acoustic elements.
Generally performed within five meters of a female, the male junco's
display includes ptiloerection, spreading of the tail feathers, fanning the
wings, hopping, bill wiping, and carrying nesting material (Nolan et al.
2002), all while singing highly variable, low-amplitude song (Reichard
et al. 2011; Titus 1998). Many of these courtship behaviors appear to function to display plumage ornaments.

B.1. Visual Cues

In general, adult juncos have gray or brown body feathers on the head,
back, and breast and a white belly. In contrast to the mostly dark tail,
the outer tail feathers are conspicuously white ("tail white"). Subspecies
differ markedly in plumage (see color plates 4 and 5). For example, the
Oregon junco (*J. h. thurberi*) has brown body feathers, with a black or
gray hood, compared to the slate-colored junco (*J. h. hyemalis*), which has
mostly gray body feathers over the back and head. Other plumage ornaments include the white wing bars of *J. h. aikeni* and the pinkish sides of
J. h. mearnsi. While the sexes are morphologically very similar, across subspecies males are generally slightly larger and have darker plumage on
the head and more tail white than females (see Milá et al., chapter 8, this
volume for more detail).

Tail white has long been suspected to be a sexually selected signal in
juncos because males have greater amounts than females and they spread
their tails during courtship revealing the outer rectrices. Female juncos
find tail white attractive: in a captive two-way mate preference study, females spent more time near and directed more sexual displays towards
males with experimentally augmented tail white (Hill et al. 1999). There
is some evidence that tail white may be condition-dependent, as birds fed

Box 12.1
What about Male Preferences for Females?

Most discussions of mate choice focus on female preferences for male traits. Even though male birds generally have the showiest plumage ornaments, in many species, females also display the same ornaments, although they are often less elaborate than the males. Do males prefer more highly ornamented females—and do these traits correlate with quality in females?

Data from choice trials are mixed, with males in some species showing a preference for more ornamented females and other showing no preference (reviewed in Amundsen 2000). In juncos, males given the choice between females with natural levels of tail white and females with experimentally increased tail white showed no preference (Wolf et al. 2004). When given the choice between the odor of females from their own population and the odor of females from a different population, male juncos again showed no preference (Whittaker et al. 2011).

While male preferences for the traits listed above are as yet unclear, males may still be making mate choice decisions based on visual and chemical cues in female juncos that may indicate fecundity. Females with smaller wings and longer tails had higher annual reproductive success (Gerlach et al., in prep) and fecundity (McGlothlin et al. 2005), suggesting that these traits may signal reproductive potential or parental ability. However, despite their effect on fecundity, these morphological features had only a weak effect on mating success in females (McGlothlin et al. 2005). Females with more "female-like" preen oil volatile profiles (that is, with lower proportions of volatile compounds that are typically found in higher proportions in male preen oil) produced more offspring, suggesting that odor may also signal fecundity in female juncos (Whittaker et al. 2013). However, male preferences for these traits are not yet known.

References

Amundsen, T. 2000. Why are female birds ornamented? *Trends in Ecology & Evolution* 15:149–155.

McGlothlin, J. W., P. G. Parker, V. Nolan, Jr., and E. D. Ketterson. 2005. Correlational selection leads to genetic integration of body size and an attractive plumage trait in dark-eyed juncos. *Evolution* 59:658–671.

Whittaker, D. J., N. M. Gerlach, H. A. Soini, and M. V. Novotny. 2013. Bird odour predicts reproductive success. *Animal Behaviour* 86:697–703.

Whittaker, D. J., K. M. Richmond, A. K. Miller, R. Kiley, C. Bergeon Burns, J. W. Atwell, and E. D. Ketterson. 2011. Intraspecific preen oil odor preferences in dark-eyed juncos (*Junco hyemalis*). *Behavioral Ecology* 22:1256–1263.

Wolf, W. L., J. M. Casto, V. Nolan, Jr., and E. D. Ketterson. 2004. Female ornamentation and male mate choice in dark-eyed juncos. *Animal Behaviour* 67:93–102.

on an enriched, high-protein diet grew larger and brighter white patches (McGlothlin et al. 2007a). Tail white, as well as plumage darkness on the head, is also associated with dominance in winter flocks (Holberton et al. 1989).

In males, the amount of tail white correlated positively with short-term increases in testosterone production in response to a physiological challenge (McGlothlin et al. 2008), and these short term increases were also correlated with levels of aggression in response to a territorial intruder (McGlothlin et al. 2007b), suggesting that tail white may be an honest signal of a male's ability to defend a territory (see chapter 6). In addition to preferring males with increased tail white, females also preferred males with experimentally elevated testosterone in mate choice trials, most likely due to the fact that these males increased their courtship intensity (Enstrom et al. 1997). When given a choice between a male with increased tail white and a male with increased testosterone, females showed individual preferences for one or the other, but neither appeared more important than the other for females as a whole, demonstrating that both plumage characteristics and courtship intensity are likely taken into account when assessing a potential mate (Hill et al. 1999).

Although females show preferences for high levels of both tail white and testosterone in laboratory preference trials (Enstrom et al. 1997; Hill et al. 1999), and though males with experimentally elevated levels of testosterone have higher levels of extrapair paternity (EPP) (Raouf et al. 1997; Reed et al. 2006), these traits do not show evidence of strong directional selection in the natural population (McGlothlin et al. 2010; Gerlach et al., in prep.). Instead, testosterone levels in males appear to be under strong stabilizing selection (McGlothlin et al. 2010). One explanation for the lack of directional selection for testosterone is that there are trade-offs inherent in high levels of testosterone—males with high levels of testosterone put more effort into mating at the expense of parenting (McGlothlin et al. 2007b), and males with experimentally elevated testosterone have lower survival rates (Reed et al. 2006). A second explanation is that females with high levels of testosterone have reduced reproductive success, suggesting a constraint on the evolution of testosterone levels in the species, resulting in stabilizing selection that represents a compromise between the two sexes (Clotfelter et al. 2004; Gerlach and Ketterson 2013; O'Neal et al. 2008; see chapter 4, this volume).

But importantly, there is no evidence that females preferentially reproduce with males with larger plumage ornaments or that they increase their investment in the offspring of such males. In *J. h. thurberi*, females did not invest more in broods of males with experimentally increased tail white; in fact, females mated to enhanced males decreased their incubation time while those males increased their nestling feeding effort (Ferree 2007a). Data from *J. h. carolinensis* in Virginia showed that although while tail white is heritable, there was no directional sexual selection on the size of this ornament (McGlothlin et al. 2005). In eighteen years of data from Virginia, tail white did not predict any aspect of male reproductive success, nor did it differ in group comparisons of successful versus unsuccessful males or between social males and the extrapair males to whom they lost paternity (Gerlach et al., in prep.). The reason for the disconnect between apparent female preference for tail white in experimental trials and the lack of evidence for mate choice based on tail white in free-living populations is not clear. However, all of the field data are concerned with genetic reproductive success; tail white could be used by females in making decisions about social rather than genetic mates. In many species, exaggerated plumage ornaments frequently correspond to reduced investment in parental care (Burley 1988; de Lope and Møller 1993; Qvarnström 1997), potentially making them less desirable in a social mate choice context when male care is important for offspring survival. Because tail white is correlated with testosterone response to a territorial challenge (McGlothlin et al. 2008), which is in turn correlated with aggressive response to that challenge (McGlothlin et al. 2007b), it may be that a large tail white ornament signals a male's potential to acquire and defend a high-quality territory. In this way, tail white may be analogous to carotenoid-based ornaments in other species, which are typically thought to signal the availability of high-quality food resources (Hill 1992; Hill & Montgomerie 1994; Wolfenbarger 1999; Casagrande 2006). Recent data reveal that tail white varies more across junco subspecies than do proxies for body size, suggesting that selection strength on this trait varies geographically, which could mean that any patterns related to tail white may be population specific (Ferree 2013).

While junco populations vary in the proportion of their tails that are covered by this white ornament, the differences between populations do not appear to correspond to equally large differences in the degree of sexual selection in these populations (as estimated by frequency of EPP). For example, *J. h. thurberi* in an urban population in San Diego County, California, had smaller amounts of tail white but similar rates of EPP compared to the *J. h. carolinensis* population in Virginia (26 percent in San Diego, compared to 28 percent in Virginia) (Atwell et al. 2014; Gerlach et al. 2012a), while EPP rates in *J. h. aikeni* in South Dakota were lower but amount of tail white was much higher (Bergeon Burns and Gerlach, unpublished data). Other conspicuous plumage ornaments, such as the white wing bars of *J. hyemalis aikeni* and the pink sides of *J. hyemalis mearnsi*, have not yet been investigated with respect to their influence on female mate choice. However, divergence in plumage among closely related populations is often inferred to be related to sexual selection (West-Eberhard 1983).

Aside from plumage, another potentially important visual cue is body size. Indices such as ratio of body mass to tarsus length are frequently used as indicators of condition, though this practice has been called into question (Schamber et al. 2009). McGlothlin et al. (2005) found that males with longer wings and shorter tails had higher mating success (as did males with longer wings and whiter tails), suggesting that a combination of morphological traits, rather than any one measurement, may be important. Data from a larger data set suggest that the increased mating success of males with longer wings resulted in increased numbers of extrapair offspring (Gerlach et al., unpublished data). Females with the opposite combination of traits—shorter wings and longer tails—had higher fecundity (McGlothlin et al. 2005) and overall reproductive success (Gerlach et al., unpublished data).

Visual cues are generally much easier to measure than traits from other sensory modalities, which is evident from the focus on visual traits in the literature. However, these traits do not appear to adequately explain variation in reproductive success in juncos, which is unsurprising given the fact that songbirds and other animals also employ other cues, such as sounds or odors, to attract potential mates.

B.2. Acoustic Cues

The role of junco song in mate choice is not well understood. Males sing two types of song: long-range song (LRS), which, as its name implies, can be heard over a long distance and serves for both mate attraction and territoriality (Titus 1998); and short-range song (SRS), which is typically sung at a low amplitude during courtship (Titus 1998) and elicits an aggressive response from males with fertile mates (Reichard et al. 2011). No tests have yet been conducted on female preference for different quality aspects of LRS, though some measurements of performance quality have been suggested (Cardoso et al. 2007). SRS is highly variable, and it is not yet known what aspects of it may be attractive to females. Given the range of transmission, it is likely that LRS is the first cue that a female uses to assess a male and decide whether or not to approach more closely. Once a female is within range, assessment of visual cues, such as plumage or courtship intensity, as well as of short-range acoustic cues like SRS, can begin. Chapter 13 by Gonçalo Cardoso and Dustin Reichard provides more details about junco song.

B.3. Use of Multiple Cues in Avian Mate Choice

In closely related sympatric bird species, the species are typically divergent in both male plumage patterns and the corresponding female preferences for conspecific plumage, thus reducing the frequency of hybridization (Sætre et al. 1997). Plumage is often more divergent than song in closely related species (Uy et al. 2009) and is thought to be more "reliable" than song as a species recognition cue because it is inherited genetically rather than culturally (Baker and Baker 1990). However, after testing females from two hybridizing bunting species, Baker and Baker (1990) concluded that visual and vocal traits each explained about half of female mate preferences. The need for multiple signals suggests that plumage ornamentation may be less reliable as a species recognition cue than predicted, perhaps because female preferences are not always divergent. In allopatric populations, females often share the same preferences for ornamentation, so that females from different populations may all prefer the

more extreme males, even if the more extreme males are heterospecifics (Collins and Luddem 2002; Price 1998). As noted previously, captive
female juncos demonstrated a preference for males with experimentally
increased tail white—even when the amount of tail white was far larger
than what is typically found in most male juncos from any subspecies (Hill
et al. 1999). Thus, female preferences for tail white suggest that this plumage trait is not likely used as a recognition cue in juncos.

Relying on multiple cues during mate choice can increase accuracy
in recognizing a conspecific (Ward and Mehner 2010) and decrease the
time and energy spent mate searching (Candolin 2003). Since vertebrate
chemical signals show signs of adaptation to reduce energetic costs (Alberts 1992), olfaction could be an inexpensive and efficient way to improve mate recognition and assessment. In the next section, we explore
the use of chemical communication in avian behavior and mate choice.

C. The Role of Olfaction in Avian Mate Choice

The two most obvious features of most songbirds to a human observer
are the bright colors and the distinctive songs of the males. Perhaps for
this reason, nearly all analyses of avian mate choice have focused on the
two senses of sight and hearing, to the exclusion of other senses such as
olfaction. Smell has long been thought to be unimportant in avian behavior and ecology, as most birds have very small olfactory bulbs relative to brain size (Bang and Cobb 1968) and all birds completely lack a
vomeronasal organ, which is an important part of the accessory olfactory
system in other taxa (Balthazart and Taziaux 2009). However, olfaction
is known to play an important role in mate attraction, recognition, and
choice in most other taxa, and recent analyses of odors produced by birds
and birds' abilities to detect them have begun to suggest that smell plays
an important role in avian mate choice.

Vertebrates have a variety of glands that may secrete potentially odor-
ous substances, such as the femoral pores of lizards or the sweat glands
of mammals. However, most birds have only one large exocrine gland,
the uropygial or "preen" gland, which produces a thick waxy substance
known as preen oil (Jacob and Ziswiler 1982). Birds use their bills to
stimulate the release of oil from this gland, and they then spread the oil
over their feathers while preening. The oil helps protect the feathers from
degradation due to environmental exposure (heat, sun, cold) and also is

thought to help deter ectoparasites and feather-eating bacteria (Douglas 2008; Moyer et al. 2003).

C.1. Odor Production

Preen oil gives off volatile, airborne compounds that lend a bird its scent. These volatile compounds vary qualitatively among bird species—in general, while there is some overlap, especially among closely related species, different species of birds produce a different range of volatile compounds (Haribal et al. 2009; Haribal et al. 2005; Mardon et al. 2010; Soini et al. 2013). Within species, most individuals appear to have the same set of compounds but vary quantitatively in the amounts and proportions of those compounds that make up an individual's unique or signature blend (Whittaker et al. 2010; Zhang et al. 2010). These compounds typically include alcohols, aldehydes, methylketones, and fatty acids, similar to volatile compounds produced by many mammalian scent glands (Burger 2005; Soini et al. 2007).

Dark-eyed junco preen oil contains nineteen volatile organic compounds that have been found to increase significantly in breeding-condition birds compared to winter-condition birds, suggesting that these compounds may play an important role in sexual signaling and breeding ecology (Soini et al. 2007). These compounds include nine linear alcohols, five methyl ketones, and five carboxylic acids (Soini et al. 2007; Whittaker et al. 2010). Several of these compounds differ in relative concentration between the sexes. Male juncos typically have a much higher proportion of the methyl ketones 2-tridecanone and 2-pentadecanone, while females have higher proportions of the linear alcohols 1-decanone and 1-undecanone and the carboxylic acids dodecanoic acid, tetradecanoic acid, and hexadecanoic acid (Whittaker et al. 2013; Whittaker et al. 2010). Individual volatile profiles were repeatable, suggesting that individuals can be reliably distinguished on the basis of odor (Whittaker et al. 2010). Closely related junco populations were significantly different with respect to volatile profiles, suggesting that odor may have a genetic basis (Whittaker et al. 2010). In another bird species, odor similarity was found to reflect genetic similarity (Leclaire et al. 2012).

As stated, these compounds in preen oil are typically present in very low concentrations in birds in winter condition but increase when birds go into breeding condition (Soini et al. 2007). At the long-term study site at Mountain Lake Biological Station in Virginia, male volatile compound concentrations increased steadily over the first four weeks of the breeding season (Whittaker et al. 2011b). In females, the concentrations increased quickly between week one and week three and then dropped in week four (figure 12.1). The peak date for first egg date that year was also in week three, suggesting a correspondence between preen oil volatile compound concentration and egg laying in females (Whittaker et al. 2011b). Circulating testosterone levels are known to follow a similar seasonal pattern (Ketterson et al. 2005), but while concentrations of some volatile compounds are responsive to experimentally increased testosterone levels, exogenous testosterone treatment failed to replicate breeding season concentrations (Whittaker et al. 2011b).

C.2. Odor Detection

Birds, particularly seabirds, are known to be capable of distinguishing odors and using olfactory cues to navigate and find food and nesting sites (Bonadonna and Bretagnolle 2002; Nevitt et al. 2008; Wallraff 2004). Evidence for intraspecific chemical communication has also been found in several seabird species (Bonadonna and Nevitt 2004; Bonadonna and Sanz-Aguilar 2012; Hagelin 2007) and in a parrot (Zhang et al. 2010). Passerine songbirds were previously thought to be the least likely group of birds to have strong olfactory capabilities due to the fact that they have the smallest olfactory bulbs relative to brain size among all birds (Bang and Cobb 1968), but genomic-level studies have revealed that songbirds have olfactory receptor repertoires of comparable size to that of seabirds (Steiger et al. 2008; Steiger et al. 2009). Furthermore, passerines have been shown to use odor as a navigational cue (Wallraff et al. 1995), to detect predator odor (Amo et al. 2008; Roth et al. 2008), and to find and replace antiparasitic herbs in the nest (Petit et al. 2002).

Songbirds, including juncos, have also been demonstrated to use and respond to odor in the context of reproduction. Zebra finches (*Taeniopygia guttata*) recognized the odor of their natal nest (Caspers and Krause 2010) and distinguished between kin and nonkin on the basis of odor (Krause et al. 2012). Nesting female juncos in Virginia responded to heterospecific (mockingbird) and unfamiliar conspecific preen oil applied to the nest by temporarily reducing the length of incubation time but showed no change in incubation in response to their own preen oil or a vehicle-only control (Whittaker et al. 2009).

Junco odor preferences in captive choice experiments also give some

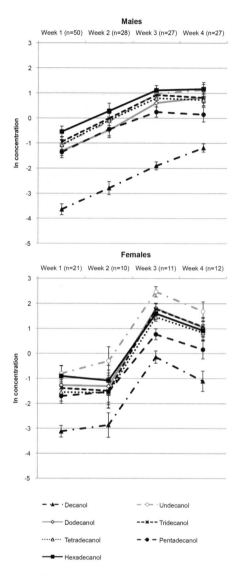

FIGURE 12.1. Mean ln-transformed preen oil linear alcohol relative concentrations (relative amounts in 100 μl of preen oil) in juncos, for each week during early season 2008, in males (top) and females (bottom). Sample sizes for each week are indicated in parentheses. Error bars indicate standard error of the mean. Adapted from Whittaker et al. 2011b.

clues as to how odor may play a role in mate preferences and assessment. In Y-maze trials, both male and female juncos spent more time with preen oil from males than females (Whittaker et al. 2011a). This preference was interpreted as possible territorial behavior from males and mate attraction for females. In the same study, when females were given the choice between the preen oil from males of their own population or subspecies or males from a different population or subspecies, females showed no preference for their own group. They also appeared to show no preference when given a choice between the odor of males (from their own population) with low or high tail white. Instead, a post hoc analysis showed that females tended to prefer the odor of males with smaller wings, regardless of population or subspecies (Whittaker et al. 2011a). Odor did correlate with wing length in juncos: birds with longer wings had a more "male-like" odor profile, with higher relative proportions of the methyl ketones 2-undecanone through 2-pentadecanone and lower proportions of 1-decanol and 1-undecanol (Whittaker et al. 2011a). Thus, when given a choice between the odor of two males, females preferred the more "female-like" odor of males with smaller wings and avoided the scent of males with larger wings. Because males with longer wings have previously been shown to have higher reproductive success (as measured by number of mates, McGlothlin et al. 2005), this result was unexpected. Males with less attractive visual traits may invest in a more attractive odor. On the other hand, in an enclosed setting such as a Y-maze, females may prefer to avoid the scent of an unseen larger and potentially more aggressive male, and this preference may not be very informative with respect to mate choice (Whittaker et al. 2011a).

C.3. Odor as a Signal of Quality

Studies from mammals and other taxa suggest that odor may signal the individual's parasite load, diet, or other aspects of condition (Liang and Silverman 2000; Penn and Potts 1998; Zala et al. 2004). Little is known about the relationship between avian odor and condition, though preen gland size is correlated with body condition and immunocompetence (Moreno-Rueda 2010). With the exception of wing length (see previous section) (Whittaker et al. 2011a), few morphological, physiological (Whittaker et al. 2011b), or behavioral traits have been found to covary with preen oil volatile composition. However, one study suggests that avian odor may predict individual reproductive success and thus may be a reliable cue for birds assessing the quality of potential mates.

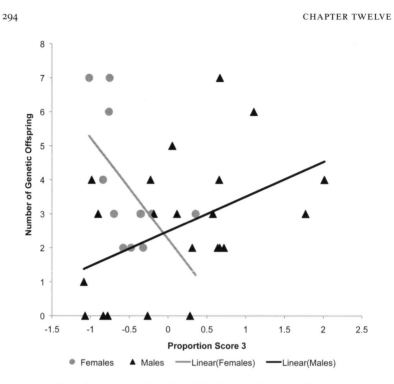

FIGURE 12.2. Reproductive success in male and female juncos is predicted by the relative concentration of preen oil volatile compounds (proportion score). Source: Whittaker et al. 2013

In a study of juncos in Virginia over one breeding season, volatile compound concentrations were strongly correlated with genetic reproductive success (Whittaker et al. 2013). Males with higher proportions of compounds that are associated with "male-like" odor (2-tridecanone, 2-tetradecanone, and 2-pentadecanone) sired significantly higher numbers of offspring (figure 12.2, reproduced from Whittaker et al. 2013). Females showed the opposite relationship, such that females with lower proportions of these same compounds produced higher numbers of offspring—thus, females with a more "female-like" odor had higher fecundity (figure 12.2). In males, the concentrations of a number of linear alcohols were significantly positively correlated with the number of social offspring (including their own offspring and extrapair offspring in the home nest) that successfully fledged from the nest. This correlation suggests that males with higher concentrations of these compounds may also have provided good paternal care, for example increased rates of offspring feeding or more effective defense against nest predators. Alternatively, these males may have been high-quality in some way unrelated to parental ability, leading to increased maternal investment in the offspring and resulting in higher reproductive success (differential allocation hypothesis, Burley 1988; Sheldon 2000). Finally, males with higher concentrations of compounds associated with a female-like profile (the carboxylic acids dodecanoic acid, tetradecanoic acid, and hexadecanoic acid) lost more paternity in their home nest to extrapair fertilizations, suggesting that females may be able to evaluate potential social and extrapair mates on the basis of different aspects of odor (Whittaker et al. 2013).

Odor is often thought to contain information about an individual's genetic quality, in particular their major histocompatibility complex (MHC) genotype. The MHC gene family plays a major role in mobilizing the adaptive immune response in vertebrates (Klein 1986) and may be important in mate choice (Penn and Potts 1999). MHC genes are highly variable, particularly at the region that interacts with peptides from pathogens, and heterozyogsity at MHC loci is thought to increase an individual's ability to combat a greater variety of diseases (Penn et al. 2002). Evidence from many species, including birds, suggests that animals prefer mates with different MHC alleles from their own, thereby increasing the heterozygosity of their offspring as well as avoiding breeding with relatives (Jordan and Bruford 1998; Landry et al. 2001; Setchell and Huchard 2010; Zelano and Edwards 2002). Odor appears to vary with MHC heterozygosity or genotype in mice and other mammals (Novotny et al. 2007; Willse et al. 2006) but such evidence is not yet available for most birds. The MHC in passerines studied to date—including juncos—is very complex with extensive gene duplication, making it difficult to draw connections between odor and genotype (Bollmer et al. 2010; Whittaker et al. 2012; Zagalska-Neubauer et al. 2010).

C.4. Potential Olfactory Displays During Courtship

Bill wiping is the act of rubbing both sides of the bill against a branch or other substrate and in most contexts functions for cleaning or honing the bill, especially after eating (Cuthill et al. 1992). Bill wiping is also observed during social interactions, especially competitive interactions or courtship. Early studies examined this behavior and dismissed it as irrelevant or "displacement behavior" (Andrew 1956; Clark 1970). However, given that birds use the bill to spread preen oil on their feathers, and that trace amounts of substances from the preen gland may remain on the bill

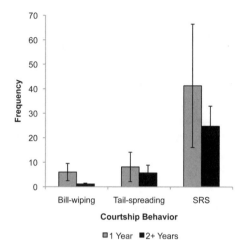

FIGURE 12.3. Courtship behaviors by young versus older males. Only bill-wiping differs significantly between the two age classes (p = 0.028). Adapted from Whittaker et al. 2015.

after preening (Whittaker, pers. obs.), one hypothesis suggests that bill wiping during social interactions could function as an olfactory display (Whittaker et al. 2015).

Preliminary results from a study of bill wiping during social interactions, induced by ten-minute simulated territorial intrusions with a caged conspecific paired with a playback, suggest that bill wiping is likely an integral part of the junco courtship ritual (Whittaker et al. 2015). Male subjects bill wiped significantly more frequently in response to a female conspecific and playback of a female precopulatory trill than to a male conspecific and LRS playback (p = 0.012). In response to a female, bill wiping was strongly positively correlated with frequency of other courtship-related behaviors, including tail spreading and singing short-range song. Younger males, and males with shorter wings, bill wiped more frequently than older males and males with longer wings—but did not differ in the frequency of tail spreading or SRS, suggesting that the difference was not due to an overall increase in courtship intensity (figure 12.3).

In conjunction with the captive evidence that female juncos preferred the odor of males with shorter wings, these data suggest that shorter-winged males and younger males (note that wings tend to get longer with age, Nolan et al. 2002) may be investing more into an alternative attractive olfactory ornament, while larger or older males may invest more in visual ornaments such as attractive plumage (Whittaker et al. 2015).

D. Mate Choice in Context

D.1. Within-Pair vs. Extrapair Mate Choice

The benefits of choosing one male over another have been explored in depth elsewhere (Andersson 1994) but can in general be divided into two main categories: direct benefits, which increase an individual's reproductive success by increasing the number of offspring they are able to produce; and indirect benefits, which increase reproductive success by increasing offspring quality. In socially monogamous mating systems, a female's choice of social mate is most likely to have an effect on her direct reproductive success. A social partner that provides high levels of parental care, or can defend a high-quality territory, will likely increase the number of offspring a female is able to raise.

However, most discussions of female mate choice focus on her choice of genetic sires and how the selected male will contribute to her offspring's quality. Females may select their mates based on some trait that indicates physical attractiveness, general longevity, or health (Forstmeier et al. 2002; Hasselquist et al. 1996; Houtman 1992; Kempenaers et al. 1992; Møller and Tegelstrom 1997), thus securing these high-quality genes for their offspring. Other data suggest that instead of an absolute scale of quality, sires may be chosen for their genetic compatibility with the female—in other words, to avoid inbreeding and/or increase offspring heterozygosity (Eimes et al. 2005; Foerster et al. 2003; Freeman-Gallant et al. 2003).

In a socially monogamous mating system, the choice of social partner can be dissociated from the choice of the genetic sire, allowing females to base these decisions on different criteria. Females do not necessarily choose different males for social versus genetic mate; indeed, in most species it is typical for the majority of offspring to be sired by the female's social mate (Griffith et al. 2002). However, some females may pair with a social partner that would be a suboptimal genetic sire, either because the preferred male(s) already had a social partner or because females may select social mates for traits (such as parental care) that do not correspond to genetic quality or compatibility. In either case, these females may improve the quality of their offspring by mating with an extrapair male.

D.2. Limits of Female Mate Choice

In this chapter, we have focused on female mate choice, but there are other factors to consider. Forced copulation is rarely observed in birds (with the exception of ducks and geese, reviewed in Gowaty and Bus-chhaus 1998). However, males may limit female choice in a number of ways, including mate guarding to prevent females from engaging in extrapair copulations (e.g. in Seychelles warblers, Komdeur et al. 1999) and sperm competition (Birkhead and Møller 1992).

Males may also limit female choice in less direct ways through intrasexual competition. Larger male juncos are more successful in competing for food during the winter (Balph et al. 1979; Holberton et al. 1989; Ketterson 1979), which may affect their ability to survive through the winter. In junco populations that migrate, males generally arrive on the breeding grounds a week or two before females (Nolan et al. 2002) and compete for territories. In this way, the males that a female encounters—and the traits that a female may choose among—may be limited by male-male competition.

D.3. Do Experimental "Preferences" Tell Us Anything about Real Mate Choice?

Two-way choice trials are a common way to test for individual preferences for particular traits. They have been used in birds to understand sexual selection for auditory (e.g., Neubauer 1999), visual (e.g., Hill et al. 1999), and even chemical (Whittaker et al. 2011a; Zhang et al. 2010) traits. In a typical choice scenario, a female in full breeding condition is given the choice between two unfamiliar males, and time spent near each male is interpreted as a measure of preference. In captive zebra finches, a female's preference during these trials has been shown to be predictive of her eventual social mate choice in a free-flight mixed-sex aviary (Clayton 1990; Rutstein et al. 2007), but similar tests of this assumption in other avian species are extremely rare (although it has been repeatedly demonstrated for fish; see Walling et al. 2010 and references therein). Furthermore, two-choice tests do not fully account for variation in female mating preferences, nor do they necessarily predict mate choice in natural populations (Wagner 1998).

The relevance of mate preference trials is further complicated by the fact that females in socially but not genetically monogamous species may make multiple mate choice decisions over the course of a breeding season, with very different contexts and different implications for fitness. As noted above, females choosing social mates may rely on cues that indicate parental care ability or investment, while females choosing extrapair partners may prefer males that are attractive or genetically compatible. In two-choice mate preference trials, it is difficult if not impossible to determine the context in which females are evaluating males; are they choosing a potential social partner, a potential sire for their offspring, both, or neither? For example, most preference trials are conducted on females in full breeding condition, yet in the Virginia population of juncos, social pair formation typically begins as early as March, shortly after birds return to the breeding grounds and before females display signs of reproductive readiness (Nolan et al. 2002). Thus, two-choice trials may be more likely to reflect female preference for extrapair mates when the breeding season is already underway. However, as discussed above, preferences shown in these experimental trials do not always correspond to realized patterns of male mating success. For this reason, it is important to understand that female trait preferences are not the only factor in real-world mate choice decisions and to validate whether preferences displayed in these trials are actually associated with increased mating success in the wild.

Mate choice is an outcome resulting from not only a female's mate preference but also from other aspects about that female, including her sampling strategy, as well as environmental factors (Wagner 1998) and the phenotypes and preferences of the males with which that female interacts (see box 12.1). While extrapair offspring are often sired by neighboring males (Gerlach, unpublished data), suggesting that a mated female has a limited radius in which to search for an extra-pair mate, we know very little about how a female samples the available males at the beginning of a mating season. The extent of an individual female's sampling strategy will affect the bias shown in mate preferences—the wider the sample of potential mates, the greater the bias that can be observed (Wagner 1998). Since female juncos gain nongenetic benefits from pair bonding, including access to resources and territorial defense, and since females who mate with multiple males have higher fitness (Gerlach et al. 2012a; Gerlach et al. 2012b) it may be that extrapair mate choice is the main arena in which a female can improve her fitness.

D.4. Sexual Selection and Junco Evolution

Female choice for male traits has been recognized as one of the two major drivers of sexual selection since Darwin first proposed the idea. However, in terms of their effects on sexual selection, not all mate choice decisions are created equal. Bateman's principle states that males who acquire more mates should produce more offspring (Bateman 1948), and therefore any traits that covary with mate acquisition should be favored by selection. However, in this context, "mate" refers to a female with which a male has successfully sired an offspring; social partners are not necessarily included. Thus, a species with relatively little skew in social mating success may still be experiencing strong sexual selection on male traits if female choice for genetic mates results in a more skewed distribution of male reproductive success.

Cues that are used by females to identify high-quality social mates may not be the same cues used to identify high-quality sires, as discussed above. If they do differ, only the traits that are used to choose genetic partners can truly be considered to be under sexual selection. However, in most socially monogamous species (including the junco), the majority of offspring are sired by their mother's social mate (Griffith et al. 2002). In these cases, cues used in female social mate choice decisions will also affect differential male reproductive success and will therefore be subject to some degree of sexual selection pressure. Indeed, in the junco, the ability to acquire and sire offspring with multiple social partners has a stronger effect on male fitness than does the ability to acquire extrapair partners, suggesting that social as well as extrapair mate choice has a strong influence on sexual selection (figure 12.4, Gerlach et al. 2012b).

Sexual selection is often thought to play a role in speciation, as shifts in traits and preferences can reinforce premating isolation. However, in allopatric populations, while males traits may vary geographically, female preferences may stay the same, especially if ornamentation is an indicator of quality (Hill 1994; Price 1998). With the exception of a few hybrid zones, dark-eyed junco populations are allopatric, and thus premating isolation is generally achieved through geographic barriers or distance rather than via mate preference. Indeed, the mate choice trials that have been conducted in juncos suggest that females show a directional preference for plumage ornamentation (Hill et al. 1999) and odor (Whittaker et al. 2010). Borja Mila and coauthors comment on the role of sexual selection in diversification of the *Junco* lineage in chapter 8, and Trevor Price and Daniel Hooper discuss speciation in juncos more thoroughly in chapter 9.

E. Conclusions and Future Directions

Mate choice is far from straightforward. Not only do the choosing individuals assess multiple traits in different sensory modalities (visual, auditory,

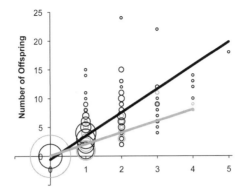

FIGURE 12.4. Bateman gradients for lifetime reproductive and mating success of male juncos, separated into within-pair (black) and extrapair (gray) mates and offspring. Size of the circle indicates the number of individuals at a point. Source: Gerlach et al. 2012b

and chemical) simultaneously but also their choices may differ depending on the context (choosing a social or extrapair mate), timing (early in the season or later), availability of other mates, or the chooser's own phenotype or genotype. To properly assess the role of sexual selection in the evolution and diversification of a species, we first need to understand the range of actual mate preferences and choices that are found in different populations. Two-way choice tests with captive birds can lead to testable hypotheses but should not be considered a substitute for studies in natural populations.

Additional research is especially needed on chemical signaling in mate choice, as our understanding of communication in this modality in birds lags far behind what is known about mammals and other taxa. In addition to continuing to decipher the information content of these signals, future work should investigate the neurobiological, endocrinological, and behavioral effects on the receivers. The role of the avian microbiome in producing these chemosignals is another important area of research. Work is already underway (Whittaker and Theis, in prep.) to test whether symbiotic bacteria in the preen gland are responsible for volatile compounds used in chemical communication, similar to what has already been described in mammals (Archie and Theis 2011).

Major gaps exist in our understanding of junco mate choice. In particular, the process of territory acquisition by males and the subsequent sampling, assessment, and choice of social mates by females is an area that needs study. Extrapair mate choice should be examined not only in light of the female's social mate but also in comparison to other neighboring males that may have been available. Future studies should attempt to collect data on multiple cues (plumage, song quality, preen oil volatile composition) and other traits (e.g., relatedness to the chooser, MHC genotype) in order to examine the relative importance of multiple traits and how they may be integrated into each mate choice decision.

Acknowledgments

We are grateful to our many colleagues who have contributed to the work in this chapter, including Jonathan Atwell, Kathryn Battle, Christine Bergeon Burns, Amy Dapper, Marine Drouilly, Ellen Ketterson, Ryan Kiley, Abby Kimmitt, Joel McGlothlin, Allison Miller, Milos Novotny, Patty Parker, Dustin Reichard, Kaitlin Richmond, Elizabeth Schultz, Samuel Slowinski, and Helena Soini. We thank Mountain Lake Biological Station, Mountain Lake Hotel, the Dolinger family, Grand Teton National Park, and University of Wyoming-National Park Service Research Station for allowing and supporting research on their grounds.

References

Alberts, A. C. 1992. Constraints on the design of chemical communication systems in terrestrial vertebrates. *American Naturalist* 139:S62–S89.

Amo, L., I. Galván, G. Tomás, and J. J. Sanz. 2008. Predator odour recognition and avoidance in a songbird. *Functional Ecology* 22:289–93.

Andersson, M. B. 1994. *Sexual Selection*. Princeton: Princeton University Press.

Andrew, R. J. 1956. Normal and irrelevant toilet behavior in *Emberiza* spp. *British Journal of Animal Behavior* 4:85–91.

Archie, E. A., and K. R. Theis. 2011. Animal behaviour meets microbial ecology. *Animal Behaviour* 82:425–36.

Atwell, J. W., G. C. Cardoso, D. J. Whittaker, T. D. Price, and E. D. Ketterson. 2014. Hormonal, behavioral and life-history traits exhibit correlated shifts in relation to population establishment in a novel environment. *American Naturalist* 184: E147–60.

Baker, M. C., and A. E. M. Baker. 1990. Reproductive behavior of female buntings: Isolating mechanisms in a hybridizing pair of species. *Evolution* 44:332–38.

Balph, M. H., D. F. Balph, and C. Romesburg. 1979. Social status signaling in winter flocking birds: An examination of current hypotheses. *Auk* 96:78–93.

Balthazart, J., and M. Taziaux. 2009. The underestimated role of olfaction in avian reproduction? *Behavioral Brain Research* 200:248–59.

Bang, B. G., and S. Cobb. 1968. The size of the olfactory bulb in 108 species of birds. *Auk* 85:55–61.

Bateman, A. J. 1948. Intra-sexual selection in *Drosophila*. *Heredity* 2:349–68.

Birkhead, T. R., and A. P. Møller. 1992. *Sperm Competition in Birds: Evolutionary Causes and Consequences*. San Diego: Academic Press.

Bollmer, J. L., P. O. Dunn, L. A. Whittingham, and C. Wimpee. 2010. Extensive MHC Class II B gene duplication in a passerine, the common yellowthroat (*Geothlypis trichas*). *Journal of Heredity* 101:448–60.

Bonadonna, F., and V. Bretagnolle. 2002. Smelling home: A good solution for burrow-finding in nocturnal petrels? *Journal of Experimental Biology* 205: 2519–23.

Bonadonna, F., and G. A. Nevitt. 2004. Partner-specific odor recognition in an Antarctic seabird. *Science* 306:835.

Bonadonna, F., and A. Sanz-Aguilar. 2012. Kin recognition and inbreeding avoidance in wild birds: The first evidence for individual kin-related odour recognition. *Animal Behaviour* 84:509–13.

Burger, B. V. 2005. Mammalian semiochemicals. *Topics in Current Chemistry* 240: 231–78.

Burley, N. T. 1988. The differential-allocation hypothesis: An experimental test. *American Naturalist* 132:611–28.

Candolin, U. 2003. The use of multiple cues in mate choice. *Biological Review* 78: 575–95.

Cardoso, G. C., J. W. Atwell, E. D. Ketterson, and T. D. Price. 2007. Inferring performance in the songs of dark-eyed juncos (*Junco hyemalis*). *Behavioral Ecology* 18:1051–1057.

Caspers, B. A., and E. T. Krause. 2010. Odour-based natal nest recognition in the zebra finch (*Taeniopygia guttata*), a colony-breeding songbird. *Biology Letters* 7:184–86.

Clark, G. A. 1970. Avian bill-wiping. *Wilson Bulletin* 82:279–88.

Clayton, N. S. 1990. Mate choice and pair formation in Timor and Australian mainland zebra finches. *Animal Behavior* 39:474–80.

Clotfelter, E. D., D. M. O'Neal, J. M. Gaudioso, J. M. Casto, I. M. Parker-Renga, E. Snajdr, D. L. Duffy, et al. 2004. Consequences of elevating plasma testosterone in females of a socially monogamous songbird: Evidence of constraints on male evolution? *Hormones and Behavior* 46:171–78.

Collins, S. A., and S. T. Luddem. 2002. Degree of male ornamentation affects female preference for conspecific versus heterospecific males. *Proceedings of the Royal Society B-Biological Sciences* 269:111–17.

Cuthill, I. C., M. Witter, and L. Clarke. 1992. The function of bill-wiping. *Animal Behaviour* 43:103–15.

de Lope, F., and A. P. Møller. 1993. Female reproductive effort depends on the degree of ornamentation of their mates. *Evolution* 47:1152–160.

Douglas, H. D., III. 2008. Prenuptial perfume: Alloanointing in the social rituals of the crested auklet (*Aethia cristatella*) and the transfer of arthropod deterrents. *Naturwissenschaften* 95:45–53.

Eimes, J. A., P. G. Parker, J. L. Brown, and E. R. Brown. 2005. Extrapair fertilization and genetic similarity of social mates in the Mexican jay. *Behavioral Ecology* 16:456–60.

Enstrom, D. E., E. D. Ketterson, and V. Nolan, Jr. 1997. Testosterone and mate choice in the dark-eyed junco. *Animal Behaviour* 54:1135–46.

Ferree, E. D. 2007a. Tail white and the influence of male attractiveness on maternal investment in dark-eyed juncos. PhD dissertation, University of California, Santa Cruz.

———. 2007b. White tail plumage and brood sex ratio in dark-eyed juncos (*Junco hyemalis thurberi*). *Behavioral Ecology and Sociobiology* 62:109–17.

———. 2013. Geographic variation in dark-eyed junco morphology and implications for population divergence. *Wilson Journal of Ornithology* 125:454–70.

Foerster, K., K. Delhey, A. Johnsen, J. T. Lifjeld, and B. Kempenaers. 2003. Females increase offspring heterozygosity and fitness through extra-pair matings. *Nature* 425:714–17.

Forstmeier, W., B. Kempenaers, A. Meyer, and B. Leisler. 2002. A novel song parameter correlates with extra-pair paternity and reflects male longevity. *Proceedings of the Royal Society B-Biological Sciences* 269:1479–85.

Freeman-Gallant, C. R., M. Meguerdichian, N. T. Wheelwright, and S. Sollecito. 2003. Social pairing and female mating fidelity predicted by restriction fragment length polymorphism similarity at the major histocompatibility complex in a songbird. *Molecular Ecology* 12:3077–83.

Gerlach, N. M., and E. D. Ketterson. 2013. Experimental elevation of testosterone lowers fitness in female dark-eyed juncos. *Hormones and Behavior* 63:782–90.

Gerlach, N. M., J. W. McGlothlin, P. G. Parker, and E. D. Ketterson. 2012a. Promiscuous mating produces offspring with higher lifetime fitness. *Proceedings of the Royal Society B-Biological Sciences* 279:860–66.

———. 2012b. Reinterpreting Bateman gradients: Multiple mating and selection in both sexes of a songbird species. *Behavioral Ecology* 23:1078–88.

Gowaty, P. A., and N. Buschhaus. 1998. Ultimate causation of aggressive and forced copulation in birds: Female resistance, the CODE hypothesis, and social monogamy. *American Zoologist* 38:207–25.

Griffith, S. C., I. P. Owens, and K. A. Thuman. 2002. Extra pair paternity in birds: A review of interspecific variation and adaptive function. *Molecular Ecology* 11:2195–212.

Hagelin, J. C. 2007. The citrus-like scent of crested auklets: Reviewing the evidence for an avian olfactory ornament. *Journal of Ornithology* 148:S195–S201.

Haribal, M., A. A. Dhondt, and E. Rodriguez. 2009. Diversity in chemical compositions of preen gland secretions of tropical birds. *Biochemical Systematics and Ecology* 37:80–90.

Haribal, M., A. A. Dhondt, D. Rosane, and E. Rodriguez. 2005. Chemistry of preen gland secretions of passerines: Different pathways to the same goal? Why? *Chemoecology* 15:251–60.

Hasselquist, D., S. Bensch, and T. von Schantz. 1996. Correlation between male song repertoire, extra-pair paternity and offspring survival in the great reed warbler. *Nature* 381:229–32.

Hill, G. E. 1994. Geographic variation in male ornamentation and female mate preference in the house finch: A comparative test of models of sexual selection. *Behavioral Ecology* 5:64–73.

Hill, J. A., D. E. Enstrom, E. D. Ketterson, V. Nolan, Jr., and C. Ziegenfus. 1999. Mate choice based on static vs. dynamic secondary sexual traits in the dark-eyed junco. *Behavioral Ecology* 10:91–96.

Holberton, R. L., K. P. Able, and J. C. Wingfield. 1989. Status signalling in dark-eyed juncos, *Junco hyemalis*: Plumage manipulations and hormonal correlates of dominance. *Animal Behaviour* 37:681–89.

Houtman, A. 1992. Female zebra finches choose extra-pair copulations with genetically attractive males. *Proceedings of the Royal Society B-Biological Sciences* 249:3–6.

Jacob, J. P., and V. Ziswiler. 1982. The uropygial gland. In *Avian Biology*, edited by D. S. Farner, J. R. King, and K. C. Parkes, 199–324. New York: Academic Press.

Johansson, B. G., and T. M. Jones. 2007. The role of chemical communication in mate choice. *Biological Review* 82:265–89.

Johnstone, R. A. 1997. The evolution of animal signals. In *Behavioural Ecology: An Evolutionary Approach*, edited by J. R. Krebs and N. B. Davies, 155–78. Oxford: Blackwell Science.

Jordan, W. C., and M. W. Bruford. 1998. New perspectives on mate choice and the MHC. *Heredity* 81:127–33.

Kempenaers, B., G. R. Verheyen, M. Vandenbroeck, T. Burke, C. Vanbroeckhoven, and A. A. Dhondt. 1992. Extra-pair paternity results from female preference for high-quality males in the blue tit. *Nature* 357:494–96.

Ketterson, E. D. 1979. Aggressive behavior in wintering Dark-eyed Juncos: Determinants of dominance and their possible relation to geographic variation in sex ratio. *Wilson Bulletin* 91:371–83.

Ketterson, E. D., V. Nolan, Jr., and M. Sandell. 2005. Testosterone in females: Mediator of adaptive traits, constraint on the evolution of sexual dimorphism, or both? *American Naturalist* 166:S85–S98.

Klein, J. 1986. *Natural History of the Major Histocompatibility Complex*. New York, NY: Wiley.

Komdeur, J., F. Kraaijeveld-Smit, K. Kraaijeveld, and P. Edelaar. 1999. Explicit experimental evidence for the role of mate guarding in minimizing loss of paternity in the Seychelles warbler. *Proceedings of the Royal Society B-Biological Sciences* 266:2075–81.

Krause, E. T., O. Krüger, P. Kohlmeier, and B. A. Caspers. 2012. Olfactory kin recognition in a songbird. *Biology Letters* 8:327–29.

Lack, D. 1968. *Ecological Adaptations for Breeding in Birds.* London: Methuen Ltd.

Landry, C., D. Garant, P. Duchesne, and L. Bernatchez. 2001. "Good genes as heterozygosity": The major histocompatibility complex and mate choice in Atlantic salmon (*Salmo salar*). *Proceedings of the Royal Society B-Biological Sciences* 268:1279–85.

Leclaire, S., T. Merkling, C. Raynaud, H. Mulard, J.-M. Bessière, É. Lhuillier, S. A. Hatch, et al. 2012. Semiochemical compounds of preen secretion reflect genetic make-up in a seabird species. *Proceedings of the Royal Society B-Biological Sciences* 279:1185–93.

Liang, D., and J. Silverman. 2000. "You are what you eat": Diet modifies cuticiular hydrocarbons and nestmate recognition in the Argentine ant, *Linepithema humile. Naturwissenschaften* 87:412–16.

Mardon, J., S. M. Saunders, M. J. Anderson, C. Couchoux, and F. Bonadonna. 2010. Species, gender, and identity: Cracking petrels' sociochemical code. *Chemical Senses* 35:309–21.

McGlothlin, J. W., D. L. Duffy, J. L. Henry, and E. D. Ketterson. 2007a. Diet quality affects an attractive white plumage pattern in dark-eyed juncos (*Junco hyemalis*). *Behavioral Ecology and Sociobiology* 61:1391–99.

McGlothlin, J. W., J. Jawor, T. J. Greives, J. M. Casto, J. L. Phillips, and E. D. Ketterson. 2008. Hormones and honest signals: Males with larger ornaments elevate testosterone more when challenged. *Journal of Evolutionary Biology* 21:39–48.

McGlothlin, J. W., J. M. Jawor, and E. D. Ketterson. 2007b. Natural variation in a testosterone-mediated trade-off between mating effort and parental effort. *American Naturalist* 170:864–75.

McGlothlin, J. W., P. G. Parker, V. Nolan, Jr., and E. D. Ketterson. 2005. Correlational selection leads to genetic integration of body size and an attractive plumage trait in dark-eyed juncos. *Evolution* 59:658–71.

McGlothlin, J. W., D. J. Whittaker, S. E. Schrock, N. M. Gerlach, J. M. Jawor, E. A. Snajdr, and E. D. Ketterson. 2010. Natural selection on testosterone production in a wild songbird population. *American Naturalist* 175:687–701.

Mock, D. W. 1985. An introduction to the neglected mating system. *Ornithological Monographs* 37:1–10.

Møller, A. P., and H. Tegelstrom. 1997. Extra-pair paternity and tail ornamentation in the barn swallow *Hirundo rustica. Behavioral Ecology and Sociobiology* 41:353–60.

Moreno-Rueda, G. 2010. Uropygial gland size correlates with feather holes, body condition, and wingbar size in the house sparrow *Passer domesticus. Journal of Avian Biology* 41:229–36.

Moyer, B. R., A. N. Rock, and D. H. Clayton. 2003. Experimental test of the importance of preen oil in rock doves (*Columba livia*). *Auk* 120:490–96.

Neubauer, R. L. 1999. Super-normal length song preferences of female zebra finches (*Taeniopygia guttata*) and a theory of the evolution of bird song. *Evolutionary Ecology* 13:365–80.

Nevitt, G. A., M. Losekoot, and H. Weimerskirch. 2008. Evidence for olfactory search in wandering albatross, *Diomedea exulans*. *Proceedings of the National Academy of Sciences of the United States of America* 105:4576–81.

Nolan, V., Jr., E. D. Ketterson, D. A. Cristol, C. M. Rogers, E. D. Clotfelter, R. Titus, S. J. Schoech, et al. 2002. Dark-eyed Junco (*Junco hyemalis*). In *The Birds of North America*, edited by A. Poole and F. Gill, no. 716. Philadelphia: The Birds of North America, Inc.

Novotny, M. V., H. A. Soini, S. Koyama, D. Wiesler, K. E. Bruce, and D. Penn. 2007. Chemical identification of MHC-influenced volatile compounds in mouse urine. I: Quantitative proportions of major chemosignals. *Journal of Chemical Ecology* 33:417–34.

O'Neal, D. M., D. G. Reichard, K. Pavlis, and E. D. Ketterson. 2008. Experimentally-elevated testosterone, female parental care, and reproductive success in a songbird, the dark-eyed junco (*Junco hyemalis*). *Hormones and Behavior* 54: 571–78.

Penn, D. J., K. Damjanovich, and W. K. Potts. 2002. MHC heterozygosity confers a selective advantage against multiple-strain infections. *Proceedings of the National Academy of Sciences of the United States of America* 99:11260–64.

Penn, D., and W. K. Potts. 1998. Chemical signals and parasite-mediated sexual selection. *Trends in Ecology & Evolution* 13:391–96.

———. 1999. The evolution of mating preferences and major histocompatibility complex genes. *American Naturalist* 153:145–64.

Petit, C., M. Hossaert-McKey, P. Perret, J. Blondel, and M. M. Lambrechts. 2002. Blue tits use selected plants and olfaction to maintain an aromatic environment for nestlings. *Ecology Letters* 5:585–89.

Price, T. 1998. Sexual selection and natural selection in bird speciation. *Philosophical Transactions of the Royal Society B-Biological Sciences* 353:251–60.

Ptacek, M. B. 2000. The role of mating preferences in shaping interspecific divergence in mating signals in vertebrates. *Behavioural Processes* 51:111–34.

Qvarnström, A. 1997. Experimentally increased badge size increases male competition and reduces male parental care in the collared flycatcher. *Proceedings of the Royal Society B-Biological Sciences* 264:1225–31.

Raouf, S. A., P. G. Parker, E. D. Ketterson, V. Nolan, Jr., and C. Ziegenfus. 1997. Testosterone affects reproductive success by influencing extra-pair fertilizations in male dark-eyed juncos (Aves: *Junco hyemalis*). *Proceedings of the Royal Society B-Biological Sciences* 264:1599–603.

Reed, W. L., M. E. Clark, P. G. Parker, S. A. Raouf, N. Arguedas, D. S. Monk, E. Snajdr, et al. 2006. Physiological effects on demography: A long-term experimental study of testosterone's effects on fitness. *American Naturalist* 167:667–83.

Reichard, D. G., R. J. Rice, C. C. Vanderbilt, and E. D. Ketterson. 2011. Deciphering information encoded in birdsong: Male songbirds with fertile mates respond most strongly to complex, low-amplitude songs used in courtship. *American Naturalist* 178:478–87.

Roth, T. C., II, J. G. Cox, and S. L. Lima. 2008. Can foraging birds assess predation risk by scent? *Animal Behaviour* 76:2021–27.

Rutstein, A. N., J. Brazill-Boast, and S. C. Griffith. 2007. Evaluating mate choice in the zebra finch. *Animal Behaviour* 74:1277–84.

Sætre, G.-P., T. Moum, S. Bures, M. Král, M. Adamjan, and J. Moreno. 1997. A sexually selected character displacement in flycatchers reinforces premating isolation. *Nature* 387:589–92.

Schamber, J. L., D. Esler, and P. L. Flint. 2009. Evaluating the validity of using unverified indices of body condition. *Journal of Avian Biology* 40:49–56.

Setchell, J. M., and E. Huchard. 2010. The hidden benefits of sex: Evidence for MHC-associated mate choice in primate societies. *Bioessays* 32:940–48.

Sheldon, B. C. 2000. Differential allocation: Tests, mechanisms, and implications. *Trends in Ecology & Evolution* 15:397–402.

Soini, H. A., S. E. Schrock, K. E. Bruce, D. Wiesler, E. D. Ketterson, and M. V. Novotny. 2007. Seasonal variation in volatile compound profiles of preen gland secretions of the dark-eyed junco (*Junco hyemalis*). *Journal of Chemical Ecology* 33:183–98.

Soini, H. A., D. J. Whittaker, D. Wiesler, E. D. Ketterson, and M. Novotny. 2013. Chemosignaling diversity in songbirds: Chromatographic profiling of preen oil volatiles in different species. *Journal of Chromatography* A 1317:186–92.

Steiger, S. S., A. E. Fidler, M. Valcu, and B. Kempenaers. 2008. Avian olfactory receptor gene repertoires: Evidence for a well-developed sense of smell in birds? *Proceedings of the Royal Society B-Biological Sciences* 275:2309–17.

Steiger, S. S., V. Y. Kuryshev, M. C. Stensmyr, B. Kempenaers, and J. C. Mueller. 2009. A comparison of reptilian and avian olfactory receptor gene repertoires: Species-specific expansion of group γ genes in birds. *BMC Genomics* 10:446.

Titus, R. C. 1998. Short-range and long-range songs: Use of two acoustically distinct song classes by dark-eyed juncos. *Auk* 115:386–93.

Uy, J. A. C., R. G. Moyle, and C. E. Filardi. 2009. Plumage and song differences mediate species recognition between incipient flycatcher species of the Solomon Islands. *Evolution* 63:153–64.

Wagner, W. E. 1998. Measuring female mating preferences. *Animal Behaviour* 55:1029–42.

Walling, C. A., N. J. Royle, J. Lindstrom, and N. B. Metcalfe. 2010. Do female association preferences predict the likelihood of reproduction? *Behavioral Ecology and Sociobiology* 64:541–48.

Wallraff, H. G. 2004. Avian olfactory navigation: Its empirical foundation and conceptual state. *Animal Behaviour* 57:189–204.

Wallraff, H. G., J. Kiepenheuer, M. F. Neumann, and A. Streng. 1995. Homing experiments with starlings deprived of the sense of smell. *Condor* 97:20–26.

Ward, A. J. W., and T. Mehner. 2010. Multimodal mixed messages: The use of multiple cues allows greater accuracy in social recognition and predator detection decisions in the mosquitofish, *Gambusia holbrooki. Behavioral Ecology* 21: 1315–20.

West-Eberhard, M. J. 1983. Sexual selection, social competition, and speciation. *Quarterly Review of Biology* 58:155–83.

Westneat, D. F., P. W. Sherman, M. L. Morton, D. Power, and R. Johnston. 1990. The ecology and evolution of extra-pair copulations in birds. *Current Ornithology* 7:331–69.

Whittaker, D. J., A. L. Dapper, M. P. Peterson, J. W. Atwell, and E. D. Ketterson. 2012. Maintenance of MHC Class IIB diversity in a recently established songbird population. *Journal of Avian Biology* 43:109–18.

Whittaker, D. J., N. M. Gerlach, H. A. Soini, and M. V. Novotny. 2013. Bird odour predicts reproductive success. *Animal Behaviour* 86:697–703.

Whittaker, D. J., D. G. Reichard, A. L. Dapper, and E. D. Ketterson. 2009. Behavioral responses of nesting female dark-eyed juncos *Junco hyemalis* to hetero- and conspecific passerine preen oils. *Journal of Avian Biology* 40:579–83.

Whittaker, D., D. Reichard, M. Drouilly, K. Battle, and C. Ziegenfus. 2015. Avian olfactory displays: A hypothesis for the function of bill-wiping in a social context. *Behavioral Ecology and Sociobiology* 69:159–67.

Whittaker, D. J., K. M. Richmond, A. K. Miller, R. Kiley, C. Bergeon Burns, J. W. Atwell, and E. D. Ketterson. 2011a. Intraspecific preen oil odor preferences in dark-eyed juncos (*Junco hyemalis*). *Behavioral Ecology* 22:1256–63.

Whittaker, D. J., H. A. Soini, J. W. Atwell, C. Hollars, M. V. Novotny, and E. D. Ketterson. 2010. Songbird chemosignals: Preen oil volatile compounds vary among individuals, sexes, and populations. *Behavioral Ecology* 21:608–14.

Whittaker, D. J., H. A. Soini, N. M. Gerlach, A. L. Posto, M. V. Novotny, and E. D. Ketterson. 2011b. Role of testosterone in stimulating seasonal changes in a potential avian chemosignal. *Journal of Chemical Ecology* 37:1349–57.

Willse, A., J. Kwak, K. Yamazaki, G. Preti, J. H. Wahl, and G. K. Beauchamp. 2006. Individual odortypes: Interaction of MHC and background genes. *Immunogenetics* 58:967–82.

Zagalska-Neubauer, M., W. Babik, M. Stuglik, L. Gustafsson, M. Cichoń, and J. Radwan. 2010. 454 sequencing reveals extreme complexity of the class II Major Histocompatibility Complex in the collared flycatcher. *BMC Evolutionary Biology* 10:395.

Zala, S. M., W. Potts, and D. J. Penn. 2004. Scent-marking displays provide honest signals of health and infection. *Behavioral Ecology* 15:338–44.

Zelano, B., and S. V. Edwards. 2002. An MHC component to kin recognition and mate choice in birds: Predictions, progress, and prospects. *American Naturalist* 160:S225-S237.

Zhang, J.-X., W. Wei, J.-H. Zhang, and W.-H. Yang. 2010. Uropygial gland-secreted alkanols contribute to olfactory sex signals in budgerigars. *Chemical Senses* 35: 375–82.

CHAPTER THIRTEEN

Dark-Eyed Junco Song

Linking Ontogeny and Function with a Potential Role in Reproductive Isolation

Gonçalo C. Cardoso and Dustin G. Reichard

The lovely tinkling chorus by the juncos in early spring, as if a myriad of woodland sprites were shaking little bells in an intensive competition. —L. de K. Lawrence

A faint, whispering warble usually much broken but not without sweetness and sometimes continuing intermittently for several minutes. —E. P. Bicknell

A. Introduction

Divergence in sexual signals may serve as a reproductive barrier among populations and facilitate speciation. Chapter 12, this volume, discussed the role of junco visual and olfactory cues in mate choice, and here we will focus on how dark-eyed junco (*Junco hyemalis*) song diversifies among populations and subspecies and whether song divergence can contribute to reproductive isolation. Birdsong is often much more elaborate than required for species or individual recognition and, thus, is an acoustic equivalent to sexual ornamentation. Developing complex, varied songs is facilitated by the ability of songbirds to learn songs socially (Marler 1990). This cultural inheritance makes birdsong a special sexual trait, in that it can diverge without genetic evolution, and, consequently, songs can potentially diverge faster and mediate reproductive isolation among populations more efficiently than nonlearned traits (Lachlan and Servedio 2004).

An additional factor of interest is that dark-eyed juncos have two distinct classes of song. Long-range songs (LRS) have a simple structure and are heard across several territories, while more elaborate short-range songs (SRS) transmit only a short distance (Titus 1998). The occurrence of two song classes is not atypical among songbirds (e.g., Anderson et al. 2008; Dabelsteen et al. 1998; Hof and Hazlett 2010; Reichard and Welklin 2015), but SRS are generally less well studied than LRS. Junco SRS is especially interesting from the perspective of female preferences and reproductive isolation because dark-eyed junco males produce it to females during directed courtship (Enstrom et al. 1997; Reichard et al. 2013; Titus 1998). Given its involvement in mate choice, divergence in SRS could contribute to prezygotic isolation in juncos as well as in other species that rely on low amplitude acoustic signals for mate choice (Reichard et al. 2011).

We first focus on LRS, explaining divergence at different spatial scales, from closely located populations, to dark-eyed junco subspecies that are geographically distant, to different species of juncos, and discuss this variation in relation to LRS function and development. We then shift to the much less studied SRS, discussing its predominant function in courtship and its potential for divergence between populations and to serve as an isolating mechanism.

B. Dark-Eyed Junco LRS

B.1. Structure and Development of LRS

Dark-eyed junco males have a small repertoire of song types, each typically consisting of a single trill (i.e., the consecutive repetition of the same syllable) (figure 13.1, first row; Konishi 1964b; Newman et al. 2008; Williams and Macroberts 1977). LRS is sung mostly by perched males at high amplitude (typically eighty to eighty-five decibels at one meter) but can also be sung softly (Nolan, Jr. et al. 2002; see soft LRS below). Average song length is about 1.5 seconds (usually between 1 and less than 2 seconds; Konishi 1964b; Newman et al. 2008; Titus 1998; Williams and Macroberts 1977; Williams and Macroberts 1978), and males typically repeat a song type many times before switching to a different type (Titus 1998). Song types differ in the morphology of the trilled syllable, which can range from a single note to more complex syllables with separate notes, frequency inflections, simultaneous voices, or buzzed sounds (e.g., figure 13.1, first row; Cardoso et al. 2007), but nonetheless generally fall within a species-typical range of acoustic parameters, such as duration or frequency.

As their song develops early in life, male dark-eyed juncos go through a phase of subsong, made up of long series of variable sounds, that then

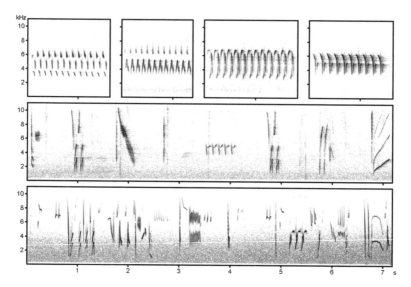

FIGURE 13.1. Examples of dark-eyed junco long-range song (LRS; upper panels), slow short-range song (SRS; middle panel), and fast SRS (lower panel). LRS is made up of a trill—consecutive repetitions of the same syllable—while SRS comprises a variety of non-trilled and sometimes-trilled syllables.

crystallizes into a few song types (Marler et al. 1962). Minor changes in their song repertoire can also happen at later stages (Marler et al. 1962; Titus et al. 1997b). Accordingly, brain nuclei involved in song learning are large during the first year of life, then small during the nonbreeding season, and increase in size seasonally during each breeding season, possibly to facilitate learning to recognize neighbors' songs (Corbitt and Deviche 2005; Gulledge and Deviche 1997).

Social stimulation is necessary for a normal song repertoire to develop in dark-eyed juncos. Young juncos raised in isolation without a tutor or deafened before song learning begins develop mostly atypical songs (e.g., with excessively long syllables, or lacking trilled syntax; Konishi 1964a; Marler et al. 1962). Nonetheless, among the songs of birds raised in isolation there are some with species-typical acoustic parameters, showing the ability of dark-eyed juncos to create normal song de novo (Marler et al. 1962). When young junco males are raised in small groups (i.e., with social stimulation but still without a tutor), they develop normal song repertoires, including some songs imitated from their peers and also several novel song types unique to the individual (Marler et al. 1962). These novel song types may be modifications of models heard from their peers (called improvisation) or develop without reference to outside models (called invention; Beecher and Brenowitz 2005). The development of novel song types during the song ontogeny of dark-eyed juncos has implications for understanding song diversity within populations and the mode of divergence across populations and subspecies.

B.2. Song Diversity within Populations

Each male dark-eyed junco sings a repertoire of two to eight song types (Newman et al. 2008; Williams and Macroberts 1977), some of which are shared with other males in the population. Sharing of song types among males decreases with the distance between their territories (Newman et al. 2008). Overall song sharing is uncommon, with most neighboring males not sharing songs (Newman et al. 2008), and in all populations studied to date the majority of song types were sung by only one male (Cardoso and Atwell 2011a; Konishi 1964b; Newman et al. 2008; Williams and Macroberts 1977; Williams and Macroberts 1978). Thus, despite the small repertoire size of individual males, dark-eyed junco populations harbor a large diversity of song types. This diversity could be explained either by early song learning followed by dispersal to other populations or by the development of novel song types by young males within the population.

Dispersal is unlikely to completely account for the low song sharing among males and high song diversity within populations because, to some degree, dark-eyed juncos are philopatric to both natal and breeding grounds (Nolan, Jr. et al. 2002; also see chapter 3, this volume). Furthermore, even in a small and isolated population that experiences limited immigration (Yeh and Price 2004; also see chapter 10, this volume), levels of song sharing remain low and song diversity remains high (Cardoso and Atwell 2011a; Newman et al. 2008). Therefore, many song types sung by only one male most likely result from improvisation or invention during song development in the wild, similar to what is observed in the lab (Marler et al. 1962).

The acoustic properties of songs (e.g., duration and frequency of syllables) differ among song types but are conserved within each song type, even for shared song types that are sung by different males (Cardoso et al. 2009). Therefore, the acoustic diversity in a population is a function of the number of song types it harbors, rather than the number of individual males singing. We next tackle divergence of these two aspects of song diversity—the song types and their acoustic properties—at increasing levels of geographic and phylogenetic separation, from nearby populations toward dark-eyed junco subspecies to full junco species (see chapter 8, this volume, for taxonomic and phylogenetic information).

B.3. Mechanisms of Song Divergence across Closely Located Populations

Song learning may lead to fast population divergence even without genetic changes, for example due to cultural drift or to cultural mutations (e.g., improvisation, invention) appearing in different populations (Podos and Warren 2007; Slabbekoorn and Smith 2002). One extreme case of rapid change through drift occurs during population bottlenecks in which a severe reduction in population size accentuates random loss of phenotypes (in this case, song types). Dark-eyed junco males have small repertoires, and small groups of juncos going through a bottleneck will retain only a very minor sample of the previous diversity of song types in their population. This initial lack of diversity could suggest high potential for song divergence, but song diversity in a recently founded population of juncos indicates otherwise. An isolated junco population established itself recently on the campus of the University of California at San Diego (UCSD), likely by wintering birds that ceased to migrate and became sedentary (Yeh and Price 2004; chapter 10, this volume). Some twenty years after establishment this population still had reduced genetic diversity, indicating that it underwent a severe bottleneck (Rasner et al. 2004; Whittaker et al. 2012), but population-level song diversity was as high as in the native range (Newman et al. 2008). In other species, reduced song diversity often persists after bottlenecks for longer periods (Parker et al. 2012; Potvin and Clegg 2015; Price 2008), which means that in juncos some mechanism rapidly restores natural levels of song diversity. Some immigration of foreign males to this population, or learning songs from the occasional wintering male that begins to sing before migrating away, may have contributed to the recovery of song diversity (Newman, et al. 2008). But the simplest explanation for the recovery of song type diversity is the juncos' aforementioned tendency to create novel song types during development.

Bottlenecks might also reduce diversity in the acoustic properties of song if the properties of the few song types surviving the bottleneck do not represent the range of the previous pool of songs. Whether creating novel song types also buffers against this type of reduction in diversity depends on how juncos improvise or invent novel songs. If the acoustic properties of songs that young birds hear during development influences the properties of the novel song types they will create, then novel song types could perpetuate differences that originated by bottlenecks or other type of random changes. If not, then the novel song types will fill in species-typical acoustic space that could have been left blank by random loss of phenotypes, and will have a homogenizing effect on the acoustic properties of song across populations. The latter seems likely in juncos because, as in other species, modifications that individuals impart on song tend to cause convergence towards the species norm rather than promote divergence. For example, young canaries shift towards species-typical syntax after having learned songs with atypical syntax (Gardner et al. 2005), and young zebra finches progressively change atypical song syllables to resemble typical ones (Fehér et al. 2009).

Thus, random cultural divergence in the acoustic properties of LRS may be limited in dark-eyed juncos, as the input of novel songs should buffer against divergence by cultural drift. Song differences in acoustic properties across junco populations, therefore, are more likely attributable to other mechanisms, such as behavioral plasticity or genetic evolution of morphological and neuronal traits subjacent to song production. We illustrate this by returning to the UCSD junco population, whose songs did diverge appreciably from the native range: minimum song frequency is on average about 500 hertz higher at UCSD than in the native range, peak frequency (frequency with highest sound amplitude) is about 150 hertz higher (Cardoso and Atwell 2011a, 2011b), and maximum frequency does not differ significantly (unpublished result with data from Cardoso and Atwell 2011a). The increase in minimum frequency in the UCSD population is substantial and is best explained as a consequence of urban noise. While juncos in the native range live in quiet forest habitat, UCSD is an urbanized area with variable levels of anthropogenic noise and other differences in the acoustic environment (Slabbekoorn et al. 2007). The amplitude of anthropogenic noise (mostly automobile traffic) is highest at lower frequencies, masking the lower frequencies of songs more severely, and, similarly to the junco, many bird species living in urban habitats have increased minimum song frequency relative to their nonurban populations (e.g., Dowling et al. 2012; Hu and Cardoso 2012; Ríos-Chelén et al. 2012). Thus, the divergence in song frequency in this population of dark-eyed juncos is not attributable to random processes such as drift but instead fits their peculiar ecological conditions.

B.4. Divergence across Dark-Eyed Junco Subspecies

If song divergence in the UCSD dark-eyed junco population had been attributable to drift or other random processes, then, extrapolating to the large geographic area occupied by dark-eyed juncos in North America, we would predict large song differences. Instead, song divergence at UCSD matches changes in ecological conditions, suggesting that the extent to which song differs across dark-eyed junco subspecies should largely depend on whether the ecological conditions that affect song also differ.

Multiple ecological factors can affect birdsong. For example, in addition to the environmental noise discussed above, changes in habitat type (Boncoraglio and Saino 2007), avian community composition (e.g., Doutrelant and Lambrechts 2001; Seddon 2005), bill adaptations to different foods (e.g., Badyaev et al. 2008), or strength of sexual selection (e.g., Price and Lanyon 2004) all can have effects. Most of these factors vary little across dark-eyed junco subspecies: all subspecies are socially monogamous, omnivorous, and typically inhabit forested or densely vegetated habitats, and most populations are migratory (though to different extents; chapter 3, this volume). Differences in body size can also cause changes in vocalizations, with larger taxa producing low-frequency vocalizations more efficiently than smaller ones (Bradbury and Vehrencamp 2011; Fletcher 2004). But body size differs little across junco subspecies, with average tarsus length of males only 6 percent larger in the largest relative to the smallest dark-eyed junco subspecies (Miller 1941). This variation is comparable to the range of variation found within populations: in two California populations, for example, the average tarsus lengths of males plus one standard deviation (SD) is 8 percent larger than the average minus one SD, and these differences were insufficient to cause a relation between song frequency and body size (Cardoso et al. 2008).

Given this ecological and morphological homogeneity, it is not surprising that LRS differs little across dark-eyed junco subspecies, despite their very large geographical distribution. Figure 13.2 compares basic acoustic properties of song for six populations where these have been studied: four populations of *J. h. thurberi* along the west coast of North America (including the two close populations discussed in the previous section) and two populations of *J. h. hyemalis* near the east coast. The small differences across populations in the mean values do not coincide with the subspecies boundary and are small by comparison to variation within populations (figure 13.2). Although information on acoustic properties of the different

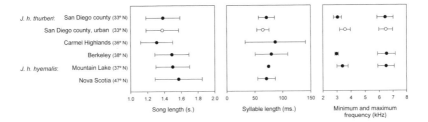

FIGURE 13.2. Average song and syllable lengths, minimum and maximum song frequencies for six populations of dark-eyed juncos near the west (*J. h. thurberi*) and east (*J. h. hyemalis*) coast of North America. Where information is available, standard deviations of the mean are indicated. Datasets or sources are: San Diego, Cardoso and Atwell (2011b); Carmel Highlands,Williams and MacRoberts (1977); Berkeley, Konishi (1964a); Mountain Lake, Titus (1998); Nova Scotia, Williams and MacRoberts (1978).

subspecies is still limited, this suggests a mosaic of limited geographic variation, not related to divergence between subspecies. Similarly to the acoustic properties of song, syntax is also uniform across dark-eyed junco subspecies. All subspecies sing simple trilled songs, and infrequent deviations to this, such as multisyllable song types (two or more trills in a single song), have been described for all populations studied (Konishi 1964b; Newman et al. 2008; Titus 1998; Williams and Macroberts 1977; Williams and Macroberts 1978; Reichard, unpublished data).

Consistent with the weak divergence of song across subspecies, playback of *J. h. thurberi* LRS from the West coast elicits typical agonistic territorial responses from *J. h. hyemalis* males from the East coast (Reichard 2014), indicating that songs of even the most geographically distant populations are recognized as conspecific. Nonetheless, the intensity of agonistic responses of *J. h. hyemalis* males is on average lower to playbacks of *J. h. thurberi* than to local *J. h. hyemalis* song (Reichard 2014). Minor differences in acoustic properties or familiarity with local song types thus cause differential responsiveness to song by male dark-eyed juncos. Discrimination of local versus foreign song exists in many songbird species, and females also often prefer local songs (reviewed in Podos and Warren 2007), likely due to song recognition being at least in part socially learned in songbirds (reviewed in Riebel 2003). If this explanation applies to the discrimination of local versus foreign song by dark-eyed juncos, then the basic mechanism for reproductive isolation due to song learning is in place. But since LRS divergence among dark-eyed junco populations is modest, and the range of acoustic properties overlaps largely even among distant subspecies, differences in LRS are likely an ineffective isolating mechanism.

B.5. Comparison among Junco Species

Once we cross the species boundary away from dark-eyed juncos, song and the pattern of song divergence become quite different. The yellow-eyed junco (*J. phaeonotus*) is the closest relative to dark-eyed juncos and has a widespread distribution across Central America (chapter 8, this volume). Like dark-eyed juncos, male yellow-eyed juncos have more than one song type and songs are of roughly the same length (one to two seconds), but yellow-eyed junco song is made of the concatenation of trilled and nontrilled syllables (Marler and Isaac 1961; Pieplow and Francis 2011). Unlike dark-eyed juncos, deafened yellow-eyed juncos are not able to develop normal song syntax (Konishi 1964a), suggesting that song development is more dependent on social learning than in dark-eyed juncos. Divergence in song traits has been documented among yellow-eyed junco subspecies *J. p. phaeonotus* and *J. p. palliatus* (Pieplow and Francis 2011), and these differences are larger than the small differences among dark-eyed junco subspecies.

Related to dark- and yellow-eyed juncos are the Guatemala and the Guadalupe juncos (*J. alticola* and *J. insularis*, the latter traditionally classified close to *J. hyemalis* because of its dark iris), then Baird's junco (*J. bairdi*, traditionally classified within *J. phaeonotus*), and then the more distantly related volcano junco (*J. vulcani*) from mountains in Costa Rica (see chapter 8, this volume, for phylogenetic relations). Guadalupe and Baird's juncos have even higher syllable diversity within songs than the yellow-eyed junco, with trills within the song being short and less frequent (Mirsky 1976; Pieplow and Francis 2011). Guatemala and volcano junco songs have not been studied in detail, but written descriptions indicate that they too have songs with more syllable diversity than dark-eyed junco LRS (Del Hoyo et al. 2011; Howell and Webb 1995). Since these among-species differences in song are much larger than the ones discussed earlier within dark-eyed juncos, it is likely that they are genetically ingrained and that they can effectively mediate species recognition and perhaps reproductive isolation.

The evolutionary diversification of the genus *Junco* was characterized by a northward expansion accompanied by isolation and speciation, with the dark-eyed junco radiation being the most recent (chapter 8, this

volume). Together with the above species differences in song, this evolutionary history indicates that the ancestral state of LRS in dark-eyed
juncos, before the split with yellow-eyed juncos, would have been more
elaborated than currently, raising the question of why dark-eyed juncos
evolved simpler LRS than their relatives. Another outstanding question,
possibly related to this, is why they create novel songs during development. In the next section we discuss these questions in relation to the
functions of dark-eyed junco LRS.

C. Functional Design of Dark-Eyed Junco LRS

C.1. Functions of Dark-Eyed Junco LRS

As in most passerines from temperate regions, only adult male dark-eyed
juncos normally sing LRS (Catchpole and Slater 2008). Female dark-eyed
juncos have the ability to sing and can be induced to sing LRS with testosterone treatment (Konishi 1964a) but only rarely do so in the wild.
The functions of male song in passerines are often twofold: an intrasexual function of territory and mate defense and an intersexual function of
female attraction and stimulation (Catchpole and Slater 2008). In dark-
eyed juncos the intrasexual function is evident, as males sing from within
their territories throughout the breeding season and countersing against
approaching males (Titus 1998). An intersexual function of female attraction is likely because unmated males remain singing well after territory establishment and sing more LRS than mated males (Ketterson et al.
1992). An intersexual function of female stimulation is also possible, as
male dark-eyed juncos continue to sing after pairing and in many species females require continuing sexual signaling to invest in reproduction
(Servedio et al. 2013). However, female stimulation does not appear to
be the primary function of LRS because males do not increase singing
rates during the fertile period of females (Titus et al. 1997a) and generally
switch from LRS to short-range song (SRS) during courtship (see below).

C.2. Unexpected Reduction in Song Complexity

Complex songs, with high syllable diversity, are thought to evolve mainly
by sexual selection (Searcy and Nowicki 2005). For example, females prefer larger repertoires in many, though not in all, passerine species (Byers
and Kroodsma 2009; Soma and Garamszegi 2011). Higher absolute lati-

tude of the breeding range (i.e., less equatorial latitudes) is an ecological trait that has been associated with increased sexual selection in socially monogamous birds (Albrecht et al. 2007; Spottiswoode and Møller 2004), putatively due to some of several possible mechanisms: breeding synchrony at higher latitudes increasing the opportunity for extrapair matings (e.g., Albrecht et al. 2007; Hammers et al. 2009; Stutchbury and Morton 1995; but also see discussion in LaBarbera et al. 2010); higher climate instability promoting female extrapair matings to increase offspring genetic diversity or to correct for suboptimal mate choice in unpredictable environments (Botero and Rubenstein 2012); migration causing mate choice based on early arrival and consequent establishment of good quality territories (Spottiswoode et al. 2006), and (together with year-to-year fluctuation in climate) increasing the variance in male genetic quality for optimal migration time (Fitzpatrick 1994, 1998). Ecological gradients, such as lower noise and lower species diversity at higher latitudes, may also promote the evolution of complex songs at higher latitudes even in the absence of increased strength of sexual selection (reviewed in Singh and Price 2015). Accordingly, latitudinal gradients of increasing song complexity have been found within species (Irwin 2000; Singh and Price 2015) and across related species (Botero et al. 2009; Cardoso et al. 2012b; Greig et al. 2013; Mountjoy and Leger 2001).

Contrary to the findings reviewed above, the dark-eyed junco is the highest latitude species of its genus but evolved simpler, not more complex, songs than the yellow-eyed junco and the remaining *Junco* species. We cannot yet explain this unexpected reduction in song complexity, but here we suggest four possible solutions, all of which need further research. First, latitude is only an indirect cue to the strength of sexual selection, and confidently establishing differences in the strength of sexual selection among junco species may require more direct assessment (e.g., prevalence of extrapair paternity). Second, the evolution of dark-eyed junco song may have been influenced by selective pressures other than sexual selection. For example, the acoustic transmission properties of different habitats influence birdsong evolution (Boncoraglio and Saino 2007), with forested habitats selecting for lower frequencies that suffer less reverberation from vegetation (Wiley and Richards 1978; Wiley and Richards 1982). While dark-eyed, yellow-eyed, and the remaining junco species all inhabit forests, research on the acoustic differences between their habitats could suggest some selective pressures for song divergence, which perhaps constrain song diversity in dark-eyed juncos.

Third, some of the sexual functions of LRS may have been transferred to other sexual signals. In certain circumstances, sexual selection acting strongly on one out of multiple coexisting ornaments can lead to decreased elaboration of the others (Shutler 2011). Notably, dark-eyed juncos use a different category of song, short-range song (SRS), for courtship and possibly also for escalated agonistic interactions (discussed in detail below), which might have reduced the strength of sexual selection for elaborate LRS. This hypothesis predicts that reduced complexity of dark-eyed junco LRS, relative to other junco species, would have been accompanied by an increase in the elaboration or the usage of other sexual signals. These could include plumage ornamentation, which varies between junco species (chapter 8, this volume), and SRS, which yellow-eyed juncos and possibly other junco species also sing (Moore 1972).

Fourth, although sexual selection can explain the evolution of complex song, it does not select for song complexity in all species (Byers and Kroodsma 2009; Garamszegi and Møller 2004; Soma and Garamszegi 2011). Other song traits are potential targets of sexual selection (Gil and Gahr 2002), some of which trade off with song complexity. For example, singing fast and wide frequency modulations can be implicated in female choice and aggressive communication (e.g., Ballentine et al. 2004; De Kort et al. 2009; Drăgănoiu et al. 2002; Dubois et al. 2009, 2011) and, in another group of birds (the wood warblers, family Parulidae), species with faster and wider frequency modulation have simpler songs with lower syllable diversity and more trills (Cardoso and Hu 2011). This appears strategic, since trills likely facilitate the assessment of the speed and breadth of frequency modulations by receivers. Advertising the performance of fast and wide frequency modulations might therefore contribute to explain the evolution of simple trilled songs in dark-eyed juncos. While there is no evidence that this aspect of trill performance indicates male quality in juncos (Cardoso et al. 2012a), it nonetheless appears meaningful for communication, since higher performance song types are used during more motivated singing (Cardoso et al. 2009). Similarly, by making it easier to assess by receivers, these simple trilled songs could be a means to advertise song consistency (Sakata and Vehrencamp 2012) or resistance to occasional mistakes (Cardoso 2013), both of which convey information on individual quality in dark-eyed juncos (Ferreira et al., submitted).

Yet another trait that may compromise song complexity is sound amplitude. High sound amplitude may be preferred by females (Pasteau et al. 2009; Ritschard et al. 2010; Searcy 1996) and be involved in aggressive intrasexual interactions (reviewed in Brumm and Ritschard 2011; Ritschard et al. 2012). Amplitude also trades off with several aspects of song complexity. For example, short or multinote syllables are generally sung with lower amplitude than longer or single-note syllables (dark-eyed juncos: Cardoso et al. 2007; other species: Cardoso and Mota 2009; Ritschard and Brumm 2011), likely because the air-sac pressure required for high sound amplitude (Suthers et al. 1999) is easier to mount gradually than in very short bursts. Also, in other songbird taxa, trills have been found to have higher sound amplitude than nontrilled syllables, both within (Cardoso and Mota 2009) and across species (Cardoso 2010). Thus, if the dark-eyed junco evolved louder LRS than its ancestors, this could help to explain its simple, trilled songs. This hypothesis predicts that song amplitude is higher in dark-eyed than yellow-eyed and other junco species and that trilled syllables are sung with higher amplitude. The latter prediction cannot be tested directly on dark-eyed junco LRS (which lacks nontrilled portions for comparison), but it might be tested in SRS (which contains an assortment of trilled and nontrilled syllables) or in the LRS of yellow-eyed and related junco species.

C.3. Unexpectedly Low Song Sharing

Song sharing—identical songs in the repertoires of different individuals— allows males to direct singing to a particular opponent in a vocal interaction by matching its song, which is often an aggressive signal (reviewed in Searcy and Beecher 2009). Aggressive signaling benefits signalers and receivers, as it helps to negotiate conflicts without a physical fight (Searcy and Nowicki 2005), and several species have learning mechanisms that facilitate song sharing and song matching with neighbors (e.g., open-ended learning, or selective retention of songs that best match neighbors', i.e., "selective attrition"; e.g., Nelson 2000). High song sharing and song matching are expected to evolve in species in which males have a relatively stable set of neighbors, while in species in which neighbors change frequently it may not be possible to continuously learn their songs (Kroodsma 1996). Song sharing is more likely in sedentary than migratory populations (Handley and Nelson 2005), supporting the association between philopatry and sharing. Although most subspecies of dark-eyed junco are migratory to some extent, males also show some philopatry and often return to their previous year's territory (Nolan et al. 2002). Therefore, notwithstanding wintering mortality, neighborhoods exhibit some stability, and we would expect dark-eyed juncos to have a repertoire of songs that allowed opportunities for matching neighbors. But, on the contrary, males do not share songs with most neighbors (Newman et al. 2008) and frequently have song types in their repertoires that are not sung by other males in the population (Cardoso and Atwell 2011a; Newman et al. 2008). Thus, dark-eyed juncos are unable to match most song types in their neighborhood.

This unexpectedly low song sharing should be related to the high input of novel song types during dark-eyed junco song development. Therefore, understanding why dark-eyed juncos create novel song types would help to explain their low song sharing. It would also help to explain their very low geographic divergence in song (since novel songs should buffer against divergence by random cultural processes; see above). One piece for this puzzle is that novel song types of dark-eyed juncos are on average of higher performance (have longer syllables, with shorter intersyllable intervals and larger frequency bandwidth) than the song types learned by imitation of social models (Cardoso and Atwell, submitted). Some of those performance traits are relevant for dark-eyed junco communication (Cardoso et al. 2009), suggesting that creating novel song types is a functional behavior in juncos. This helps to explain why juncos do not exhibit higher levels of song sharing. But it is likely an incomplete explanation, since we know yet little about the development of novel song types (after over fifty years, Marler et al.'s 1962 work remains the only descriptive study on song development in juncos). More research on how novel song types are developed and on which social and ecological circumstances they are employed could give new insight on the functions of the dark-eyed junco mode of song development.

D. Low-Amplitude Songs of the Dark-Eyed Junco

In addition to high-amplitude, long-range songs, many avian (Dabelsteen et al. 1998; Reichard and Welklin 2015) and nonavian species (bats: Behr and Von Helversen 2004; fish: Simões et al. 2008; insects: Nakano et al. 2008; Sueur and Aubin 2004; Zuk et al. 2008) produce low-amplitude (whispered) songs during close-proximity interactions. Unlike long-range songs, low-amplitude songs are sung at reduced amplitude and, depending on the species, these songs can differ markedly in structure and function (Reichard and Welklin 2015). Male dark-eyed juncos produce two distinct categories of low-amplitude song: (1) soft LRS, which does not differ structurally from high-amplitude LRS, and (2) short-range song (SRS), which is structurally distinct from LRS. The remainder of the chapter will focus primarily on junco SRS, which functions predominantly in courtship and may be relevant for reproductive isolation among juncos.

D.1. The Structure of Low-Amplitude Songs in Dark-Eyed Juncos

Dark-eyed junco SRS differs substantially in structure from LRS, to the point where these two song classes seem like they could be sung by different species (figure 13.1). While junco LRS is a simple trill, junco SRS is a continuous and variable song of indeterminate length, produced with brief pauses in between syllables. Each male has a large repertoire of SRS syllable types of varying complexity (on average twenty-four different syllable types were observed within two minutes of singing; Titus 1998). SRS syllable types are much more diverse than those sung in LRS (figure 13.1), with the frequency bandwidth of SRS (1.1—11.4 kilohertz) being much larger than that of LRS (2.4—8.2 kilohertz; Titus 1998). Male juncos sing SRS at two distinct tempos, slow and fast (figure 13.1, second and third rows). The average length of the pauses separating each syllable is much larger in slow SRS (0.84 ± 0.27 seconds) than in fast SRS (0.12 ± 0.14 seconds), suggesting that these two tempos are discrete song categories (Reichard et al. 2011). It remains to be determined whether the slow and fast SRS sung by individual males contain similar or distinct syllable types.

D.2. The Function of Low-Amplitude Songs in Juncos

Male juncos typically produce soft LRS and SRS during close-range interactions with conspecifics, including rival males, their social mates or potential extrapair mates (Nolan, Jr. et al. 2002; Reichard et al. 2011, 2013; see chapter 12, this volume, for information on extrapair behavior). The social contexts in which males sing soft LRS, slow SRS and fast SRS differ, suggesting that these low-amplitude songs have somewhat different functions (Reichard et al. 2013; Reichard et al. 2011). Soft LRS is often produced when a male encounters another male or a female conspecific on his territory and near his social mate while she is actively building a nest (Nolan, Jr. et al. 2002; Reichard et al. 2013). Research in other avian species has shown that singing soft LRS indicates aggressive intent (Ballentine et al. 2008; Hof and Hazlett 2010; Ręk and Osiejuk 2011; Searcy et al. 2006); however, male juncos will also sing soft LRS near their mate or other female conspecifics, which raises the possibility that, similar to LRS, soft LRS has a dual role in both territoriality and mate attraction or stimulation (Reichard et al. 2013). During nest building, for example, soft LRS may serve to stimulate the male's mate while she is fertile or as an "all-clear" signal to indicate to the female that she can decrease her vigilance (Wingelmaier et al. 2007). In summary, soft LRS appears to serve multiple functions depending on the social context.

In contrast to soft LRS, numerous lines of evidence suggest that male juncos produce SRS almost exclusively in the presence of females. Free-living males presented with a live male or female conspecific produce slow SRS only to females, never to males (Reichard et al. 2013). Similarly, in a study of captive juncos, male juncos sang predominantly LRS before a female was introduced to the test apparatus, and afterwards males sang predominantly SRS (Enstrom et al. 1997). In the field, production of SRS by males is higher during the courtship and fertile periods (seven to ten days preceding laying of the penultimate egg) than during incubation and nestling care, while production of LRS does not differ detectably across the nesting cycle (Titus 1998; Titus et al. 1997a). Based on these data, SRS appears to function predominantly as a female-directed, courtship signal.

In addition to observations of the social context in which juncos produce low-amplitude songs, playback experiments have tested whether territorial males respond differently to playbacks of LRS, soft LRS, slow SRS, or fast SRS at different stages of their mates' fertility cycle (Reichard et al. 2011). During a mate's fertile period, male juncos are at highest risk for losing paternity to intruders and should respond more aggressively to songs that signal courtship. Consistent with this prediction, males responded more aggressively to playbacks of slow SRS when their mates were fertile, suggesting that slow SRS is perceived as a courtship signal. In contrast, males responded similarly to playback of soft LRS irrespective of their mate's fertility status. Male response to fast SRS was tested only during the nonfertile period, and males responded more aggressively to fast SRS than soft LRS. Thus, fast SRS is a potent elicitor of aggression from male juncos, but whether fast SRS is perceived as aggressive, intense courtship, or some combination remains unknown.

In summary, the different categories of dark-eyed junco song appear to play complementary roles in attracting and stimulating potential mates. In making their mate choice decisions, female juncos may first choose which males to approach based predominantly on differences in male LRS and territory quality. After approaching a specific male, females are exposed to that male's SRS in addition to other visual and olfactory signals before making a decision of whether or not to pair or copulate with that male (see chapter 12, this volume). In the lab, mate choice experiments support the importance of SRS in courtship as female juncos prefer males that sing more SRS, in addition to differences in visual displays (Enstrom et al. 1997). Because of these functions in courtship and mate choice, we hypothesize that SRS is a likely target of intersexual selection and that, if it diverges between populations, SRS has the potential to serve as a reproductive isolating mechanism. We next evaluate the potential of SRS to diverge between populations and to play a role as a driver of speciation in dark-eyed juncos.

D.3. The Development of SRS

Very little is currently known about the development of SRS, and it has not been tested whether or not SRS follows a similar developmental trajectory to LRS, including social learning, socially stimulated improvisation or invention of novel syllable types, and song crystallization. The SRS repertoires of individual males are very diverse and include some syllable types that are shared with multiple males and other syllable types that are shared at low rates or possibly not at all (Reichard, unpublished data). SRS bouts also often include the production of junco-specific call notes (Nolan et al. 2002; Reichard, pers. obs.). Species-specific call notes are thought to be innate (inherited with little to no influence from the social environment; Catchpole and Slater 2008) and the same may be true of the junco, which suggests that SRS may include unlearned components. Additionally, SRS can include syllable types that are variants or exact copies of LRS syllables and syllable types that appear to be mimicked sounds of heterospecific species (Nolan et al. 2002). Thus, it appears that social learning and improvisation or invention are involved in the development of a male's SRS repertoire, similarly to the development of LRS.

During the process of song learning, juvenile songbirds can eavesdrop and learn the LRS of many males in the local population because LRS transmits over long distances (Templeton et al. 2009). SRS, in contrast, does not transmit over long distances, and learning low-amplitude songs requires juvenile songbirds to be close to the singer to accurately overhear the song. Approaching a vocalizing, territorial male can be a challenging situation for a juvenile songbird as males are often aggressive towards intruders on their territory, particularly during their mate's fertile period when SRS is most often produced (Reichard et al. 2011; Ritschard et al. 2012; but see Templeton et al. 2012). Consequently, juveniles may be limited in the available tutors for SRS, perhaps leaving the social father to serve as the primary tutor. As juncos have multiple broods in a breeding season, females begin building the next nest before the current offspring reach independence (Nolan et al. 2002), and during this period, males care for fledglings from the previous nest and court and copulate with their mate. These courtship events may provide prime opportunities for juveniles of both sexes to learn SRS and other courtship behaviors.

If young males and females learn SRS from their social fathers and a few other neighboring males, then SRS may be a reliable indicator of natal origin. In this scenario, female preferences for local versus foreign SRS (Podos and Warren 2007; Riebel 2003) may then contribute to reproductive isolation among populations. Alternatively, the more restrictive set of social tutors and large repertoire sizes of SRS suggest that improvisation or invention of novel syllables may play an important role in SRS development. In that case, similarly to the above discussion for LRS, the input of novel syllables should buffer against cultural divergence. We still lack detailed analyses of the development and mode of transmission of SRS between generations, and so it remains unclear to which extent SRS may serve as an indicator of population origin.

D.4. Potential Mechanisms of Divergence in SRS

In addition to the possibility that SRS diverges by learning from a restricted number of local males, there are numerous selective factors that may contribute to the divergence of acoustic signals between populations. The strength of those selective factors may be different for low-amplitude songs than they are for high-amplitude songs. Perhaps the greatest difference lies in the effect of the local acoustic environment. High-amplitude songs, such as LRS, experience degradation as they transmit through the environment that can blur and mask the signal's structure and information content. Due to selection for efficient sound transmission over long distances, the structure of high-amplitude songs is typically restricted to a frequency range and level of complexity that facilitates accurate long-distance communication (Wiley and Richards 1978; Wiley and Richards 1982). Low-amplitude songs likely experience this constraint to a negligible or reduced degree, given that receivers are in very close proximity to the singer (Balsby et al. 2003; Dabelsteen et al. 1993).

With reduced or absent transmission constraints, one might predict that low-amplitude songs need not be restricted to the frequency range and structure that transmits most effectively over long distances. Indeed, we have already noted that dark-eyed junco SRS covers a much broader frequency range and is substantially more complex than LRS. Similarly to juncos, Eurasian blackbirds (*Turdus merula*) and whitethroats (*Sylvia communis*) sing SRS that is more complex and has a broader frequency range than their LRS, and their SRS has been found to degrade more rapidly than LRS as distance from the singer increases (Balsby et al. 2003; Dabelsteen et al. 1993). Since the structure of SRS appears less constrained by transmission characteristics of the local environment, then other factors might more easily drive population or subspecies divergence in SRS.

In contrast, divergence in morphology or body size may have a stronger effect on SRS than LRS. The minimum frequency of SRS is much lower than LRS and thus more likely to be limited by size-related production constraints (Hall et al. 2013). If dark-eyed juncos produce SRS near the limit of their production capabilities in terms of minimum or maximum frequency, then differences in morphology or size across subspecies could affect the structure of SRS more than LRS. Preliminary data from white-winged (*J. h. aikeni*) and slate-colored (*J. h. carolinensis*) dark-eyed juncos suggests that these two subspecies differ in the minimum frequency of SRS in the direction predicted by their differences in body size (*J. h. aikeni* is larger and has lower minimum frequency; Reichard, unpublished data). Thus, despite small differences in body size among subspecies (Miller 1941), it is possible that changes in morphology have a stronger effect on junco SRS than LRS.

Since SRS plays an integral role in junco courtship behavior (Reichard et al. 2011, 2013), it may be particularly responsive to selection driven by divergence in female mating preferences. The strength and direction of sexual selection for male ornaments is often unstable and can shift rapidly among ornaments or different aspects of the same ornament across time and space (Bro-Jørgensen 2010; Cornwallis and Uller 2010; Schluter and Price 1993). Thus, divergence in female mating preferences could select for different traits of SRS in different populations, such as the presence or absence of specific note types, repetition rate, repertoire size, or frequency bandwidth. If SRS and female preference diverge substantially between populations of juncos, then SRS will have the potential to act as a reproductive isolating mechanism.

E. Main Conclusions and Open Questions

Male dark-eyed juncos sing simple, trilled LRS, with much lower syllable diversity than in other junco species. Development of LRS is peculiar in that some song types are socially learned and shared with other males in the population while others are improvised or invented during development. The input of these novel songs explains the low levels of song sharing among males and buffers against geographic differentiation by cultural mechanisms. In fact, while dark-eyed junco LRS can diverge rapidly, for example under changing acoustical conditions of habitats, LRS differs minimally across the large geographic distribution of subspecies in North America, indicating little potential for mediating reproductive isolation. Simpler LRS than in related species and the use of improvised or invented songs, with the consequent low song sharing among males, are functionally unexpected for a species with the ecological characteristics of dark-eyed juncos. Yet these song traits seem responsible for the low song divergence across subspecies. Explaining those traits functionally is a challenge for future research.

In contrast, SRS of dark-eyed juncos is a quiet, complex vocalization that differs substantially from LRS and is particularly important during courtship. Owing to the fact that SRS is directed to a close receiver, it is likely less affected by selection for efficient sound transmission than LRS, which may facilitate divergence between populations, for example in response to differences in female preferences. Similarly to LRS, the likelihood that SRS will diverge rapidly and serve as a reliable indicator of population origin will depend heavily on how much of SRS is socially learned, an important topic for future study. It is unlikely that SRS is the sole determinant of mating success in juncos, and thus the relative importance of junco courtship signals in each modality (acoustic, visual, olfactory) must be considered to fully understand the potential for premating isolation through divergent courtship signals.

Acknowledgments

We thank Ellen Ketterson, Jonathan Atwell, and Trevor Price for discussions and comments on earlier versions of this chapter. GCC was supported by fellowship SFRH/BPD/46873/2008 from the Fundação para a

Ciência e a Tecnologia; DGR was supported by an NSF graduate research fellowship and DDIG IOS-1011145.

References

Albrecht, T., J. Schnitzer, J. Kreisinger, A. Exnerová, J. Bryja, and P. Munclinger. 2007. Extrapair paternity and the opportunity for sexual selection in longdistant migratory passerines. *Behavioral Ecology* 18:477–86.

Anderson, R. C., W. A. Searcy, S. Peters, and S. Nowicki. 2008. Soft song in song sparrows: Acoustic structure and implications for signal function. *Ethology* 114: 662–76.

Badyaev, A. V., R. L. Young, K. P. Oh, and C. Addison. 2008. Evolution on a local scale: Developmental, functional, and genetic bases of divergence in bill form and associated changes in song structure between adjacent habitats. *Evolution* 62:1951–64.

Ballentine, B., J. Hyman, and S. Nowicki. 2004. Vocal performance influences female response to male bird song: An experimental test. *Behavioral Ecology* 15:163–68.

Ballentine, B., W. A. Searcy, and S. Nowicki. 2008. Reliable aggressive signalling in swamp sparrows. *Animal Behaviour* 75:693–703.

Balsby, T. J. S., T. Dabelsteen, and S. B. Pedersen. 2003. Degradation of whitethroat vocalisations: Implications for song flight and communication network activities. *Behaviour* 140:695–719.

Beecher, M. D., and E. A. Brenowitz. 2005. Functional aspects of song learning in songbirds. *Trends in Ecology & Evolution* 20:143–49.

Behr, O., and O. Von Helversen. 2004. Bat serenades: Complex courtship songs of the sac-winged bat (*Saccopteryx bilineata*). *Behavioral Ecology and Sociobiology* 56:106–15.

Boncoraglio, G., and N. Saino. 2007. Habitat structure and the evolution of birdsong: A meta-analysis of the evidence for the acoustic adaptation hypothesis. *Functional Ecology* 21:134–42.

Botero, C. A., N. J. Boogert, S. L. Vehrencamp, and I. J. Lovette. 2009. Climatic patterns predict the elaboration of song displays in mockingbirds. *Current Biology* 19:1151–55.

Botero, C. A., and D. R. Rubenstein. 2012. Fluctuating environments, sexual selection, and the evolution of flexible mate choice in birds. *PLoS One* 7: e32311.

Bradbury, J. W., and S. L. Vehrencamp. 2011. *Principles of Animal Communication.* Sunderland, MA: Sinauer Associates.

Bro-Jørgensen, J. 2010. Dynamics of multiple signalling systems: Animal communication in a world in flux. *Trends in Ecology & Evolution* 25:292–300.

Brumm, H., and M. Ritschard. 2011. Song amplitude affects territorial aggression of male receivers in chaffinches. *Behavioral Ecology* 22:310–16.

Byers, B. E., and D. E. Kroodsma. 2009. Female mate choice and songbird song repertoires. *Animal Behaviour* 77:13–22.

Cardoso, G. C. 2010. Loudness of birdsong is related to the body size, syntax and phonology of passerine species. *Journal of Evolutionary Biology* 23:212–19.

———. 2013. Sexual signals as advertisers of resistance to mistakes. *Ethology* 119:1035–43.

Cardoso, G. C., and J. W. Atwell. 2011a. Directional cultural change by modification and replacement of memes. *Evolution* 65:295–300.

Cardoso, G. C., and J. W. Atwell. 2011b. On the relation between loudness and the increased song frequency of urban birds. *Animal Behaviour* 82:831–36.

Cardoso, G. C., and J. W. Atwell. Submitted. Shared songs are of lower performance in the dark-eyed junco. Manuscript.

Cardoso, G. C., J. W. Atwell, Y. Hu, E. D. Ketterson, and T. D. Price. 2012a. No correlation between three selected trade-offs in birdsong performance and male quality for a species with song repertoires. *Ethology* 118:584–93.

Cardoso, G. C., J. W. Atwell, E. D. Ketterson, and T. D. Price. 2007. Inferring performance in the songs of dark-eyed juncos (*Junco hyemalis*). *Behavioral Ecology* 18:1051–57.

———. 2009. Song types, song performance, and the use of repertoires in dark-eyed juncos (*Junco hyemalis*). *Behavioral Ecology* 20:901–7.

Cardoso, G. C., and Y. Hu. 2011. Birdsong performance and the evolution of simple (rather than elaborate) sexual signals. *American Naturalist* 178:679–86.

Cardoso, G. C., Y. Hu, and P. G. Mota. 2012b. Birdsong, sexual selection, and the flawed taxonomy of canaries, goldfinches and allies. *Animal Behaviour* 84:111–19.

Cardoso, G. C., A. T. Mamede, J. W. Atwell, P. G. Mota, E. D. Ketterson, and T. D. Price. 2008. Song frequency does not reflect differences in body size among males in two oscine species. *Ethology* 114:1084–93.

Cardoso, G. C., and P. G. Mota. 2009. Loudness of syllables is related to syntax and phonology in the songs of canaries and seedeaters. *Behaviour* 146:1649–63.

Catchpole, C. K., and P. J. B. Slater. 2008. *Bird Song: Biological Themes and Variations*. 2nd edition. Cambridge: Cambridge University Press.

Corbitt, C., and P. Deviche. 2005. Age-related difference in size of brain regions for song learning in adult male dark-eyed juncos (*Junco hyemalis*). *Brains, Behavior and Evolution* 65:268–77.

Cornwallis, C. K., and T. Uller. 2010. Towards an evolutionary ecology of sexual traits. *Trends in Ecology & Evolution* 25:145–52.

Dabelsteen, T., O. N. Larsen, and S. B. Pedersen. 1993. Habitat-induced degradation of sound signals: Quantifying the effects of communication sounds and bird location on blur ratio, excess attenuation, and signal-to-noise ratio. *Journal of the Acoustical Society of America* 93:2206–20.

Dabelsteen, T., P. K. Mcgregor, H. M. Lampe, N. E. Langmore, and J. Holland. 1998. Quiet song in song birds: An overlooked phenomenon. *Bioacoustics* 9:89–105.

De Kort, S. R., E. R. B. Eldermire, E. R. A. Cramer, and S. L. Vehrencamp. 2009. The deterrent effect of bird song in territory defense. *Behavioral Ecology* 20: 200–206.

Del Hoyo, J., A. Elliot, and D. A. Christie. 2011. *Handbook of the Birds of the World*, vol. 16. Barcelona: Lynx Edicions.

Doutrelant, C., and M. M. Lambrechts. 2001. Macrogeographic variation in song: A test of competition and habitat effects in blue tits. *Ethology* 107:533–44.

Dowling, J. L., D. A. Luther, and P. P. Marra. 2012. Comparative effects of urban development and anthropogenic noise on bird songs. *Behavioral Ecology* 23: 201–9.

Drăgănoiu, T. I., L. Nagle, and M. Kreutzer. 2002. Directional female preference for an exaggerated male trait in canary (*Serinus canaria*) song. *Proceedings of the Royal Society B-Biological Sciences* 269:2525–31.

Dubois, A. L., S. Nowicki, and W. A. Searcy. 2009. Swamp sparrows modulate vocal performance in an aggressive context. *Biology Letters* 5:163–65.

———. 2011. Discrimination of vocal performance by male swamp sparrows. *Behavioral Ecology and Sociobiology* 65:717–26.

Enstrom, D. A., E. D. Ketterson, and V. Nolan Jr. 1997. Testosterone and mate choice in the dark-eyed junco. *Animal Behaviour* 54:1135–46.

Fehér, O., H. Wang, S. Saar, P. M. Mitra, and O. Tchernichovski. 2009. De novo establishment of wild-type song culture in the zebra finch. *Nature* 459:564–68.

Ferreira, A. C., J. W. Atwell, D. J. Whittaker, E. D. Ketterson, and G. C. Cardoso. Submitted. Communication value of mistakes in dark-eyed junco song. Manuscript.

Fitzpatrick, S. 1994. Colourful migratory birds: Evidence for a mechanism other than parasite resistance for the maintenance of "good genes" sexual selection. *Proceedings of the Royal Society B-Biological Sciences* 257:155–60.

———. 1998. Intraspecific variation in wing length and male plumage coloration with migratory behaviour in continental and island populations. *Journal of Avian Biology* 29:248–56.

Fletcher, N. H. 2004. A simple frequency-scaling rule for animal communication. *Journal of the Acoustical Society of America* 115:2334–38.

Garamszegi, L. Z., and A. P. Møller. 2004. Extrapair paternity and the evolution of bird song. *Behavioral Ecology* 15:508–19.

Gardner, T. J., F. Naef, and F. Nottebohm. 2005. Freedom and rules: The acquisition and reprogramming of a bird's learned song. *Science* 308:1046–49.

Gil, D., and M. Gahr. 2002. The honesty of bird song: Multiple constraints for multiple traits. *Trends in Ecology & Evolution* 17:133–41.

Greig, E. I., J. J. Price, and S. Pruett-Jones. 2013. Song evolution in Maluridae: Influences of natural and sexual selection on acoustic structure. *Emu* 113:270–81.

Gulledge, C. C., and P. Deviche. 1997. Androgen control of vocal control region volumes in a wild migratory songbird (*Junco hyemalis*) is region and possibly age dependent. *Journal of Neurobiology* 32:391–402.

Hall, M. L., S. A. Kingma, and A. Peters. 2013. Male songbird indicates body size with low-pitched advertising songs. *PLoS One* 8:e56717.

Hammers, M., N. Von Engelhardt, N. E. Langmore, J. Komdeur, S. Griffith, and M. J. L. Magrath. 2009. Mate-guarding intensity increases with breeding synchrony in the colonial fairy martin, *Petrochelidon ariel*. *Animal Behaviour* 78: 661–69.

Handley, H. G., and D. A. Nelson. 2005. Ecological and phylogenetic effects on song sharing in songbirds. *Ethology* 111:221–38.

Hof, D., and N. Hazlett. 2010. Low-amplitude song predicts attack in a North American wood warbler. *Animal Behaviour* 80:821–28.

Howell, S. N. G., and S. Webb. 1995. *A Guide to the Birds of Mexico and Northern Central America*. New York: Oxford University Press.

Hu, Y., and G. C. Cardoso. 2012. Which birds adjust the frequency of vocalizations in urban noise? *Animal Behaviour* 79:863–67.

Irwin, D. E. 2000. Song variation in an avian ring species. *Evolution* 54:998–1010.

Ketterson, E. D., V. Nolan, L. Wolf, and C. Ziegenfus. 1992. Testosterone and avian life histories: Effects of experimentally elevated testosterone on behavior and correlates of fitness in the dark-eyed junco (*Junco hyemalis*). *American Naturalist* 140:980–99.

Konishi, M. 1964a. Effects of deafening on song development in two species of juncos. *Condor* 66:85–102.

———. 1964b. Song variation in a population of Oregon juncos. *Condor* 66:423–36.

Kroodsma, D. E. 1996. Ecology of passerine song development. In *Ecology and Evolution of Acoustic Communication in Birds*, edited by D. E. Kroodsma and E. H. Miller, 3–19. Ithaca: Cornell University Press.

Labarbera, K., P. E. Llambías, E. R. A. Cramer, T. D. Schaming, and I. J. Lovette. 2010. Synchrony does not explain extrapair paternity rate variation in northern or southern house wrens. *Behavioral Ecology* 21:773–80.

Lachlan, R. F., and M. R. Servedio. 2004. Song learning accelerates allopatric speciation. *Evolution* 58:2049–63.

Marler, P. 1990. Song learning: The interface between behaviour and neuroethology. *Philosophical Transactions of the Royal Society B-Biological Sciences* 329: 109–14.

Marler, P., and D. Isaac. 1961. Song variation in a population of Mexican juncos. *Wilson Bulletin* 73:193–206.

Marler, P., M. Kreith, and M. Tamura. 1962. Song development in hand-raised Oregon juncos. *Auk* 79:12–30.

Miller, A. H. 1941. Speciation in the avian genus *Junco*. *University of California Publications in Zoology* 44:173–434.

Mirsky, E. N. 1976. Song divergence in hummingbird and junco populations on Guadalupe Island. *Condor* 78:230–35.

Moore, N. J. 1972. Ethology of the Mexican junco (*Junco phaenotus palliatus*). PhD dissertation, University of Arizona, Tucson.

Mountjoy, D. J., and D. W. Leger. 2001. Vireo song repertoires and migratory distance: Three sexual selection hypotheses fail to explain the correlation. *Behavioral Ecology* 12:98–102.

Nakano, R., N. Skals, T. Takanashi, A. Surlykke, T. Koike, K. Yoshida, H. Maruyama, S. Tatsuki, and Y. Ishikawa. 2008. Moths produce extremely quiet ultrasonic courtship songs by rubbing specialized scales. *Proceedings of the National Academy of Sciences of the United States of America* 105:11812–17.

Nelson, D. A. 2000. Song overproduction, selective attrition and song dialects in the white-crowned sparrow. *Animal Behaviour* 60:887–98.

Newman, M. M., P. J. Yeh, and T. D. Price. 2008. Song variation in a recently founded population of the dark-eyed junco (*Junco hyemalis*). *Ethology* 114: 164–73.

Nolan, V., Jr., E. D. Ketterson, D. A. Cristol, C. M. Rogers, E. D. Clotfelter, R. C. Titus, S. J. Schoech, and E. Snajdr. 2002. Dark-eyed junco (*Junco hyemalis*). In *The Birds of North America*, edited by A. Poole and F. Gill, no. 716. Ithaca: Cornell Lab of Ornithology.

Parker, K. A., M. J. Anderson, P. F. Jenkins, and D. H. Brunton. 2012. The effects of translocation-induced isolation and fragmentation on the cultural evolution of bird song. *Ecology Letters* 15:778–85.

Pasteau, M., L. Nagle, and M. Kreutzer. 2009. Preferences and predispositions of female canaries (*Serinus canaria*) for loud intensity of male sexy phrases. *Biological Journal of the Linnean Society* 96:808–14.

Pieplow, N. D., and C. D. Francis. 2011. Song differences among subspecies of yellow-eyed juncos (*Junco phaeonotus*). *Wilson Journal of Ornithology* 123: 639–49.

Podos, J., and P. S. Warren. 2007. The evolution of geographic variation in birdsong *Advances in the Study of Behavior* 37:403–58.

Potvin, D. A., and S. M. Clegg. 2015. The relative roles of cultural drift and acoustic adaptation in shaping syllable repertoires of island bird populations change with time since colonization. *Evolution* 69:368–80.

Price, J. J., and S. M. Lanyon. 2004. Patterns of song evolution and sexual selection in the oropendolas and caciques. *Behavioral Ecology* 15:485–97.

Price, T. D. 2008. *Speciation in Birds*. Greenwood Village, CO: Roberts and Company.

Rasner, C. A., P. Yeh, L. S. Eggert, K. E. Hunt, D. S. Woodruff, and T. D. Price. 2004. Genetic and morphological evolution following a founder event in the dark-eyed junco, *Junco hyemalis thurberi*. *Molecular Ecology* 13:671–81.

Reichard, D. G. 2014. Male dark-eyed juncos (*Junco hyemalis*) respond differentially to playback of local and foreign song. *Wilson Journal of Ornithology* 126: 605–11.

Reichard, D. G., R. J. Rice, E. M. Schultz, and S. E. Schrock. 2013. Low-amplitude songs produced by male dark-eyed juncos (*Junco hyemalis*) differ when sung during intra- and inter-sexual interactions. *Behaviour* 150:1183–202.

Reichard, D. G., R. J. Rice, C. C. Vanderbilt, and E. D. Ketterson. 2011. Deciphering information encoded in birdsong: Male songbirds with fertile mates respond most strongly to complex, low-amplitude songs used in courtship. *American Naturalist* 178:478–87.

Reichard, D. G., and J. F. Welklin. 2015. On the existence and potential functions of low-amplitude vocalizations in North American birds. *Auk* 132:156–66.

Ręk, P., and T. S. Osiejuk. 2011. Nonpasserine bird produces soft calls and pays retaliation cost. *Behavioral Ecology* 22:657–62.

Riebel, K. 2003. The "mute" sex revisited: Vocal production and perception learning in female songbirds. *Advances in the Study of Behavior* 33:49–86.

Ríos-Chelén, A. A., C. Salaberria, I. Barbosa, C. Macías Garcia, and D. Gil. 2012. The learning advantage: Bird species that learn their song show a tighter adjustment of song to noisy environments than those that do not learn. *Journal of Evolutionary Biology* 25:2171–80.

Ritschard, M., and H. Brumm. 2011. Effects of vocal learning, phonetics and inheritance on song amplitude in zebra finches. *Animal Behaviour* 82:1415–22.

Ritschard, M., K. Riebel, and H. Brumm. 2010. Female zebra finches prefer high-amplitude song. *Animal Behaviour* 79:877–83.

Ritschard, M., K. Van Oers, M. Naguib, and H. Brumm. 2012. Song amplitude of rival males modulates the territorial behaviour of great tits during the fertile period of their mates. *Ethology* 118:197–202.

Sakata, J. T., and S. L. Vehrencamp. 2012. Integrating perspectives on vocal performance and consistency. *Journal of Experimental Biology* 215:201–9.

Schluter, D., and T. D. Price. 1993. Honesty, perception and population divergence in sexually selected traits. *Proceedings of the Royal Society B-Biological Sciences* 253:117–22.

Searcy, W. A. 1996. Sound-pressure levels and song preferences in female red-winged blackbirds (*Agelaius phoeniceus*) (aves, Emberizidae). *Ethology* 102: 187–96.

Searcy, W. A., R. C. Anderson, and S. Nowicki. 2006. Bird song as a signal of aggressive intent. *Behavioral Ecology and Sociobiology* 60:234–41.

Searcy, W. A., and M. D. Beecher. 2009. Song as an aggressive signal in songbirds. *Animal Behaviour* 78:1281–92.

Searcy, W., and S. Nowicki. 2005. *The Evolution of Animal Communication: Reliability and Deception in Signaling Systems*. Princeton: Princeton University Press.

Seddon, N. 2005. Ecological adaptation and species recognition drives vocal evolution in neotropical suboscine birds. *Evolution* 59:200–215.

Servedio, M. R., T. D. Price, and R. Lande. 2013. Evolution of displays within the pair bond. *Proceedings of the Royal Society B-Biological Sciences* 280:20123020.

Shutler, D. 2011. Sexual selection: When to expect trade-offs. *Biology Letters* 7: 101–4.

Simões, J. M., P. J. Fonseca, G. F. Turner, and M. C. P. Amorim. 2008. African cichlid *Pseudotropheus* spp. males moan to females during foreplay. *Journal of Fish Biology* 72:2689–94.

Singh, P. and T. D. Price. 2015. Causes of the latitudinal gradient in birdsong complexity assessed from geographical variation within two Himalayan warbler species. *Ibis* 157:511–27.

Slabbekoorn, H., and T. B. Smith. 2002. Bird song, ecology and speciation. *Philosophical Transactions of the Royal Society B-Biological Sciences* 357:493–503.

Slabbekoorn, H., P. Yeh, and K. Hunt. 2007. Sound transmission and song divergence: A comparison of urban and forest acoustics. *Condor* 109:67–78.

Soma, M., and L. Z. Garamszegi. 2011. Rethinking birdsong evolution: Meta-analysis of the relationship between song complexity and reproductive success. *Behavioral Ecology* 22:363–71.

Spottiswoode, C. N., and A. P. Møller. 2004. Extrapair paternity, migration, and breeding synchrony in birds. *Behavioral Ecology* 15:41–57.

Spottiswoode, C. N., A. P. Tøttrup, and T. Coppack. 2006. Sexual selection predicts advancement of avian spring migration in response to climate change. *Proceedings of the Royal Society B-Biological Sciences* 273:3023–29.

Stutchbury, B. J., and E. S. Morton. 1995. The effect of breeding synchrony on extra-pair mating systems in songbirds. *Behaviour* 132:675–90.

Sueur, J., and T. Aubin. 2004. Acoustic signals in cicada courtship behaviour (order hemiptera, genus *Tibicina*). *Journal of Zoology* 262:217–24.

Suthers, R. A., F. Goller, and C. Pytte. 1999. The neuromuscular control of birdsong. *Philosophical Transactions of the Royal Society B-Biological Sciences* 354: 927–39.

Templeton, C. N., Ç. Akçay, S. E. Campbell, and M. D. Beecher. 2009. Juvenile sparrows preferentially eavesdrop on adult song interactions. *Proceedings of the Royal Society B-Biological Sciences* 277:447–53.

Templeton, C. N., S. E. Campbell, and M. D. Beecher. 2012. Territorial song sparrows tolerate juveniles during the early song-learning phase. *Behavioral Ecology* 23:916–23.

Titus, R. C. 1998. Short-range and long-range songs: Use of two acoustically distinct song classes by dark-eyed juncos. *Auk* 115:386–93.

Titus, R. C., C. R. Chandler, E. D. Ketterson, and V. Nolan. 1997a. Song rates of dark-eyed juncos do not increase when females are fertile. *Behavioral Ecology and Sociobiology* 41:165–69.

Titus, R. C., E. D. Ketterson, and V. Nolan. 1997b. High testosterone prior to song crystallization inhibits singing behavior in captive yearling dark-eyed juncos (*Junco hyemalis*). *Hormones and Behavior* 32:133–40.

Whittaker, D. J., A. L. Dapper, M. P. Peterson, J. W. Atwell, and E. D. Ketterson. 2012. Maintenance of MHC Class IIB diversity in a recently established songbird population. *Journal of Avian Biology* 43:109–18.

Wiley, R. H., and D. G. Richards. 1978. Physical constraints on acoustic communication in atmosphere—implications for evolution of animal vocalizations. *Behavioral Ecology and Sociobiology* 3:69–94.

———. 1982. Adaptations for acoustic communication in birds: Sound transmission and signal detection. In *Acoustic Communication in Birds*, edited by D. E. Kroodsma and E. H. Miller, 131–81. New York: Academic Press.

Williams, L., and M. H. Macroberts. 1977. Individual variation in songs of dark-eyed juncos. *Condor* 79:106–12.

———. 1978. Song variation in dark-eyed juncos in Nova Scotia. *Condor* 80: 237–40.

Wingelmaier, K., H. Winkler, and E. Nemeth. 2007. Reed bunting (*Emberiza schoeniclus*) males sing an "all-clear" signal to their incubating females. *Behaviour* 144:195–206.

Yeh, P. J., and T. D. Price. 2004. Adaptive phenotypic plasticity and the successful colonization of a novel environment. *American Naturalist* 164:531–42.

Zuk, M., D. Rebar, and S. P. Scott. 2008. Courtship song is more variable than calling song in the field cricket *Teleogryllus oceanicus*. *Animal Behaviour* 76: 1065–71.

CHAPTER FOURTEEN

Standing on the Shoulders

Agendas for Future Research Addressing Evolutionary and Integrative Biology in a Rapidly Evolving Songbird

Ellen D. Ketterson amd Jonathan W. Atwell

Whither thou goest I shall go.
—Ruth 1:16, King James Bible

A. *Birds of North America* Accounts of the Dark-Eyed and Yellow-Eyed Junco

L ooking back twelve to fifteen years to the priorities for future research described in the *Birds of North America* species accounts of the dark-eyed and yellow-eyed junco (Nolan, Jr. et al. 2002; Sullivan 1999), we happily see that some of the gaps in knowledge identified in those accounts have been filled. Others remain as enticing challenges, and new priorities have emerged.

A.1. The Dark-eyed Junco

Priorities for Future Research (text in italics from Nolan, Jr. et al. 2002)

1. *Coordinated studies of Dark-eyed Junco breeding populations at northern (especially Canada) and southern sites would throw light on geographic variation in population dynamics, annual cycle, response to initial predictive and supplemental regulatory factors, and related endocrinological mechanisms. Probability*

that global climate change will affect reproductive schedules (Visser et al. 1998)
makes such studies even more desirable.

Ongoing research addresses this priority. Chapters 10 and 11 by Atwell
et al. and Bergeon Burns et al., respectively, report on geographic variation
in reproductive schedules by comparing junco populations in California
and South Dakota to each other and to the previously studied population
in Virginia. A more thorough and systematic set of comparisons could be
highly informative. Current research on hormonal mediation of migratory
behavior and reproductive timing in migrant and resident groups is being
conducted by T. Greives of North Dakota State University in collaboration
with A. Fudickar and E. Ketterson of Indiana University.

2. *No molecular study of traditional questions regarding taxonomy of junco has*
   appeared. Questions about speciation and subspecific taxonomy remain unset-
   tled despite thorough study by Miller (1941). Ecological speciation of juncos
   would be interesting to study. Are hybrid zones between subspecies stable? Note
   that new molecular methods to study hybridization are constantly being devel-
   oped.

Chapters 8 and 9 specifically address progress towards these goals, and
ongoing research in the laboratories of B. Milá and J. McCormack are employing the most recent molecular techniques to further resolve the relationships among the taxa of juncos. Milá's group is sampling specifically
in hybrid zones and is collaborating with Ketterson's group to study geographic variation in gene expression related to plumage coloration. On
the research agenda are studies of female mate choice in relation to seasonality and plumage differences.

3. *Other desiderata are studies of form and function of winter flocks. A winter study*
   of J. h. carolinensis would be welcome, especially one that established winter lo-
   cation and habitat of adult females (see Ketterson et al. 1991; Rabenold and Ra-
   benold 1985). Use of stable isotopes (see Marra et al. 1998) may enable tracing of
   populations from breeding to wintering ranges.

Currently researchers are making exciting progress towards these goals.
Leah Wilson, inspired by J. L. Goodson of Indiana University, is investi-
gating the neurobiology of flocking behavior using the junco as a model.
Bird species vary greatly in the degree to which they are attracted to and

form bonds with conspecifics. These species differences have been attributed to the action of nonapeptide hormones in the brain (e.g., arginine vasopressin [AVP]) (Goodson 2008). Because juncos form flocks during winter and are territorial during breeding, it is possible to ask whether the nonapeptides that give rise to species differences in attraction to conspecifics also mediate seasonal differences in flocking behavior within a species. Atwell and A. Fudickar and E. Bridge are undertaking a project to use stable isotopes, genomics, and geolocators to establish connectivity between wintering and breeding locations of migratory dark-eyed juncos in the eastern United States, Canada, and Alaska (see chapter 3, this volume) (Mckinnon et al. 2013). The winter locations preferred by females in the partially migratory population of juncos in the Appalachian Mountains are still not known and remain a question seeking an answer.

4. *Only the most rudimentary facts are known about* J. h. insularis; *any detailed information about any aspect of its life history would be a contribution.*

Considerable progress has been made on this front and continues. The film *Ordinary Extraordinary Junco* includes a segment filmed on Guadalupe Island, and chapter 8 (Milá et al., this volume) reports that the Guadalupe junco, despite its dark eye, is more closely related to the yellow-eyed junco than the dark-eyed junco, a conclusion that differs from all previously published taxonomies (Aleixandre et al. 2013). Importantly, this research resulted directly in the recent decision by the American Ornithologists' Union to recognize the Guadalupe junco as a unique full species, *Junco insularis*.[1] Milá's research group is currently conducting studies of the life history of the Guadalupe junco, whose breeding dates are still imperfectly known. Interestingly, the Guadalupe junco's bill differs in shape from mainland forms, and the cones of the Guadalupe cypress tree (*Cupressus guadalupensis*) may provide an explanation (compare chapters 8 and 9, this volume). Similarly, the song of *J. insularis* is also quite different from mainland forms and would be interesting to study (compare chapter 13, this volume). Another wide-open opportunity for research is the volcano junco (*Junco vulcani*) in the highlands of Costa Rica. Many opportunities for comparative research remain.

---

1. See AOU Classification Committee's *North and Middle America Proposal Set 2014-A* (no. 6, page 23).

A.2. The Yellow-Eyed Junco

Sullivan (2002) also noted the need to establish relationships among the juncos using molecular methods, and these are ongoing (see chapters 8 and 9, this volume). Sullivan listed attributes of the yellow-eyed junco that make it easy to study, including that it is abundant, sedentary, easy to catch, forages on the ground, and is quite tame. She wrote as follows:

> This species provides an unusual opportunity to study the behavioral and population ecology of juvenile passerines because once the young can fly, they are not secretive, they are easily observed and followed, they remain near their natal territory for 23 months after fledging, and they are philopatric, breeding near their natal territory. The site fidelity shown by returning and first-year breeders makes it possible to measure individual variation in reproductive success as the number of young recruited into the population and to estimate gene flow in an avian population. To date, all studies of behavior, ecology, and physiology have been carried out in southern Arizona at the extreme northern extent of the range of this species. It is not known whether these populations are representative of populations throughout the range (Sullivan 2002).

The opportunity to study lifetime reproductive success and short-distance natal dispersal is rare in songbirds and, to our knowledge, has not been capitalized upon (compare dispersal in juncos from a Virginia population that is partially migratory) (Liebgold et al. 2013). Lack of knowledge of juvenile behavior persists in ornithology and represents an important life stage for study (Price et al. 2008). Some progress has been made in acquiring comparative knowledge of yellow-eyed junco behavior and ecology as reported in chapter 8 (see also Pieplow and Francis 2011), but Sullivan's recommendation for further study remains on target.

B. Recommendations for Further Study, Parts 2–4

Each chapter in this volume concludes with priorities for future research, and we add some of our own here.

B.1. Part 2: Hormones, Phenotypic Integration, and Life Histories:
An Endocrine Approach

Chapter 4 summarizes findings of Reed et al. (2006) that male juncos with experimentally elevated testosterone had higher fitness than control males primarily because testosterone-treated males were more successful at extrapair fertilizations (EPFs). Data in Reed et al. (2006) and earlier papers could be used to model how a "mutant with elevated testosterone" might spread. If males with chronically elevated testosterone were to exist, their EPF advantage over males with lower testosterone would necessarily decline as their relative abundance increased and they began to steal copulations from one another's mates. The model's predictions could be tested if relative abundance of testosterone-treated males and controls in the field were varied and fitness compared. The relative advantages of a chronically high T phenotype should also vary with the environment, and this too could be modeled to learn whether hormone-mediated traits favored under high density would be disfavored at low density or during years of high or low nest predation, cold or mild winters, or early or late springs. High-T males that allocate more time and energy to mating as opposed to parental effort might, for example, fare well in years of high food abundance or low density when parental care is not as beneficial as having more mates. The reverse might hold true if food were in short supply or nest predators are abundant. In general, experimental studies of life history evolution need to go on for longer and be conducted in multiple populations because some years and environments will favor traits that are disfavored in other years or environments.

Other experimental studies related to hormones that are wanting include manipulations to reduce the levels of testosterone reaching target tissues, as might be accomplished with compounds that block hormone receptors. Many traits mediated by testosterone involve conversion of testosterone to estradiol, but no studies have asked how the phenotype might vary if the enzyme that converts testosterone to estradiol (aromatase) were disabled. Given recent findings about seasonal variation in the impact of aromatase on aggression in the song sparrow (Heimovics et al. 2012), this approach holds promise. Another unexplored avenue is testosterone in juveniles. Researchers have not yet looked for organizational as opposed to activational effects of testosterone on phenotype in the wild.

Other intriguing questions include why, in a mechanistic sense, males with experimentally elevated testosterone are poorer providers of food to their young. What sensory, neural, and peripheral changes lead such males to sing rather than to feed offspring, to patrol large home ranges as opposed to participate in nest defense? Do they not hear their offspring beg? Are they less sensitive to the pitch of begging calls? Is their own hunger suppressed so they are less inclined to forage in general? What exactly is different about a high-testosterone male as compared to an unmanipulated male? Similar unanswered questions regarding females, testosterone, and other endocrine signals abound.

Chapter 5 by McGlothlin and Ketterson calls for quantitative genetic studies of selection on correlated traits and the degree to which hormone-mediated traits may facilitate or retard adaptation and the resolution of sexual conflict. To what extent do genetic correlations predict responses to correlated selection? How weak or strong is the response to simultaneous selection on traits that are positively and negatively correlated with fitness? Answering these questions might be too challenging in the junco and most other songbirds if they cannot be induced to breed in captivity. However, as more is learned about the neuroendocrine basis of reproduction, it may be possible to induce more species of captive songbirds to breed at times or in environments where they currently do not, perhaps by blocking mechanisms such as GnIH (gonadotropin inhibiting hormone) that suppress breeding (Tsutsui et al. 2012). And as methods for assessing relatedness within populations improve, it may become possible to relate genetic similarity to correlated traits without doing crosses or having detailed pedigrees of marked animals.

Chapter 6 by Cain, Jawor, and McGlothlin reports on natural variation in the ability of individual males and females to elevate testosterone in response to a hormonal challenge and how that variation relates to physiology, behavior, morphology, and fitness. These promising studies are based on just one or two seasons and there is more to be learned.

The GnRH challenge method has been informative, but more could be learned about how best to apply it and how best to interpret the findings. For example, the method was used to measure the ability of an individual to produce testosterone after thirty minutes. Rosvall, Bergeon Burns, and Atwell are currently asking whether individuals or populations vary in the timing of peak testosterone after a challenge (Bergeon Burns et al. 2013). Would different conclusions be reached if T in response to GnRH were measured at fifteen minutes or forty-five minutes? We also do not know why some traits correlate significantly with the maximum T generated by GnRH (e.g., aggression and parental behavior); others correlate more strongly with testosterone prior to a GnRH challenge (e.g., immune function); and still others with the difference in testosterone as measured before and after a challenge (e.g., plumage coloration) (see chapter 6, figure 1, this volume). Females elevate testosterone in response to GnRH during some stages of reproduction but not during others, and the reasons for this are not known.

Other unanswered questions include the role of hormones in accounting for variation among individuals in suites of traits, including personality (Wolf et al. 2007). To date, response to a GnRH challenge has been related to immune function, behavior, and morphology, but not when all were measured in the same individuals. Further, researchers are only beginning to investigate plasticity in individual response to hormonal challenges. For example, to date no study has assessed whether T in response to a GnRH challenge would differ before and after an immune challenge (would response decline?), or before and after experimentally increasing brood size by adding nestlings to a nest (would response decline?), or before and after a manipulation of sex ratio that would reduce the availability of mates (would response increase?). Also not known is whether an environmental manipulation that increases one trait via a change in hormone levels will simultaneously drive down another (e.g., if brood size is increased, does T decline and immune function improve?). Study of hormonal interactions may also be highly informative. Reproductive hormones and hormones that respond to stressors are linked via the HPG and HPA axes, and each axis is likely to affect plasticity (M. Abolins-Abols, pers. comm.). Rich opportunities for measuring reaction norms in hormonal responses exist, and results can be applied to fundamental questions in adaptation and response to environmental change.

In chapter 7, Rosvall et al. recommend an individual-based approach to identifying how hormone-mediated behaviors such as courtship or aggression might evolve by adaptive and tissue-specific changes in hormone sensitivity. For example, do differences in steroid sensitivity between females and males permit divergent sex-specific responses to similar circulating levels of hormone? How rapidly can selection favor dissociation of correlated traits by adjusting hormonal sensitivity in different tissues? With respect to parenting, Rosvall et al. write "... we might expect selection to favor downregulated (or completely eliminated) sensitivity to T in areas of the brain controlling parental care, particularly in females or species with extensive parental care (Lynn 2008)." This prediction can be tested and resonates with the suggestion in chapter 5 that we apply concepts from quantitative genetics to questions of correlated traits, adaptation, and constraint. Chapter 7 also moves beyond hormone levels and targets and opens up fascinating questions of hormone-influenced gene expression where truly much remains to be learned (Peterson et al. 2014).

B.2. Part 3: Evolutionary Diversification in the Avian Genus Junco: Pattern and Process

Chapters 8 and 9 present numerous questions for future research. In chapter 8 Milá et al. consider the perennial question "what is a species?" and conclude that for the dark-eyed junco, "the designation of species is particularly difficult." They predict that the "ongoing quest for additional molecular data will help resolve some relationships and patterns of gene flow among hybridizing junco forms in mainland North America." As for the "yellow-eyed junco," their research has led to a clear designation of three phylogenetic species: one to be found on Guadalupe Island, one in southern Baja, and one in Guatemala. Their chapter emphasizes that understanding lineage divergence in the junco will require greater knowledge of the causes of phenotypic variation within and between junco forms, and they call for studies of song and odor variation among populations (see chapters 12 and 13, this volume), quantitative analyses of plumage color across the range, and clinal analyses of molecular markers. They also emphasize the need "to unveil the genetic basis of adaptive variation. Identifying the genes that code for the traits that selection acts upon will allow us to associate polymorphism with levels of divergence and gene flow, and infer strength and modes of selection, the number of loci involved, and many other fascinating aspects of this complex evolutionary process."

Milá et al. conclude that "The genus *Junco* provides the possibility of investigating old speciation events as well as striking cases of 'speciation-in-action,' and we are confident that juncos will remain an illuminating and challenging system in the study of vertebrate diversification" (chapter 8, this volume).

Price and Hooper also call for exploration of how color differences evolve and note how little we understand about the evolutionary processes that determine color patterns, whether they have been convergent or parallel, and how they might reflect local adaptation (see section C.1, "Ecology," below). They also appeal for efforts to identify the loci underlying color differences as a critical step "towards an understanding of mechanisms of divergence at these loci . . . ." They predict that the loci coding for color differences will prove to be much older (more divergent) than the genetic differences at mitochondrial and nuclear loci that have been studied thus far.

Price and Hooper also comment on the impact of climate on junco divergence and attribute the greater genetic divergence seen in southern forms as described by Milá et al. to greater climatic stability in Meso-america as compared to North America. In contrast, they are less ready to attribute the lack of genetic divergence in northern forms to recent post-glacial events and more inclined to attribute it to periodic mixing during periods of glacial advance and retreat, or to ongoing gene mixing that accompanies migratory behavior. Thus they view parts of the dark-eyed junco genome as older than the last glacial maxima, a view that contrasts strongly with that of Milá et al. Once again, resolution will require more complete genomic data than are currently available and greater knowledge regarding the genetic contributions to phenotypic variation within *Junco hyemalis.*

Particularly exciting are Price and Hooper's speculations about a potential role for chromosomal rearrangements in maintaining complexes of locally adapted alleles in the face of gene flow (Kirkpatrick and Barton 2006). They suggest that chromosomal inversions might explain why juncos look so different from place to place, despite displaying so little overall genetic divergence. If genes within the junco's inversion on chromosome 5 were to code for color or migratory behavior and not recombine during hybridization, then genomes measured by molecular markers outside the inversions could be quite similar even when phenotypic differences are significant. Interestingly, the frequency of an inversion on chromosome 5 in the junco varies geographically and is more common in migratory groups (contrast to a recent report of lack of covariation between allelic variation in clock genes and migratory behavior in the junco) (Peterson et al. 2013). They recommend comparing subspecies and predict greater genetic divergence within the inversion than outside of it.

A new and very exciting paper (Horton et al. 2014) is highly relevant and will serve as a prime example of the potential for synergy between integrative and evolutionary biology. The white-throated sparrow, *Zono-trichia albicollis*, has two morphs, tan-striped and white-striped, that differ in aggression, plumage, and parental behavior. Mating between the morphs is disassortative, and they differ in the presence/absence of a chromosomal inversion. The white-striped morph carries the inversion;

the tan-striped morph does not. The gene that codes for estrogen receptor (ERα) maps to the inversion and has at least two alleles that differ by sequence in the promoter region. The allele on the chromosome with the inversion is transcribed more rapidly than the allele found in the chromosome that lacks the inversion. That is, the more rapidly transcribed form of the gene is found in the more aggressive white-striped morph, not in the tan-striped morph. The difference in transcription rate predicts greater abundance of estrogen receptor and potentially stronger response to a given level of circulating or locally produced hormone (compare Rosvall et al. in chapter 7). In short, white-striped males may be the more aggressive morph because they bear a chromosomal arrangement that includes a copy of a gene that is transcribed more rapidly, giving rise to greater sensitivity to sex steroids and thus more aggressive behavior. This study is the first to establish a connection between a chromosomal rearrangement, alleles for sensitivity to hormones, and behavior, and is precisely the kind of alteration that Price and Hooper would predict for the junco, except that the genes they expect to find within the junco's inversion would relate to plumage color or migratory behavior (i.e., the traits that are associated with the different subspecies of dark-eyed junco).

Finally, both Milá et al. and Price and Hooper emphasize the need for studies of mating preferences. Before long it might be possible to alter eye color through developmental genetic manipulation to assess the importance of this trait to juncos or manipulate volatiles emanating from the preen gland (Whittaker et al. 2011). A. Kimmett is working with Ketterson to assess the role of timing mechanisms in mate choice. Might, for example, ecologically induced variation in migratory behavior inhibit interbreeding between migratory and sedentary forms during seasons of sympatry, while sexual selection acting during breeding might foster divergence in plumage during seasons of allopatry (Winker 2010)?

B.3. Part 4: Mechanisms of Divergence among Populations

Chapter 10 by Atwell et al. considers the junco's recent colonization of urban environments. A nagging question that cannot be answered with certainty is whether the colonists responded rapidly to the altered selection regime of the city, as seems most likely, at least with respect to some traits, or whether they were preadapted in the sense that the trait that favored establishment (cessation of migration) tended to be coexpressed with other traits that fare well in urban environments. A third possibility—that colonists happened to express traits that fare well in urban environments by chance (founder effect)—is deemed highly unlikely (Yeh and Price 2004).

What can be addressed is the potential role of phenotypic plasticity in maintaining the phenotypic distributions that currently characterize urban and nonurban birds. To date results from common garden experiments indicate a mosaic—some traits are consistent with selection having led to rapid genetic divergence (the plumage trait tail white and a behavioral trait, boldness), while others traits remain plastic (seasonal elevation of gonadal steroids) and can be induced on a similar schedule in both populations under mild conditions in captivity. Research underway created an "urban-like common garden" with frequent disturbance. R. Hanauer and M. Abolins-Abols are currently comparing physiology and gene expression in disturbed and undisturbed birds. Population comparisons of plasticity in the field may also be quite informative. When confronted with disturbance, are city birds less perturbed hormonally and behaviorally than mountain birds (Hope et al. 2014)? More generally, much remains to be learned about the extent to which hormonal induction of multiple traits might hasten colonization of novel environments.

In chapter 11, Bergeon Burns et al. address the role of circulating hormones and target tissue sensitivity in mediating differences in life-history traits such as aggression among individuals within and between populations. They speculated that initial divergence between populations may be facilitated by shifts in hormone signal, whereas the multiple changes in hormone sensitivity might require longer divergence times (compare Ketterson et al. 2009). They concluded that the "junco species complex provides an extraordinary opportunity to understand variation in the mechanisms underlying hormone-mediated phenotypic evolution." One of those opportunities is to extend our knowledge of mechanisms of divergence in females. It is females that decide when and where to breed and who to mate with, and we know far too little about the mechanisms underlying those decisions.

In chapter 12, Whittaker and Gerlach focus directly on the topic of mate choice, warning that the study of "mate choice is far from straightforward." Females may base their choice on multiple traits using multiple sensory systems and choice may vary depending on their own condition or genotype and on what they are choosing a mate for (e.g., as a sire for their offspring or as a partner to help rear offspring). Choice may also vary depending on available alternatives, which means that choice needs to be measured in different contexts. While many studies have employed two-way choice tests, chapter 12 encourages investigators to study choice in natural populations. We have long known that females sample before they choose, but it has been hard to follow females while they choose (Neudorf et al. 2002). New technologies should make it far more feasible to study female choice in nature and to add in the complexity that is part of the natural process.

When confronted with the difficult decision of how to classify the northernmost populations of yellow-eyed juncos that occasionally interbreed with dark-eyed juncos, Milá et al. decided that differences in song between the two groups were too striking to ignore. In chapter 13, Cardoso and Reichard ask how and why songs vary and urge greater knowledge about song learning, particularly the role of social copying and improvisation in song learning. Importantly, improvisation buffers against the geographic differentiation that might occur via social copying, because a substantial number of songs are produced de novo. Dark-eyed juncos have far less diversity in song than would be expected for such a broadly distributed species, and their geographic similarity in song contrasts strongly with their divergence in plumage. Cardoso and Reichard pose the question of why dark-eyed juncos have unexpectedly simple songs, low copying, and high improvisation as "a challenge for future research."

Another rich area for future research is the junco's low amplitude songs that are used during courtship and known as short-range song (SRS) (Reichard et al. 2011). These songs are highly complex and quite difficult to study in nature because of their low amplitude. Future progress can be expected, however, because of the possibility to study free-living birds equipped with microphone transmitters (Reichard, pers. comm.). Still to be learned is whether SRS differs between populations, is learned through copying, or plays a critical role in mate choice.

C. Topics by Discipline to Which Juncos Can Contribute

In this section we describe how questions raised in the chapters can be combined to make significant contributions to the fields of ecology, speciation, microevolution, and animal behavior and physiology. We also continue to point to how methodological advances may enable studies that have not been feasible to date.

C.1. Ecology

If we are to make full use of knowledge gained to date, we need to know more about the junco's ecology, for example, what accounts for variation in its tendency to be a generalist with broad tolerances and preferences in North America or a specialist that confines itself to mountain tops in the subtropics in Mesoamerica (chapter 3, this volume). More information is needed about how it interacts with competitors, parasites, and predators (Clotfelter et al. 2007). Juncos conform to biogeographic rules, but not completely. Larger subspecies of the dark-eyed juncos do not necessarily occur in colder climates as would be predicted by Bergmann's rule (Ferree 2013). Whether larger individuals within a species are also found in cooler climates, particularly during winter, is not known despite the suspected influence of body size on distance migrated (Ketterson and Nolan, Jr. 1976). But limited data based on wing length suggest it does not (Nolan, Jr. and Ketterson 1983). Paler forms of the junco do appear to occupy dry and cool climates as Gloger's rule would predict (e.g., Baird's junco in the southern Baja peninsula), but why dry-cool climate favors dull plumages in this or other species is still unknown. A suggestion that feather-degrading bacteria might play a role in related species is worth pursuing (Burtt and Ichida 2004). The explanation for the yellow-eyed junco's yellow eye also remains to be found, and the new phylogeny that joins dark- and yellow-eyed forms in the same species suggests that an adaptive explanation may exist.

One exciting and recent advance in evolutionary ecology is niche modeling, which can be used to reconstruct species distributions based on climate. This approach is particularly interesting when applied to historical distributions as Price and Hooper describe in chapter 9. Using climate data and data on current distributions (e.g., the Breeding Bird Survey or e-bird), they mapped historical species distributions based on climates during earlier eras (e.g., the last glacial maximum approximately 15,000 years ago). The accuracy of these estimates obviously depends on how well past climates can be mapped and on the assumption that a species' climate preferences have not changed. Accuracy also depends on the importance of climate per se in determining a species' distribution. If current distributions are determined more by biotic factors (e.g., predators or parasites that respond to climate in their own independent ways), then the distributional estimates produced by niche modeling may be noisy. Thus if the junco's altitudinal distributions are strongly affected by the distribution of nest predators such as snakes or small mammals, then climate-based niche models may miss the mark. Caveats aside, however, it is fascinating to ask where juncos were during the last glacial maximum and whether or not Alaska was a major refuge as some models predict. The answer is also hugely relevant to how rapidly phenotypes can evolve (see chapters 8 and 9, this volume). In one view, junco plumage types may have evolved in the past 10,000 years. In another view, juncos may have been through repeated cycles of retreat and advance between glacial cycles and their refugia may have included Alaska as well as Mexico. New techniques for niche modeling, better estimates of present day distributions and abundance, better methods for estimating genetic divergence, and greater knowledge of the role of gene expression in the development of plumage color and eye color will all be important in answering these questions in historical and evolutionary ecology. Furthermore, the answers are sure to be informative when making predictions about the ecological consequences of climate change for juncos and other taxa.

Another incredibly hot area in ecology is the role of bacteria and the microbiomes they form in accounting for variation among populations and individuals. In the case of the junco, microbiomes may prove to play important roles in explaining variation in plumage, diet, and behavior. Also of growing interest is urban ecology. As we saw in chapter 10, juncos are invading or colonizing cities, and this new environment requires that they adjust to light at night, noise, and novel predators (e.g., house cats) and likely new diseases (Cardoso and Atwell 2011; Gil and Bruum 2013). Current research on avian personalities and response to stressors creates a bridge between ecology and animal behavior (Partecke et al. 2006) and is currently under study in the junco (M. Abolins-Abols and R. Hanauer, pers. comm.).

C.2. Divergence, Speciation, and Microevolution

We have already emphasized the importance of ever more effective methods for assessing relatedness, including next generation sequencing, single nucleotide polymorphisms (SNPs), and RAD sequencing (chapter 8) (McCormack et al. 2012). Future research needs to focus on hybrid zones and their apparent stability. We lack measures of the abundance and fitness of F2 and more introgressed individuals in hybrid zones (see *Ordinary Extraordinary Junco* for footage of a hybrid pair caring for offspring in Nevada). Until we know how hybrids perform we will not be able to properly interpret the implications of hybridization.

The need for a more complete phylogeny extends beyond proper classification. Those interested in the evolution of behavior need an accurate tree to test hypotheses. A recent study that attempted to relate variation in candidate genes to migratory behavior was limited by uncertainties regarding the junco's phylogeny (Peterson et al. 2013). Returning to hybrid zones, they offer splendid opportunities to study how the development and evolution of behavior, migration, mate choice, and song are affected by genes and environment.

Maternal and other intergenerational effects, along with the stupendous growth of epigenetics (changes in gene expression not related to changes in gene sequence) as a field, are certain to have huge effects on our understanding of how phenotypes develop and differ. To pick just one consequence that is relevant here, common gardens have long played a significant role in partitioning variation and attributing it to genes or environment, and several studies have taken this approach when studying the junco (see chapter 10). Traditionally, when comparing populations, differences observed in the wild that were not apparent in a common garden are interpreted as having been caused by differences in the environment, whereas differences that persist in a common garden are attributed to genes. We now know that mothers can pass information about the environment to their offspring that can influence development independently of variation in gene sequences. Particularly relevant to studies reviewed in this volume may be steroids and immunoglobulins deposited by females in yolk.

The chapter summaries have already highlighted the importance of hybridization, chromosomal rearrangements, and sequence divergence in the junco's evolutionary history. The charge for the future will be to complete the links relating gene sequence to gene expression to systems biology.

C.3. Organismal Biology: Animal Behavior and Physiology

A point of view that focuses on the organism as the expression of the genotype and the agent that does or does not pass that genotype along can be enlightening and provide an explicit alternative to evolution gene by gene (Noble 2013). Until recently the prevailing intellectual environment held that if variation isn't genetic, it isn't interesting. That environment is rapidly shifting towards one in which it has become critical to model and understand how environmental variation interacts with genetic mechanisms to give rise to phenotypic variation. Organismal biology can contribute to this shift and thus to the unification of behavioral ecology and molecular biology. These bold statements demand a few examples. The area of sexual conflict is one that will be advanced by identifying the impact of the external and internal environment on the enhancement or suppression of gene expression in one sex or the other (Pavitt et al. 2014). We are learning more and more about plasticity in gene expression in response to the abiotic environment. Populations exposed to cold or heat, humidity or drought, will vary in the expression of gene networks; seasonal variation in phenotype will do the same (Piersma and Van Gils 2011; Storz et al. 2010). Coexisting populations that differ in migratory behavior will differ in gene expression, and comparing populations will reveal the physiological basis for migration.

This last thread provides the strongest argument for why evolutionary biologists and organismal biologists can benefit from understanding each other's biology. We conclude with five themes relating hormones to evolution: hormonal pleiotropy, phenotypic plasticity, sexual selection, maternal effects, and population divergence.

- Owing to hormonal pleiotropy, hormones allow for coordinated phenotypic expression of multiple traits and coordinated response to environmental change.
- Owing to the role of hormones in growth, development, and sexual differentiation, hormones serve as mechanistic links on which selection can act to alter trait expression in ways that relate to viability and fecundity selection.
- Owing to the role of hormones in reproductive behavior and the production of ornaments, traits mediated by hormones are likely to be subject to sexual selection and to help explain the evolution of sex differences.
- Owing to the presence of hormones in egg yolk and other aspects of maternal physiology, hormones can serve as a medium for maternal effects that transduce the environments of the maternal generation and prepare the offspring generation to produce one or more of an array of possible phenotypes that are most suited to the actual environment the offspring will occupy. This assumes autocorrelation of environments and so does not produce a perfect match and maintains standing variation.
- Closely related groups may diverge owing to selection on hormone-mediated traits, thus hormones can play a role in speciation.

References

Aleixandre, P., J. Hernández Montoya, and B. Milá. 2013. Speciation on oceanic islands: Rapid adaptive divergence vs. cryptic speciation in a Guadalupe Island songbird (Aves: *Junco*). *PLoS ONE* 8:e63242.

Bergeon Burns, C. M., K. A. Rosvall, T. P. Hahn, G. E. Demas, and E. D. Ketterson. 2013. Examining sources of variation in HPG axis function among individuals and populations of the dark-eyed junco. *Hormones and Behavior* 65:179–87.

Burtt, E. H. J., and J. M. Ichida. 2004. Gloger's rule, feather-degrading bacteria, and color variation among song sparrows. *Condor* 106:681.

Cardoso, G. C., and J. W. Atwell. 2011. On the relation between loudness and the increased song frequency of urban birds. *Animal Behaviour* 82:831–36.

Clotfelter, E. D., A. B. Pedersen, J. A. Cranford, N. Ram, E. A. Snajdr, V. Nolan, Jr., and E. D. Ketterson. 2007. Acorn mast drives long-term dynamics of rodent and songbird populations. *Oecologia* 154:493–503.

Ferree, E. 2013. Geographic variation in dark-eyed junco morphology and implications for population divergence. *Wilson Journal of Ornithology* 125:454–70.

Gil, D. D., and H. Bruum, eds. 2013. *Avian Urban Ecology: Behavioural and Physiological Adaptations*. Oxford: Oxford University Press.

Goodson, J. L. 2008. Nonapeptides and the evolutionary patterning of sociality. *Progress in Brain Research* 170:3–15.

Heimovics, S. A., N. H. Prior, C. J. Maddison, and K. K. Soma. 2012. Rapid and widespread effects of 17β-estradiol on intracellular signaling in the male songbird brain: A seasonal comparison. *Endocrinology* 153:1364–76.

Hope, S. F., M. Abolins-Abols, and E. D. Ketterson. 2014. Stress affects the behavior of urban and montane populations differently. *Integrative and Comparative Biology* 54:E288.

Horton, B. M., W. H. Hudson, E. A. Ortlund, S. Shirk, J. W. Thomas, E. R. Young, W. M. Zinzow-Kramer, and D. L. Maney. 2014. Estrogen receptor alpha polymorphism in a species with alternative behavioral phenotypes. *Proceedings National Academy of Sciences of the United States of America* 111:1443–48.

Ketterson, E. D., J. W. Atwell, and J. W. McGlothlin. 2009. Phenotypic integration and independence: Hormones, performance, and response to environmental change. *Integrative and Comparative Biology* 49:365–79.

Ketterson, E. D., and V. Nolan, Jr. 1976. Geographic variation and its climatic correlates in sex-ratio of eastern-wintering dark-eyed juncos (*Junco-hyemalis hyemalis*). *Ecology* 57:679–93.

Kirkpatrick, M., and N. Barton. 2006. Chromosome inversions, local adaptation, and speciation. *Genetics* 173:419–34.

Liebgold, E. B., N. M. Gerlach, and E. D. Ketterson. 2013. Similarity in temporal variation in sex-biased dispersal over short and long distances in the dark-eyed junco, *Junco hyemalis*. *Molecular Ecology* 22:5548–60.

Lynn, S. E. 2008. Behavioral insensitivity to testosterone: Why and how does testosterone alter paternal and aggressive behavior in some avian species but not others? *General and Comparative Endocrinology* 157:233–40.

McCormack, J. E., J. M. Maley, S. M. Hird, E. P. Derryberry, G. R. Graves, and R. T. Brumfield. 2012. Next-generation sequencing reveals phylogeographic structure and a species tree for recent bird divergences. *Molecular Phylogenetics and Evolution* 62:397–406.

McKinnon, E. A., K. C. Fraser, and B. J. M. Stutchbury. 2013. New discoveries in landbird migration using geolocators, and a flight plan for the future. *Auk* 130: 211–22.

Neudorf, D. L., D. J. Ziolkowski, V. Nolan, Jr., and E. D. Ketterson. 2002. Testosterone manipulation of male attractiveness has no detectable effect on female home-range size and behavior during the fertile period. *Ethology* 108:713–26.

Noble, D. 2013. Physiology is rocking the foundations of evolutionary biology. *Experimental Physiology* 98:1235–43.

Nolan, Jr., V., and E. D. Ketterson. 1983. An analysis of body mass, wing length, and visible fat deposits of dark-eyed juncos wintering at different latitudes. *Wilson Bulletin* 95:603–20.

Nolan Jr., V., E. D. Ketterson, D. Cristol, C. Rogers, E. D. Clotfelter, R. C. Titus, S. Schoech, and E. Snajdr. 2002. Dark-eyed junco (*Junco hyemalis*).In *The Birds of North America*, edited by A. Poole and F. Gill, no. 716. Philadelphia: The Birds of North America, Inc.

Partecke, J., I. Schwabl, and E. Gwinner. 2006. Stress and the city: Urbanization and its effects on the stress physiology in European blackbirds. *Ecology* 87:1945–52.

Pavitt, A. T., C. A. Walling, A. S. McNeilly, J. M. Pemberton, and L. E. B. Kruuk. 2014. Variation in early life testosterone within a wild population of red deer. *Functional Ecology* 28:1224–34.

Peterson, M. P., M. Abolins-Abols, J. W. Atwell, R. J. Rice, B. Milá, and E. D. Ketterson. 2013. Variation in candidate genes CLOCK and ADCYAP1 does not consistently predict differences in migratory behavior in the songbird genus *Junco* [v1; ref status: indexed, http://f1000r.es/11p]. *F1000 Research* 2:115.

Peterson, M. P., K. A. Rosvall, C. A. Taylor, J. A. Lopez, J. H. Choi, C. Ziegenfus, H. Tang, J. K. Colbourne, and E. D. Ketterson. 2014. Potential for sexual conflict assessed via testosterone-mediated transcriptional changes in liver and muscle of a songbird. *Journal of Experimental Biology* 217:507–17.

Pieplow, N. D., and C. D. Francis. 2011. Song differences among subspecies of yellow-eyed juncos (*Junco phaeonotus*). *Wilson Journal of Ornithology* 123:464–71.

Piersma, T., and J. A. Van Gils. 2011. *The Flexible Phenotype: A Body-Centered Integration of Ecology, Physiology, and Behaviour*. New York: Oxford University Press.

Price, T. D., P. J. Yeh, and B. Harr. 2008. Phenotypic plasticity and the evolution of a socially selected trait following colonization of a novel environment. *American Naturalist* 172:S49–S62.

Reichard, D. G., R. J. Rice, C. C. Vanderbilt, and E. D. Ketterson. 2011. Deciphering information encoded in birdsong: Male songbirds with fertile mates respond most strongly to complex, low-amplitude songs used in courtship. *American Naturalist* 178:478–87.

Storz, J. F., G. R. Scott, and Z. A. Cheviron. 2010. Phenotypic plasticity and genetic adaptation to high-altitude hypoxia in vertebrates. *Journal of Experimental Biology* 213:4125–36.

Sullivan, K. A. 1999. Yellow-eyed junco (*junco phaeonotus*).In *The Birds of North America Online*, edited by A. Poole. Ithaca: Cornell Lab of Ornithology.

Tsutsui, K., T. Ubuka, G. E. Bentley, and L. J. Kriegsfeld. 2012. Gonadotropin-inhibitory hormone (GnIH): Discovery, progress, and prospect. *General and Comparative Endocrinology* 177:305–14.

Whittaker, D., H. Soini, N. Gerlach, A. Posto, M. Novotny, and E. Ketterson. 2011. Role of testosterone in stimulating seasonal changes in a potential avian che-mosignal. *Journal of Chemical Ecology* 37:1349–57.

Winker, K. 2010. On the origin of species through heteropatric differentiation: A review and a model of speciation in migratory animals. *Ornithological Monographs* 69:1–30.

Wolf, M., G. Van Doorn, O. Leimar, and F. Weissang. 2007. Life-history trade-offs favour the evolution of animal personalities. *Nature* 447:581–84.

Yeh, P. J., and T. D. Price. 2004. Adaptive phenotypic plasticity and the successful colonization of a novel environment. *American Naturalist* 164:531–42.

Glossary

accessory olfactory system. See **vomeronasal organ**

activational effect. The effect of a hormone that fluctuates with circulating hormone levels and is often reversible (i.e., not permanent, as with organizational effects).

agonistic responses. In chemical communication it refers to the process by which a chemical binds to a receptor causing a subsequent cellular responses; in animal behavior, it refers broadly to any social behaviors related to fighting such as threats, displays, retreats, etc.

alloparapatric speciation. Differentiation between populations taking place during periods when migration is absent, as well as periods when migration and, potentially, gene exchange occurs.

allopatric (distribution) / allopatry. The geographic distribution of taxa (e.g., populations or species) found in different locations; contrast **sympatric,** in which distributions of taxa overlap.

allopatric speciation. Differentiation between populations taking place in the complete absence of migration and gene exchange between them.

altitudinal (migration). A pattern of seasonal movement common in many bird populations, in which breeding typically occurs at higher elevation sites, with migration to wintering habitats at lower elevations.

androgen receptor (AR). A nuclear receptor that is activated by binding with testosterone in the cytoplasm and carrying it across the membrane into the cell nucleus.

annual cycle. The series of events or stages that consistently occur each year throughout an individual's or a population's life, including, for example in birds, behavioral and physiological changes relating to nesting, rearing young, molting, migration, and wintering.

anterior hypothalamus. A small brain region that is part of the hypothalamus and is known to regulate several environmentally sensitive physiological functions, including temperature regulation but also seasonal responses to changes in day length (photoperiod).

anthropocene. A new "geologic era" defined by some scientists to refer to the present and recent centuries in which human-generated activities such as climate change, extinctions, and habitat alteration are fundamental drivers of changes in earth's geological and ecological processes and composition.

aromatase. An enzyme that catalyzes the conversion of androgens (e.g., testosterone) into estrogens (e.g., estradiol).

asynchrony. Mismatches or incongruences in timing, for example when the timing of seasonal activities or life-history stages such as breeding or migration differ between individuals or populations.

autocrine signaling. A type of cell signaling in which a given cell secretes a chemical messenger that binds to receptors in that same cell, initiating subsequent changes; in contrast to classical endocrine signaling between different cells.

avian microbiome. The myriad of microorganisms (e.g., bacteria, fungi, archaea) that reside within or among an avian "host" individual's internal or external tissues.

Bateman's principle. An axiom that suggests that in most species, the amount of variation in reproductive success among individuals is likely to be much greater in males than in females, generally owing to the fact that a single male is capable of fertilizing many eggs from many females (or none at all), whereas females rarely produce more offspring by mating with more males.

behavioral imprinting. The process by which early (developmental) exposure to a stimulus during a critical learning period has lasting effects on behavior; the classic example involves young chicks "imprinting" on heterospecific, nonavian, or even human "mothers"; can involve multiple sensory stimuli, including sight, smell, or sound.

behavioral insensitivity. A lack of physiological responsiveness to behavioral or social stimuli that are known to often induce hormonal responses; for example, a failure to elevate testosterone levels in response to a territorial stimulus, as is often observed in the context of the challenge hypothesis.

behavioral syndrome. A suite of interrelated behavioral traits, often measured to be repeatable and generalizable across different behavioral contexts; sometimes used interchangeably with "animal personality."

Bergmann's rule. A biogeographic principle that suggests that within a given species or taxa, animals that live at higher latitudes (i.e., colder climates) should exhibit a larger body size, and hence surface-area-to-volume ratio, in contrast to those smaller-bodied populations at lower latitudes (i.e., warmer climates).

biogeographic island. An isolated patch of habitat within a "sea" of unsuitable habitat type, effectively invoking island biogeographic theory with respect to population dynamics and gene flow; examples include "sky islands" in which inhabitants of high elevation habitat types on mountaintops are isolated from other breeding populations.

biogeography. The study of how and why populations, species, and ecosystems are distributed throughout geographic space and geological time.

biological species concept. Defines species as members of actually (or potentially) interbreeding groups in nature, distinct from each other when they do not (or would likely not be able to) interbreed; one of the most commonly held views of "species" delineations, but difficult to assess for nonoverlapping (i.e., allopatric) populations.

bottlenecks. Sharp reductions in the size of a population due to environmental events or range shifts or colonization, oftentimes leading to a reduction in phenotypic or genetic variance (e.g., genetic bottleneck).

challenge hypothesis. The concept that the arrival of an aggressive intruder induces a hormonal response, often an elevation in testosterone, in the animal on which it intrudes. Similar to an immune response to a pathogen, which can prepare an organism for subsequent encounters with the pathogen, the hormonal response to an aggressive challenge is hypothesized to induce physiological preparation for future intrusions.

chromosomal inversion. A chromosomal rearrangement in which a segment of a chromosome is reversed end to end.

chromosomal rearrangements. Changes in the physical structure of chromosomes such as deletions, duplications, inversions, or translocations, generally thought to occur during meiosis but can also occur during mitosis; often involves breakage in the DNA at two different locations.

circadian rhythms. Biological changes or rhythms that happen predictably across a twenty-four-hour cycle (i.e., daily rhythms), based on endogenous timing mechanisms, generally entrained by changes in exposure to light and darkness.

common garden. A type of experiment, originally conceived in studies of plants, in which developing individuals (e.g., plant seeds, avian eggs, or nestlings) originating from different environments are raised under identical conditions; trait differences that persist are attributed to have a likely genetic basis, whereas those that converge are attributed to phenotypic plasticity.

comparative approach. Studies on the evolution of phenotypic characteristics (e.g., behavior, morphology, physiology) that involve comparing trait changes alongside reconstructions of the evolutionary histories among populations, species, families, or other taxonomic units.

connectivity. See **migratory connectivity**

constraint (evolutionary). A concept that describes how certain traits or combinations of traits are not generated by evolution despite ways in which they might be apparently advantageous; for example, limitations to available genetic variation, or genetic or developmental covariation with other traits, can help explain why certain evolutionary dimensions are "constrained."

convergence (evolutionary). Independent evolution of analogous structures or behavioral or physiological characteristics in species of different lineages in response to similar selection pressures.

correlational selection. A type of natural or sexual selection that arises when traits interact in their effects on fitness; may act over time to assemble groups of traits that work well together, including hormone-mediated suites.

cultural drift. In contrast to genetic drift, a mechanism by which behaviors can evolve randomly, particularly in small populations or in the context of bottlenecks, range expansions, or colonization events.

cultural inheritance. In contrast to genetic or epigenetic mechanisms of inheritance, the process by which behaviors or ideas (i.e., memes) can be vertically transmitted across generations.

cultural mutations. Analogous to genetic mutations of DNA, the process by which the characteristics of behavioral traits can change randomly, for example during the learning process, with the "mutated" version subsequently passed on via learning or cultural inheritance.

decapeptide. A polypeptide composed of a chain of ten amino acids (e.g., GnRH).

developmental plasticity. The general process by which an organism's characteristics (e.g., morphological, neural, behavioral, physiological, etc.) are altered or shaped during development in response to their environment; can be considered either canalized (i.e., irreversible) or reversible. See also **phenotypic plasticity**

differential migration. Describes a scenario in which cohorts of individuals within a population (e.g., males versus females, younger versus older) vary in the distances, geography, and/or timing of their migratory journeys.

direct (fitness) benefits. Factors influencing the number of offspring produced by an individual; in contrast to indirect (fitness) benefits, which include those factors related to how many relatives (e.g., nieces, nephews, grandchildren, etc.) an individual produces.

directional selection. A "mode" of natural selection in which variation in fitness favors a single "direction" of the trait distribution or a single allele combination (e.g., larger versus smaller or darker versus lighter are favored), causing a directional shift in the mean trait value or allele frequency in the subsequent generation; contrast with **disruptive selection** or **stabilizing selection**.

disruptive selection. Also known as "diversifying selection," describes a "mode" of natural selection in which extreme trait values on both ends of a distribution are favored and intermediate values are selected against; leads to an increase in trait variance and a bimodal distribution in the subsequent generation.

divergence (population). Can refer to either genetic divergence or phenotypic divergence, or both; the general process by which two or more populations become different across time, for example in response to different selection pressures or as a product of mutation and genetic drift in the absence of gene flow.

ecological speciation. Population divergence resulting from different populations or groups adapting to their local environment. Reproductive isolating barriers may arise in sympatry or allopatry but are triggered by divergent natural selection in response to the environment.

effective population size. An estimate of the number of breeding individuals contributing to a population's gene pool, or within-population genetic diversity; can be estimated by counting the number of individuals or by estimating the standing genetic variation.

epigenetics. When traits are heritable but not caused by DNA sequence differences, likely due to differences in gene expression or cellular phenotype.

estrogen receptor (ER). Formed by a group of proteins inside a cell and activated by binding with estrogen.

evolutionary potential. Contrast with evolutionary constraint; refers to the possible ways in which traits in a population might possibly evolve in the future, given the extent of genetic variation and genetic correlations, including possible developmental constraints.

extrapair paternity. Those offspring produced outside the social pair bond in socially (but not genetically) monogamous animals, as the result of "extrapair behavior" and "extrapair copulations."

facultative migration. A form of animal movement in which members of a population vary over time in whether or not they migrate in response to environmental variation; for example, migration will occur in some years but not others, based on local resource availability or climatic conditions.

femoral pores. Portions of a secretory gland found on the legs of lizards and some amphibians that secrete compounds that act in chemical communication.

fitness. A quantifiable measure of survival and reproduction in evolutionary biology. In population genetics, fitness is often described by allele frequencies.

follicle-stimulating hormone (FSH). Stimulates sperm production in the testes of male birds and initiates the development of egg follicles in female birds.

founder effect. When a small number of individuals establishes a new population, there is a loss of genetic diversity. This loss of genetic diversity can lead to distinctive differences between the new colony and the original population with respect to both genotype and phenotype.

frequency-dependent selection. When the fitness of an individual depends on the frequency of its phenotype relative to others in the population. For example, when being a rare phenotype is more attractive to mates.

gene expression. The process by which the information contained in a gene is used to synthesize functional gene products such as proteins. Evolution may occur in the timing, location, and amount of gene expression, which can change the actions of a gene or cell.

gene flow. Movement of genes from one population to the other, that is, the establishment of immigrant gene copies in a resident population.

gene networks. Also described as "gene regulatory network," is a collection of genes (i.e., DNA segments) that interact with one other in relation to gene expression levels of mRNA and proteins and, ultimately, physiological, behavioral, and morphological trait expression.

genetic assimilation. Occurs when a phenotype expressed in response to the environment later becomes genetically encoded through selection often via the canalization of new developmental pathways.

genetic correlation. When two or more traits share some proportion of variance, causing them to be more likely to be inherited or expressed together.

genetic drift. Change in the allele frequency due to the random processes of sampling in nature over time. Not a response to selection.

genetic incompatibility. When incompatibilities arise between populations or species because of mismatched genes leading to reduced viability or fertility.

Gloger's rule. A biogeographic axiom that states that within a species or taxa, more heavily pigmented (i.e., darker) forms are more likely to be found in moister or more humid environments, whereas lighter forms are more likely to occur in drier habitats; invokes the idea of crypsis, or camouflage, in which visually matching background habitat reduces predation risk.

GnRH. See **gonadotropin releasing hormone (GnRH)**

GnRH challenge. A sampling protocol that involves an intramuscular injection of GnRH, in order to measure the short term elevation of circulating (plasma) testosterone produced in response to the physiological "challenge" of the hypothalamic-pituitary-gonadal endocrine axis; allows for a more repeatable and standardized sampling of both seasonal and individual variation in testosterone levels.

gonadotropin releasing hormone (GnRH). A trophic peptide hormone responsible for the release of FSH and LH from the anterior pituitary (also known as luteinizing-hormone-releasing hormone, LHRH).

gonadotropins. Protein hormones secreted by the anterior pituitary of vertebrates, including follicle-stimulating hormone (FSH) and luteinizing hormone (LH), which lead to reproductive development.

haplotype. A cluster of tightly linked genes on a chromosome that are likely to be inherited together from one parent; also refers to a set of genetic polymorphisms that are associated statistically, although they may not be physically linked or on the same chromosome. A "hapolotype network" is a hierarchical tree (i.e., a phylogeny) that displays the relationship between haplotypes that vary between individuals from different populations and is used to map evolutionary and geographical history.

Hardy-Weinberg equilibrium. A model for calculating changes in allele frequencies in which allele frequencies remain constant over generations under the assumption that other evolutionary forces (e.g., selection, migration, mutation) are absent. Actual allele frequencies can be observed and compared to the model to analyze what forces might be affecting a population.

heritability. The proportion of variation observed for a trait among individuals in a population that is apparently due to genetic differences, and hence likely to be passed on to the subsequent generation.

heteropatric speciation. Divergence that occurs between resident and seasonally migrant populations of the same species, which may often occur in the same location for a part of the annual life cycle (Winker 2010).

heteropatry (distribution). A geographic distribution in which populations of variably migratory species or taxa are allopatric (i.e., geographically separate) during the breeding season but share habitats in common during the winter, where they are found in "sympatry" during the nonbreeding season, as well as during their migratory journeys.

hippocampus. A region in the medial temporal lobe of the vertebrate brain, which plays an important role in memory, including spatial memory and navigation in the context of migration biology and behavioral ecology.

honest signal. A costly behavior or phenotype (e.g., a morphological trait) produced by one individual (the signaler) that conveys reliable information to a receiving individual about the quality or characteristics of the signaler.

hormonal pleiotropy. Coordination of a suite of correlated or co-occurring traits by a common underlying hormonal mechanism.

hormone. A chemical messenger molecule that is released from specialized glands or cells into the circulation and regulates a biological response at target cells or tissues.

hormone-mediated suite. A group of traits that are correlated owing to the influence of a hormone.

hormone-mediated trait. A phenotype affected by the action of a hormone.

hybrid incompatibilities. Reproductive barriers between populations or species resulting in a reduction in viability or fertility.

hybrid sterility. Reduced male or female function in hybrid individuals.

hybrid zones (hybrid population, hybrid complex). Regions or areas of overlap between species or groups where individuals thought to be isolated either ecologically or geographically co-occur, typically used to describe areas where hybrids have been observed.

hyperphagia. Excessive eating, often associated with early spring restlessness in anticipation of migration.

hypophyseal. A blood vessel system in the brain that connects the hypothalamus with the anterior pituitary.

hypophysiotropic. Hormones produced by the hypothalamus that act on the pituitary gland to maintain endocrine function.

hypothalamus. Portion of the brain that links the nervous system to the endocrine system via the pituitary gland.

immigrant selection. A scenario in which nonrandom subsets of individuals are more likely to be founding colonists of a new population, or immigrants into a different population, based on their phenotypic characteristics; in some cases, can be considered as a type of nonrandom bottleneck.

immunocompetence. The ability of an organism to produce a normal immune response to an antigen; in contrast to immunodeficiency.

immunocytochemistry (ICC). A common laboratory technique that is used to localize specific proteins, for example a hormone receptor, in particular anatomical regions (e.g., a brain region), by utilizing specific antibodies that bind to the target protein and then change color in response to a secondary antibody that can be viewed under microscopy.

improvisation. In birdsong, the process by which birds spontaneously generate vocalization types during the process of repertoire development and song crystallization.

incipient speciation. The early stages of the speciation process when new groups (populations, species) are genetically or phenotypically distinct but still capable of interbreeding.

indirect (fitness) benefits. Contrast with direct (fitness) benefits; refers to those factors that lead to greater number of genetic relatives (e.g., nieces, nephews, grandchildren) produced; part of inclusive fitness calculations.

in situ **hybridization.** A laboratory protocol that uses (e.g., fluorescently) "labeled" complementary DNA or RNA strands as "probes" to localize and visualize specific DNA or RNA sequences in target cells or tissues.

introgression. The movement or spread of genetic material from one population into another through hybridization.

last glacial maximum. The period in the earth's climatic history when glacial coverages (i.e., ice sheets) were at their most recent greatest geographic extent, particularly in North America and Eurasia, estimated to be from approximately 10,000 to 30,000 years ago.

lateral septum. A brain region considered part of the "vertebrate social behavior network," which lies within the basal ("limbic") forebrain and is invoked as a regulator of aggression, stress, and social communication in a variety of avian and mammalian species.

lateral ventromedial hypothalamus. A brain region considered part of the "vertebrate social behavior network," which lies within the basal ("limbic") forebrain inside the hypothalamus and is invoked as a regulator of satiety, stress, play behavior, sexual behavior, vocal communication, and other agonistic responses.

latitudinal (migration). The classic example of avian migration, in which birds make significant north-south and south-north journeys each season to and from breeding and wintering grounds. As an example, in the northern hemisphere, many species breed at higher latitudes and migrate to and from more southerly wintering grounds.

life history. Generally refers to the nature and timing of the series of key events, changes, stages, or transitions that occur during an organism's lifetime; examples include juvenile development, age of sexual maturity and first reproduction, number of offspring and parental investment, seasonal migration or molt strategies, senescence, and life span.

life-history theory. Posits that the schedule and duration of key events across an organism's lifetime are shaped by natural selection to maximize production of (surviving) offspring and are hence predicted to vary among populations and species inhabiting different environments.

linkage disequilibrium. Nonrandom association of inherited alleles at different genetic loci.

linked. Genes that do not freely recombine. An alternative definition, not employed here, describes genes on the same chromosome.

local adaptation. Adaptation by a population or species to a region or environment that can lead to phenotypic or genetic differences between groups.

long-range song. Contrast with short-range song; the primary long-distance communication modality typically used by birds and other animals to communicate, for example to advertise territorial ownership and attract mates.

luteinizing hormone (LH). Stimulates the production of the male hormone testosterone by Leydig cells in the testes.

maternal effects. The process by which an individual organism's phenotype is influenced not only by its environment and its genotype but also by the environment and genotype of its mother; classic examples include the *in utero* environment of mammalian offspring, as well as the *in ovo* environment of birds, reptiles, and amphibians.

medial (extended) amygdala. A neural (brain) "node" in the vertebrate social behavior network, considered to be responsible for integrating hormonal and chemosensory signals, including olfactory cues, and subsequently regulating changes in social and sexual behavior.

medial bed nucleus of the stria terminalis. A brain region considered part of the "vertebrate social behavior network," which lies within the basal ("limbic") forebrain alongside the thalamus and is invoked as a regulator of stress and anxiety in response to perceived threats.

medial preoptic area. A brain region considered part of the "vertebrate social behavior network," which lies within the anterior hypothalamus and is invoked as a regulator of vocal-acoustic communication, thermoregulation, sexual behavior, and parental behavior.

median eminence. Region of the brain located at the base of the hypothalamus that serves as an interface between the neural and peripheral endocrine systems.

meme. In behavioral ecology, a behavioral performance, characteristic, idea, or style that can spread from individual to individual via learning or other cultural transmission. In birdsong, typically refers to particular song types or variants, which can be sung by one or more individuals in a population, collectively comprising the population's "meme pool."

MHC loci (major histocompatibility complex genes). A set of genes (and their cell surface protein products) that form a major component of the vertebrate immune system, as the MHCs function to bind protein fragments derived from pathogens and display them on the cell surface for recognition by T-cells (lymphocytes).

microarray. A collection of microscopic DNA or RNA fragments attached to a solid surface used in laboratory protocols to measure gene expression levels for large numbers of genes simultaneously.

migration. Movement of animals from one location to another; typically used in the context of seasonal migration, involving journeys from breeding habitats to wintering habitats and back again each year.

migratory connectivity. The geographic relationships between populations of animals (especially birds) at different time points during the year, in particular the spatial and temporal links between specific breeding and wintering sites for individuals and populations.

morphological species concept. Classifies organisms into species based on their morphology. Individuals in the same species are similar to one another in morphology; individuals in different species are different in morphology.

morphology. The study of the physical form and structure of organisms.

mtDNA. Refers to DNA found in mitochondria, used to measure the degree and timing of divergence among populations.

natural selection. Variation in fitness or variation in lifespan and reproductive success that relates to the potential to contribute genes or traits or offspring to future generations.

neophobia. Fear of novel objects.

neurosecretory. The secretion of hormones by nerve cells.

nucleus taeniae. The avian medial amygdala, a region of the brain associated with aggressive behavior.

obligate migration. A form of animal movement in which all members of a population are migratory.

open-ended learning. A form of birdsong development in which song structure can continue to change over the life span.

opportunity for sexual selection. The variance in relative (evolutionary) fitness in a population from all sources, environmental and genetic. Since the strength of natural selection on any trait depends upon the correlation between individual phenotypic value and relative fitness, the opportunity for selection places an upper limit on the total amount of adaptive evolutionary change that can be accomplished in a single generation.

organizational effect. Developmental effect of a hormone occurring early in life, including prenatal; usually irreversible and often involves a specific sensitive period.

ornaments. Structures that do not have obvious utility and may act in mate attraction and be the result of sexual selection (e.g., extreme feathers, bright colors).

paracrine signaling. When the products of a cell cause changes in nearby or adjacent cells.

parallel selection. Similar traits found in related species owing to the existence of the trait in a common ancestor and the failure of the trait to diverge since the species separated.

parapatric (distribution) / peripatry. A geographic distribution of taxa (e.g., populations or species) found in nearby or adjacent locations; in contrast to sympatric distribution, in which distributions of taxa overlap, or allopatric distribution, in which they are separate.

parapatric speciation. The evolution of reproductive isolation between two populations that may continue to exchange migrants through the whole process.

partial migration. A form of animal movement in which some members of a population are migratory and others are sedentary; less frequently used to compare populations within a species as opposed to individuals within a population.

phenology. Timing, onset, duration, and sequence of stages of the annual cycle.

phenotype. The outward appearance of an organism; the value of all traits assumed by the organism; the outcome of the interaction of genotype and environment.

phenotypic engineering. An experimental approach used to assess the adaptive value of a trait or traits by manipulating individual phenotypes (e.g., experimentally altering hormone levels and measuring behavioral, performance, and/or fitness consequences).

phenotypic integration. Patterns of correlation or interdependence among different parts of the phenotype; can be mediated by common underlying hormonal mechanisms of trait expression or development.

phenotypic plasticity. The capability to express more than one phenotype for a given genotype, often mediated by hormonal mechanisms. This may occur at many timescales.

philopatric / philopatry. Site fidelity, or the tendency for dispersal distances to be short or for migrants to return to their previous year's home range; contrast natal site fidelity, in which the return is to the birth site.

photoperiodism. Seasonal responses of organisms to annual changes in day length; a photoperiodic species is one that employs changes in day length to time reproduction, molt, migration, and other recurrent events.

photorefractoriness. A stage in the annual cycle in which birds show no immediate response to a change in photoperiod, caused by constant exposure to long days; important for preparation for fall migration.

photostimulate. The use of light to experimentally trigger preparedness for molt, migration, and/or reproduction.

phylogenetic species concept. A species delineation concept based strictly on unique shared evolutionary history; definition is not based on reproductive compatibility or morphological similarity.

plasticity. The ability of a genotype to produce more than one phenotype depending on the environment; more broadly, change in phenotype with change in the environment.

polymorphism. Two or more distinct phenotypes co-occurring in the same population.

postzygotic. After fertilization

prezygotic. Prior to fertilization.

prezygotic isolation. A form of reproductive isolation that acts prior to fertilization (e.g., mating preferences based on species recognition or female reproductive fluids that favor conspecific over heterospecific sperm).

proximate mechanisms. Causal factors such as variation in genes or physiology that relate to a phenotypic response; contrast to ultimate mechanisms that provide evolutionary explanations.

ptiloerection. A term applied to a bird that has elevated its feathers and thus appears to be bigger than it is, used by birds in thermoregulation and social display.

quantitative PCR. A molecular technique for measuring transcript abundance in tissues, where a transcript is a sequence of messenger RNA with the potential to be translated into a peptide; transcripts are isolated, amplified by polymerase chain reaction, and quantified as a measure of gene expression.

race. A classification used to distinguish among bird populations or groups based on morphology, geography, song type, and breeding preferences.

reaction norm. Describes the relationship between a genotype and its environment on a given phenotype (e.g., genotype x may grow tall in one environment and be short in another); more broadly, compares the phenotype of populations or individuals across multiple environments.

receptor. A protein on the surface or interior of a cell that binds to a hormone, leading to modulation of cellular functioning (e.g., activation/inhibition of gene transcription or second messenger networks).

recombination. Refers to the exchange of genetic material occurring during gamete formation as part of the process of sexual reproduction.

refugia. Isolated portions of a population's or species's distribution that was once more extensive; used with reference to glacial refugia.

regression (gonadal). The stage in the annual cycle when reproduction is ending and gonads shrink in size.

repeatability. A statistical description of the degree to which measurements of the same phenomenon coincide; a measure of reliability of measurements.

repertoires. Used to describe variation in the structure of songs a bird sings; multiple song types sung by an individual are referred to as the bird's repertoire; repertoire size is variable.

reproductive isolation. Barriers to hybridization between species; encompasses a variety of mechanisms including mate choice, genetic incompatibility, hybrid sterility.

response to selection. Changes in phenotype that occur when a population experiences selection (differential survival or reproduction) acting on heritable variation.

reverse speciation. Loss of biodiversity that occurs when species formed in ecological isolation collapse when conditions change and their distributions again overlap.

seasonality. Refers broadly to changes in the environment that occur on a seasonal basis or seasonal changes in the phenotype of organisms (e.g., hibernation, migration), or the study of the impact of changing seasons on organisms.

seasonal migration. Movement of individuals from one portion of their geographic range to another and back again over the course of the year, typically in response to seasonal changes in climate.

sedentary. Individuals and populations that do not migrate; contrast migratory.

selection gradient. Relationship between variation in a trait (e.g., tail length, tendency to migrate) and variation in fitness (e.g., survival or reproductive success); a steep selection gradient implies strong directional selection.

selection pressures. Environmental factors (e.g., climate, predators, disease, competitors) that determine how natural selection affects variation in traits (e.g., behavior, morphology, physiology).

selective advantage. The impact that a particular trait has on fitness, or the relative ability to survive, reproduce (e.g., the degree to which possessing longer tails makes males more attractive to females than those possessing shorter tails).

selective sweep. The reduction of molecular genetic variation (nucleotide diversity) in DNA as the result of strong natural selection.

sensitivity. The response shown by tissues that serve as targets of circulating hormones; low response is low sensitivity.

sexual conflict. Also referred to as sexual antagonism; occurs when the fitness optimum for a trait differs in males and females, resulting in selection acting in opposing directions on the two sexes.

short-range song. Refers to a complex song produced at low amplitude that is audible only when the receiver is located near the sender.

single nucleotide polymorphisms (SNPs). Variants in the DNA sequence found among individuals within populations in which a single nucleotide (e.g., A, T, C, or G) differs at a particular location in a sequence; can be used to determine level of genetic divergence between populations.

song crystallization. A term that refers to a stage of song learning in which highly variable subsong transitions into normal adult song.

speciation. The process(es) by which new species are formed (e.g., by splitting or through gradual accumulation of change over time).

species. A fundamental taxonomic category for all organisms; a group of organisms that are morphologically and/or genetically more similar to one another than to other groups, share a single lineage, and are more or less reproductively isolated.

stabilizing selection. Refers to a "mode" of natural selection in which traits or individuals that are close to the average for the population are favored and those that deviate more from the average are disfavored; acts to reduce diversity; contrast directional or disruptive selection.

stable isotope. One of multiple forms of atoms that vary in the number of neutrons; some isotopes are radioactive (emit radiation), others do not, and the latter are referred to as stable; used in animal ecology to trace food webs by determining isotopic ratios that differ geographically and by food source.

standing variation. Refers to the current level of genetic variation within a population in which high or low levels may indicate greater or lesser potential for natural selection to lead to evolutionary change.

subspecies. A taxonomic category below the level of species used to distinguish a genetically distinct set of subpopulations, each within a discrete range, and each genetically compatible with other subpopulations of that species.

sympatric (distribution) refers to the geographic distribution of taxa (e.g., populations or species) found in the same or overlapping geographic locations; contrast allopatric, in which distributions of taxa do not overlap.

systematics The term used to describe the study of classification of organisms (taxonomy) based on their evolutionary relationships.

target / target tissues. Tissues or cells whose function is influenced by a hormone owing to the presence of hormone receptor proteins.

testosterone (T). A hormone that is important for development of the testes in male birds and ovulation in female birds and is also important for secondary sexual morphology and brain function.

trade-off. The situation that occurs when two traits (often components of fitness, e.g., mating effort versus parental effort) cannot be simultaneously maximized because the expression of each comes at the expense of the other.

transcriptome / transcriptomics. A term to describe the messenger RNA (mRNA) extracted from a tissue and representing all the sequences with the potential to be translated into peptides (proteins) (compare to the genome); the study of transcriptomes.

transmission constraints. Limitations of the ability of sounds to move through an environment.

uropygial gland. A tissue also known as the preen gland, located at the base of the tail, which secretes oily compounds that can be applied to feathers for protection and are also thought to act in chemical communication or in defense from microbial parasites.

ventral medial telencephalon (VmT). A region in the rear of the forebrain that lies within the telencephalon, at the boundary of the pallidum and pallium; this region includes the medial amygdalae (or nucleus teniae), which are often studied in relation to animal social behavior.

vomeronasal organ. A component of the chemosensory system found in the nasal cavity of amphibians, reptiles, and mammals but underdeveloped or absent in birds.

Zugunruhe. A German word to describe restlessness in caged birds during migration; also called migratory restlessness.

Contributors

PAU ALEIXANDRE
National Museum of Natural Sciences
Spanish Research Council (CSIC)
Madrid 28006
Spain

SOFÍA ALVAREZ-NORDSTRÖM
National Museum of Natural Sciences
Spanish Research Council (CSIC)
Madrid 28006
Spain

JONATHAN W. ATWELL
Department of Biology
Indiana University
Bloomington, IN 47405
USA

CHRISTINE M. BERGEON BURNS
Department of Biology
Indiana University
Bloomington, IN 47405
USA

KRISTAL E. CAIN
Evolution, Ecology & Genetics
Australian National University
Canberra ACT 0200
Australia

GONÇALO C. CARDOSO
CIBIO—Centro de Investigação em Biodiversidade e Recursos
    Genéticos
Universidade do Porto
Vairão 4485-661
Portugal

NICOLE M. GERLACH
University of Florida
Gainesville, FL 32611
USA

DANIEL M. HOOPER
Department of Ecology and Evolution
University of Chicago
Chicago, IL 60637
USA

JODIE M. JAWOR
Department of Biological Sciences
University of Southern Mississippi
Hattiesburg, MS 39406
USA

ELLEN D. KETTERSON
Department of Biology
Indiana University
Bloomington, IN 47405
USA

JOHN MCCORMACK
Moore Laboratory of Zoology and Department of Biology
Occidental College
Los Angeles, CA 90041
USA

JOEL W. MCGLOTHLIN
Department of Biological Sciences
Virginia Polytechnic Institute and State University
Blacksburg, VA 24061
USA

BORJA MILÁ
National Museum of Natural Sciences
Spanish Research Council (CSIC)
Madrid 28006
Spain

DAWN M. O'NEAL
Huyck Preserve and Biological Research Station
Rensselaerville, NY 12147
USA

MARK P. PETERSON
Department of Biology
Viterbo University
Lacrosse, WI 54601
USA

TREVOR D. PRICE
Department of Ecology and Evolution
University of Chicago
Chicago, IL 60637
USA

DUSTIN G. REICHARD
University of California, Davis
Davis, CA 95616
USA

KIMBERLY A. ROSVALL
Department of Biology
Indiana University
Bloomington, IN 47405
USA

DANIELLE J. WHITTAKER
BEACON Center for the Study of Evolution in Action
Michigan State University
East Lansing, MI 48824
USA

Index

Page numbers followed by *f* or *t* indicate a figure or table, respectively.